# The Mongolian Gerbil Brain

Avraham Mayevsky

# The Mongolian Gerbil Brain

## Mitochondrial Function, Vasculature, and Pathophysiological States

Springer

Avraham Mayevsky
Faculty of Life Sciences
Bar-Ilan University
Ramat-Gan, Israel

ISBN 978-3-031-69548-3 ISBN 978-3-031-69549-0 (eBook)
https://doi.org/10.1007/978-3-031-69549-0



This Springer imprint is published by the registered company Springer Nature Switzerland AG
The registered company address is: Gewerbestrasse 11, 6330 Cham, Switzerland

# Preface

This new book is the fourth one, in my set of books, covering my studies on the Mongolian Gerbil's brain blood supply, mitochondrial function and other physiological activities *in vivo*.

At the end of 1940, I was born in Tel Aviv, Israel, about the same time that Britton Chance earned his PhD and started his interest and research on mitochondrial biochemistry. Shortly after my arrival to Philadelphia in 1972, I started my post-doctorate training with Prof. Britton Chance at the University of Pennsylvania, Philadelphia, studying brain mitochondrial function *in vivo* using NADH fluorescence in real time. A few years later, I studied the mitochondrial NADH redox state monitored *in vivo* in the Gerbil's brain and tried to test the effect of oxygen supply and demand on brain energy metabolism evaluated by the monitoring the redox state of the mitochondria.

During the early period of my stay in Philadelphia and later at my home University, Bar-Ilan University, in Israel, I continued the development of the technology enabling the monitoring of mitochondrial NADH redox state in the unanesthetized Gerbils.

In the next step, I developed a new multiparametric monitoring system in order to monitor the gerbil brain. This system enabled the monitoring of up to eight biochemical and physiological parameters from the brain of unanesthetized gerbils exposed to various experimental protocols.

Using the developed monitoring systems, I was able to collect a large volume of results and published around 30 full-length publications. The first chapter in the current book is an introduction and an historical review of the published material relevant to the Gerbil's brain. Chapter 2 provides a scientific background to the technologies used in the current studies. Chapter 3 provides the various technologies and protocols used in all our studies. Chapters 4, 5, 6, 7, 8, 9, and 10 contain the results of all the experimental protocols performed in my laboratory during the years using the gerbil brain. Chapter 11, the last chapter, contains the discussion and conclusions of the studies.

Ramat Gan, Israel — Avraham Mayevsky

# Acknowledgments

This book is dedicated to my dearest high school teacher of the sciences, Mr. Joseph Feuchtwanger from Kefar Haroeh in Israel, for his vision, patience, and support during a critical period of my life. Without his help and encouragement, my academic achievements and writing of this book were impossible to be accomplished.

Also, I dedicate this book to my wife, Zipora and four children, Amotz, Sarit, Hagit, and Shalev that understood my devotion to the unusual career and enabled me to fulfill all of my obligations and academic activities.

My deep appreciation is given to my parents, Shabtai and Rachel Mayevsky, for their devotion and support throughout my life span.

My appreciation is given to, the late, Prof. Britton Chance, an outstanding scientist, teacher, and colleague for almost four decades of fruitful collaboration.

Many thanks are given to Dr. Zipi Breuer. Her contribution to the studies of the Gerbil's brain were critical and without her involvement, the studies were less successful. Her contributions to Chaps. 8 and 10 were especially critical. A special appreciation is given to Dr. Judith Sonn, who joined my laboratory as a PhD student and contributed significantly to the studies.

Special thanks are given to Mrs. Avrille Goldreich for her dedication to help bringing this book to press in its excellent form.

During almost 60 years of research activities at Bar-Ilan University, I trained many MSc and PhD students. I want to thank all of them for their confidence and the efforts that they put during the experimental work as well as in preparation and publication of the manuscripts.

Special appreciation is given to Nili Zarchin, MSc, and Dr Efrat Barbiro-Michaeli who joined my research team after their graduation. Their assistance in running the experiments and training new undergraduate and graduate students is appreciated very much.

# Contents

# Chapter 1
# Historical Background of the Mongolian Gerbil in Research

**Abstract** This chapter introduces the main subjects that will be discussed in this book. It summarizes the knowledge of basic physiological and biochemical events developed in the brain of the Mongolian gerbil (*Meriones unguiculatus*) when exposed to various pathological situations.

1. The discovery of the Mongolian gerbil
2. The gerbil as a model to study brain Ischemia
3. The gerbil as a model to study epilepsy
4. The use of the Mongolian gerbil in medical research

**Keywords** Brain metabolism · Tissue energy balance · Oxygen supply · Brain monitoring

## 1.1 Introduction

The first description of *Meriones unguiculatus* found in China was made in 1867 by M. Milne Edwards [1], as presented in Figs. 1.1, 1.2, and 1.3. Since its discovery in 1867, more than 4000 papers were published on the gerbil, as indicated by the PMC search. In this book, I will summarize our studies related to the monitoring of brain mitochondrial NADH redox state and other physiological activities measured in the intact brain in real time.

In the 1960s, two papers that described two main events developed in the Mongolian gerbil (*M. unguiculatus*) were published.

In 1966, Levine and Payan [2] claimed that the gerbil became commercially available as follows: "Recently, a desert rodent, the Mongolian gerbil, has become commercially available. It has proved useful in infectious disease and behavioral research [3, 4]. We have attempted to determine whether this animal has any special advantage for experimental neurology." The summary of their paper was as follows:



A. Mayevsky, *The Mongolian Gerbil Brain*,
https://doi.org/10.1007/978-3-031-69549-0_1

**Fig. 1.1** The scientist who discovered (1867) the Mongolian gerbil (*Meriones unguiculatus*) in north China (Part 1) [1]

"The Mongolian gerbil was highly susceptible to cerebral ischemia induced by ligation of the common carotid arteries. Bilateral ligation was uniformly lethal. Unilateral ligation produced ischemic lesions, sometimes lethal, in the brain or the brain and eye of over half the subjects. The gerbil was also highly susceptible to water intoxication, a finding that may be related to peculiarities of its water metabolism. The gerbil had no special advantage for the study of other adverse factors explored. The gerbil deserves a place in laboratories of experimental neurology, especially those concerned with cerebrovascular disorders, the visual system, or water intoxication."

In 1968, Thiessen et al. [5] reported as follows: "After noting several convulsions of the gerbil (*Meriones unguiculatus*) in our laboratory, we attempted to clarify the conditions that evoke and suppress the response."

The final part of their conclusion was as follows: "There is an obvious familial transmission since the majority of our seizure-prone animals were derived from two breeding pairs. In one pair the male was affected and in the other pair the female

Observations on few Mammals in North China
By: Alphonse Milne Edwards

OBSERVATIONS
SUR QUELQUES MAMMIFÈRES DU NORD DE LA CHINE,

Par M. ALPHONSE MILNE EDWARDS.

(Extrait.)

Le père Armand David, missionnaire de la Congrégation des Lazaristes à Pékin, a recueilli en Chine, et particulièrement en Mongolie, des collections zoologiques importantes, comprenant un grand nombre d'espèces nouvelles. J'ai déjà fait connaître quelques-unes d'entre elles, et dans le travail dont je présente ici un extrait (1), je donne la description de plusieurs autres appartenant à l'ordre des Insectivores, à celui des Rongeurs et à celui des Ruminants.

Le père David a pris en Mongolie une Taupe qui paraît fort rare. Par son aspect extérieur, elle se rapproche beaucoup de la Taupe d'Europe ; elle est un peu plus petite, d'une couleur grisâtre foncée, brillante et miroitante; elle répand une odeur de musc des plus prononcées. Son système dentaire est très-particulier, et ne permet pas de ranger ce petit Mammifère dans le genre *Talpa;* en effet, il a pour formule :

$$\text{Inc.}\ \frac{3}{4}\ \text{can.}\ \frac{1}{1}\ \text{prém.}\ \frac{2}{2}\ \text{mol.}\ \frac{4}{4}$$

Il existe par conséquent une prémolaire de moins que chez la Taupe, et d'ailleurs la forme de chacune des dents considérée isolément est très-différente. Je crois donc devoir former pour cette espèce une division générique particulière, et je la désigne sous le nom de *Scaptochirus moschatus.*

L'ordre des Rongeurs compte en Chine de nombreux représentants, parmi lesquels je signale plusieurs petites espèces qui, par leur aspect extérieur, ressemblent assez aux Arvicoles, mais s'en distinguent par l'existence d'abajoues très-développées et par la conformation des dents molaires ; sous ces rapports, ils se rapprochent donc des Hamsters, sans cependant pouvoir prendre place dans le même genre.

(1) Ce travail paraîtra prochainement dans les *Nouvelles archives d'histoire naturelle.*

376 ALPHONSE MILNE EDWARDS.

Pour rappeler ces analogies, je leur donne le nom de *Cricetulus.* L'espèce la plus commune (*Cricetulus griseus*) est à peu près de la taille du Mulot; son pelage est fin, très-doux, d'un gris très-légèrement teinté de roux, et marqué d'une raie noire longitudinale sur le dos.

Les Rats-Taupes de Chine sont bien distincts du Zoccor de Sibérie, qui jusqu'à présent était la seule espèce connue du genre *Siphneus.* Ces Rongeurs sont très-difficiles à distinguer à l'aide de leurs caractères extérieurs ; mais on arrive à les déterminer d'une manière rapide et sûre par la considération de leur système dentaire. Les molaires qui sont à croissance continue ne changent pas de forme par les progrès de l'âge, et diffèrent beaucoup suivant les espèces. Le *Siphneus Fontanierii*, dont M. Fontanier, consul honoraire à Pékin, nous a procuré un exemplaire provenant des environs de cette ville, a été rencontré également à Si-wan par le père David. Chez ce Rat-Taupe, la région occipitale du crâne est fortement bombée en arrière ; les molaires supérieures sont très-allongées, surtout la dernière.

Chez le *Siphneus Armandii*, qui jusqu'à présent n'a été trouvé qu'en Mongolie, la région occipitale est presque verticale comme chez le Zoccor, mais on le distingue de cette dernière espèce par la petitesse de la dernière molaire supérieure, et par l'existence d'un sillon vertical unique à la face interne de la première de ces dents.

Un Spermophile (*Spermophilus Mongolicus*), très-abondant en Mongolie où on l'appelle *Hoang-chou*, diffère des nombreuses espèces du même genre qui habitent l'Europe et la Sibérie. Il est de petite taille ; sa queue est de longueur médiocre, rousse en dessous, et bordée de poils noirs à extrémité jaunâtre ; son pelage est doux et uniformément fauve-clair, tiqueté de noir.

La Gerboise (*Dipus annulatus*), que le père David a rencontrée dans les collines sablonneuses de la haute Mongolie, ne peut être confondue avec aucune de celles déjà décrites. Indépendamment des caractères fournis par le crâne et la dentition, ce *Dipus* se distingue par l'anneau de poils blancs qui surmonte la bande noire subterminale de la queue.

Il existe aussi deux espèces de Gerbilles qui vivent en troupes

**Fig. 1.2** The scientist who discovered (1867) the Mongolian gerbil (*Meriones unguiculatus*) in north China (Part 2) [1]

MAMMIFÈRES DU NORD DE LA CHINE. 377

nombreuses dans les plaines stériles et pierreuses de toute la Mongolie. La plus petite (*Gerbillus brevicaudatus*) est, ainsi que son nom l'indique, remarquable par la brièveté de sa queue très-poilue et jaune. Le corps est d'un brun jaunâtre en dessus et blanc en dessous ; ses ongles sont blancs.

La seconde espèce (*Gerbillus unguiculatus*) diffère de la précédente par sa queue plus longue, son pelage d'un brun grisâtre en dessus, d'un blanc sale en dessous ; ses ongles sont noirs et plus robustes.

The second strain(Gerbilus unguiculatus) is differ from the previous one (Gerbillus breicaudatus) in his longer tail , the fur is gray-brown on the dorsal side and dirty white in the ventral side; The nails are black and very strong. (Free Translation)

Radde, dans son voyage en Sibérie, a décrit et figuré une espèce de Chèvre-Antilope, qu'il a rapportée à l'*Antilope crispa* (Temm.), du Japon ; mais j'ai pu me convaincre, par l'examen de l'individu-type conservé au Musée de Leyde, que ce rapprochement était erroné. Cette espèce, appelée par les Chinois *Chan-iang* (Chèvre des montagnes), est parfaitement caractérisée par sa queue très-longue et descendant jusqu'aux talons, par son pelage d'un gris brunâtre et sa gorge jaune. Il est évident que le *Chan-iang* est bien distinct de l'*Antilope crispa*, et je la désigne sous le nom de *Caudata*, à raison de la conformation de sa queue.

M. Fontanier a rapporté de Chine plusieurs bois d'un Cerf de grande taille dont il n'a pu se procurer la dépouille, et qui, d'après les renseignements qui lui ont été fournis, habite la Mantchourie. Les Chinois l'appellent *Cerf-Chameau*. Ces bois diffèrent de ceux de toutes les espèces décrites jusqu'ici et rappellent un peu ceux des jeunes Élans ; ils sont peu élevés et remarquablement massifs. Le merrain, très-court, porte un andouiller basilaire antérieur, qui, chez les individus bien adultes, est presque aussi fort que la perche et se bifurque. Les empaumures sont très-larges, surtout dans le jeune âge. La perche se courbe brusquement en arrière, et donne naissance à un second andouiller dirigé presque directement en haut. Les bois que j'ai sous les yeux paraissent provenir d'un individu déjà vieux, ce qui tend à prouver que le nombre des andouillers ne doit guère être jamais plus considérable. J'ai inscrit provisoirement cette espèce, sur les catalogues du Muséum, sous le nom de *Cervus Cameloides*.

**Fig. 1.3** The scientist who discovered (1867) the Mongolian gerbil (*Meriones unguiculatus*) in north China (Part 3) [1]

was affected. Our data are most congruent with the hypothesis of a dominant autosomal condition with penetrance as a function of arousal level (64.9% of offspring derived from two pairs seized at some time). We are now involved in breeding studies to explore the mode of inheritance more fully."

This book deals mainly with the two pathologies just mentioned, namely brain ischemia and oxygen delivery as well as epileptic seizures and the concomitant event—Cortical Spreading Depression.

## 1.2 Historical Background

According to Gulotta [6], the first assignment of unguiculatus to Meriones was published in 1908 by Oldfield Thomas [7]. In 1933, Leiper described the susceptibility of the gerbil to bilharziasis [8]. In 1954, van der Hoeden described the pathogenicity of leptospiras to field rodents (including the gerbil) in Israel [9]. In 1959, Robinson described the metabolism of the gerbil *M. unguiculatus* [10]. The abstract of Robinson's paper is as follows: "The critical temperature for the Mongolian gerbil, M. unguiculatus, is determined to be 30 °C. The zone of thermal neutrality extends to 40 °C. This species possesses a high degree of tolerance for heat and has a greater capacity for temperature regulation than is reported for many of the desert rodents" [11, 12]. In 1963, Walters et al. described the gerbil as a subject in behavioral research [4].

According to Lorene Stockberger, "All of the gerbils in the United States were derived from stock bred in the laboratory since 1935, when Dr. C. Kasuga captured 20 pairs along the Amur River basin in Eastern Mongolia. In 1954, Dr. Victor Schwentker obtained 11 pairs from the colony started with Dr. Kasuga's captives and started the gerbil colony at Tumblebrook Farm" [13]. In 1978, Fisher and Llewellyn [14] described the natural history, care, and maintenance of the Mongolian gerbil.

### *1.2.1 The Gerbil as a Model to Study Brain Ischemia*

The Mongolian gerbil (*M. unguiculatus*) has become a very useful small animal model for brain ischemia since the pioneering work of Levine and Payan [2] in 1966. According to Schwentker [3], the Mongolian gerbil belongs to the genus *Meriones*, sub-family Gerbillinae, family Muridae. The common name—gerbil—applies to all species of the sub-family Gerbillinae. The most commonly used gerbil in medical research is the Mongolian gerbil, *M. unguiculatus*, although several other species are also in use as described by Robinson [15]. The Israeli gerbil, *Meriones tristrami*, which has a different origin, is also in use in medical research [16], but was not used for ischemia studies as yet. The anatomy of blood vessels supplying the brain in those two species of gerbils and their biochemical consequences under ischemia were compared [17], but it will not be discussed here.

The Mongolian gerbil was introduced as an animal model for stroke studies nearly 60 years ago, and a few hundred publications appeared during this period, as summarized by Robinson [18–20]. During the first 5 years of usage (1966–1971), three publications addressed the general description of the gerbil as a model for stroke [2, 21, 22].

In 1972, Wexler [23] reported that indeed the gerbil could be used as a model for human stroke situations. The first detailed anatomical study was published by Kahn in 1972 [24], followed by few other papers that contributed to the establishment of the gerbil as a model for stroke studies [25–32]. As discussed by Molinari and Laurent [33], the diversity of ischemic brain models made it difficult to compare results from various laboratories, and the gerbil model provided a partial solution. A large number of papers appeared from 1975 to 1980, and only a few representative papers will be mentioned and cited. Several investigators measured cerebral blood flow and microcirculation as well as energy metabolism and neuro-transmitters [34–48]. During the years, the usage continued to grow at an exponential rate and was applied to new directions of stroke research. The involvement of the hippocampus in ischemic damage and the effects of various drugs on brain responses to ischemia are two examples. Most of the other publications were listed in the three articles of Robinson [18–20].

In 1978, Van Uitert and Levy [49] measured the regional blood flow in the conscious gerbil using $^{14}$C-butanol. They concluded that "the described method gives reproducible results and provides the first non-invasive regional brain blood flow measurements in the gerbil." Payan and Conard [50] studied the effects of age, sex, and gonads on the responses to carotid ligation in gerbils. They concluded that "the young animals prior to sexual maturity and the sexually mature female are more resistant than the sexually mature male. This difference is assumed to be due to the influence of the gonads on the blood flow of the brain." A few years ago, Tuo et al. [51] reviewed the rodent models (including the Mongolian gerbil) that could be used in vascular cognitive impairment studies. In 1980, Jarrott and Domer [52] suggested that the gerbil model of cerebral ischemia is suitable for drug evaluation. A few papers discussed the various aspects of using the gerbil as a model for various levels of ischemia [32, 53–57].

### 1.2.2 The Gerbil as a Model to Study Epilepsy

Three years after the publication of Thiessen "Spontaneous seizures in the Mongolian Gerbil" [5], Goldblatt et al. presented, in 1971, an abstract in a meeting regarding seizures in gerbils [58]. Also, in 1968, Robinson assumed that the animal is stressed when placed in a new environment [59]. In 1972, Kaplan and Miezejeski [60] published a full description of the seizure in gerbils and its parameters with regard to sex and age. They suggested that "Gerbils might provide an effective model for the study of epilepsy in humans." Loskota et al. in 1974 [61] discussed the gerbil as a model for the study of the seizure patterns and ontogenesis in gerbils. They cited their previous paper published in 1972 [61], discussing the seizure habituation and seizure patterns.

In 1975, Kaplan [62] concluded that "The results of the present study indicate that novelty in the form of a strange environment is indeed the more effective means of triggering seizures in gerbils." I suggest, however, that novelty (as well as any other stimulus that triggers seizures) is effective only in proportion to its stress-inducing properties. In another study published in 1975 by Loskota and Lomax, they measured the EEG during the epileptic activity in the Mongolian gerbil [63]. They concluded that "These observations of the correlation of EEG abnormalities with the motor manifestations of seizures in the gerbil strengthen the suggestion that this animal may be a valuable preparation for the study of the epilepsies." In a letter to the editor published in 1975 [64] by McCarty, he concluded that "It would appear from the evidence presented that Mg deficiency *is* not the causative agent of seizures in gerbils. Rather, the restriction of magnesium intake may have greatly reduced the threshold for seizure activity of genetically susceptible animals placed in the open field."

In 1979, Schonfeld and Glick [65] studied the connection between cerebrovascular abnormalities and seizure susceptibility in the Mongolian gerbil and summarized that "Although it now appears that both the stroke and seizure susceptibility of the Mongolian gerbil are associated with two separate anterior cerebral arteries, the relationship between these two phenomena has not been determined. In the gerbil, seizures are commonly observed as one of the neurological symptoms of stroke. There have been several reports in the clinical literature describing an association between vascular-induced epilepsy and stroke, a syndrome known as 'precursive epileptic attacks', although this relationship has not been well established. The gerbil might be an excellent animal model with which to study the possible association between the cerebrovasculature, seizures and strokes."

In 1981, Kaplan [66] tested the effect of fostering on seizure activity in the Mongolian gerbil and summarized that "The purpose of this experiment was to ascertain whether seizure propensity in Mongolian gerbils is affected by differences in behavior and/or nurturing ability of seizure-prone (SP) and nonseizure-prone (NSP) parents. Day-old SP and NSP pups were cross-fostered, while other SP pups were fostered by *SP* parents. Matched nonfostered littermates served as controls. The *SP* gerbils fostered to NSP parents exhibited delayed onset of seizure activity whereas SP gerbils fostered to SP parents did not. Neither fostered nor nonfostered NSP gerbils ever underwent seizures. This experiment establishes that some familial defect underlies seizure-proneness in gerbils, but that expression of this tendency is affected by parental behavior and/or nurturing ability."

In 1982, Donadio et al. published a paper entitled "Brain Vasculature and Induced Ischemia in Seizure—Prone and Non seizure—Prone Gerbils" [67]. They concluded that "it appears that the physiological mechanisms underlying the propensity of the seizure-prone gerbils to undergo cortical depolarization when subjected to stress, remains to be elucidated. The present evidence, however, indicates that the mechanism underlying stroke or stroke-induced seizures produced by unilateral ligation, are not significantly related to the naturally occurring seizure of the gerbil."

In 1983, Majkowsky and Kaplan [68] discussed the value of Mongolian gerbils in antiepileptic drug evaluation. They concluded as follows: "While gerbils may not be the subject of choice for long-term evaluation of drugs, their study still poses some interesting questions. One is the specific mechanisms underlying their seizures, and another is the relationship between habituation to stress and the decrease in seizure frequency, which may provide a good model for the non-pharmacological control of seizures."

In 1995, Bertorelli et al. [69] discussed the value of the Mongolian gerbil in experimental epilepsy. They concluded that "Altogether, data available indicate that the gerbil model is interesting in ethological studies and may be useful for investigating the mechanisms underlying 'spontaneous' seizures. In Pharmacology, however, the model has some weaknesses which limit its application in the study of new and established antiepileptic drugs."

Few more papers appeared during the early 2000s discussing the phenomenon of epilepsy and the gerbil [70, 71]. In 1979, Brown et al. [72] suggested a very interesting idea, namely, that epileptic activity could be developed by common carotid artery occlusion.

## *1.2.3 The Use of the Mongolian Gerbil in Medical Research*

In 2012, Batchelder et al. [73] published a chapter in a book summarizing the use of the Mongolian gerbil as a laboratory animal. This paper provided a list of more than 230 references. Another important paper, a very long paper (272 pages) providing a brain atlas of the gerbil CT/MRI-aided stereotaxic coordinates, was published in 2016 by Radtke-Schuller et al. [74].

### 1.2.3.1 The Growth and Aging of the Gerbil

Norris and Adams [75] described the growth of the gerbil during the life span. Figure 1.4 illustrates the growth pattern of the Mongolian gerbil from birth to 150 days of age. A clear difference between the sexes in favor of the male begins to emerge about 35 days of age. By 90 days when the growth rate tends to decline irrespective of sex, the males are on average about 10% heavier than the females.

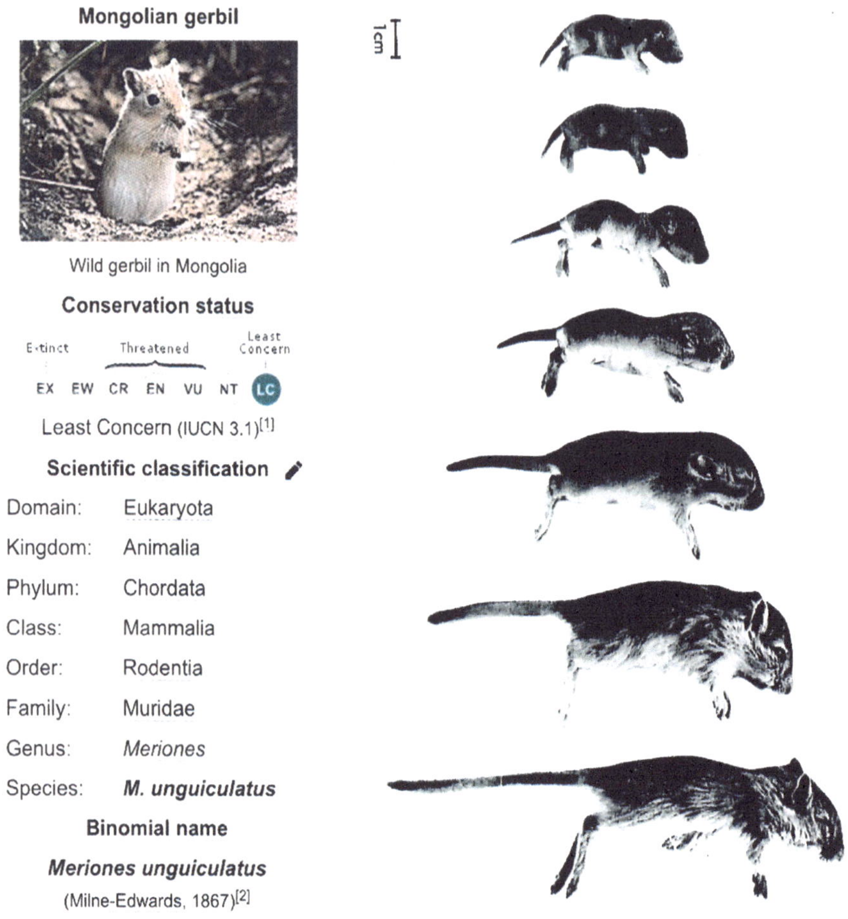


**Fig. 1.4** *Plate 1*. Growth and development of the Mongolian gerbil from birth to 32 days of age. The stages represented are Days 1, 2, 4, 8, 16, 24, and 32, when the body weights were 3.0, 3.2, 3.6, 6.7, 12.2, 18.2, and 21.8 g, respectively. The 16- and 32-day specimens were females; the remainder were males; all were derived from the same litter. (John Wiley & Sons, reprinted with permission [75])

The external development of the Mongolian gerbil from birth to weaning is depicted in Fig. 1.5. To obtain this series, a litter of eight young was selected and individuals were removed for photography at fixed intervals up to 32 days of age. Earlier reports on the development of the external features have been summarized by Marston and Chang [76]. In 1964, Kramer [77] provided the body and organ weights and linear measurements of the adult Mongolian gerbil. M. Cheal [78] described in many detail the uniqueness of the gerbil as a model for research in aging. His conclusions are as follows: "Evidence in support of the use of the Mongolian gerbil as a model for aging research has been presented. This evidence includes factors that

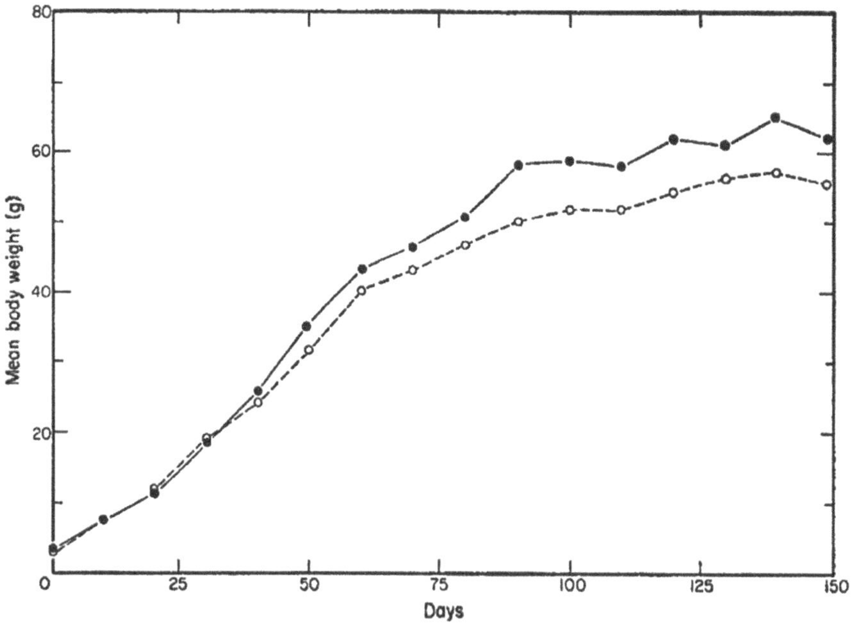


**Fig. 1.5** The growth of the Mongolian gerbil from birth to 150 days of age. ● Male, ○ Female. (John Wiley & Sons, reprinted with permission [75])

demonstrate the gerbils' suitability in fulfilling six of the practical and scientific considerations important in determining an animal model for research on aging [79]. Eight unique physiological attributes of gerbils, plus ten behavioral and nine biological processes that have been studied in gerbils and could offer valuable information for gerontology."

In 1992, Hara et al. [80] discussed the effects of aging on signal transmission and transduction in the gerbil brain. Later, in 1996 Bertrand et al. [81] tested the effect of cerebral ischemia on cholinergic metabolism, whereas Guedes et al. [82] studied the effect of aging on cortical spreading depression.

#### 1.2.3.2 Papers Related to Gerbil's Brain Vasculature

The detailed basic anatomy of the arterial blood supply to the gerbil brain will be presented in Chap. 8 of this book. Here, I will cite a few more papers that discuss various aspects of brain vasculature. Kuchinka et al. [83] described the arteries supplying the base of the brain in the gerbil. The main conclusion of this paper is that "The communicating arteries posterior between basilar artery and internal carotid arteries were not observed." In another paper published by Kuchinka in 2018 [84], he described the anatomy of the stapedial artery, which is the only source of blood supplied to the orbit.

In 1983, Yoshimine and Yanagihara [85] developed a model that created regional cerebral ischemia which is reproducible. They concluded that "Occlusion of the PCoA, MCA, or CCA also produces a model that is suitable for investigation of the postischemic period. A combination of these three experimental models is useful for investigation of regional vulnerability and for the study of regional metabolic differences in cerebral ischemia."

Hata et al. [86] studied the differential vulnerability of the hindbrain neurons during hindbrain ischemia in the gerbil. They concluded that "The present study suggests that selective vulnerability during hindbrain ischemia depends mainly on different metabolic characteristics inherent to various neurons in the hindbrain." Ito et al. [87] adopted the unilateral carotid occlusion as a model of brain ischemia. They found that in the bilateral carotid occlusion model, great variation was observed. Xiao Yun Du [88] defined two aims to their study "(1) study the anatomy of the brain vasculature in our gerbil population and identify heritable characteristics of CoWs and (2) use the results of this study to selectively breed an ischemia-sensitive gerbil group." They concluded as follows: "There are eight types of patterns of CoWs in Mongolian gerbils and they are inherited genetically (in particular, the ACoAs of mothers and their offspring are closely related). Post-UCO neurological symptoms correlate with CoW patterns: only gerbils with (1) both incomplete PCoAs and ACoAs or (2) smaller branches of ACoAs show moderate to severe neurological symptoms. With this knowledge we successfully established focal ischemia-prone strains of gerbils: the number of incomplete CoWs and the incidence of UCO ischemia in these animals were clearly and significantly higher than those in the F1 generation."

The *same* group published two more papers in 2017 discussing the ischemia-prone Mongolian gerbils [89, 90]. Very recently, in 2022 Abe et al. [91] reported that the commercially available strain lacking the posterior communicating artery may not be a strain appropriate for ischemia studies. They concluded that "the commercially available MON/Jms/GbsSlc gerbil can easily regain PcomA, and we established a new gerbil strain (MON/Jms/SlcMz) displaying non-PcomA."

#### 1.2.3.3 Effects of Ischemia of the Hippocampus and Brain Physiology

The effects of cerebral ischemia on neuronal death in the hippocampus were reported by several groups of scientists [92–97]. In 1975, Levy and Duffy [98] tested the effects of ischemia on the energy metabolism-high energy phosphates in the gerbil brain. Martin et al. [99] measured the disappearance of oxygen during stop flow in the gerbil brain. They found that after bilateral carotid artery occlusion "The fall in $PO_2$ was found not to be totally linear to zero in many cases, but instead consisted of a linear phase followed by a period of decreasing slope. We believe this phase of changing slope represents a diminishing oxygen consumption rate. The exact nature of this decrease is not known but perhaps is an inhibitory response to the accumulation of metabolites as a result of the circulation arrest."

The effects of carotid ligation on CBF and other parameters were tested in two publications [100, 101]. In 1993, Colbourne et al. [102] measured the temperature changes after forebrain ischemia.

#### 1.2.3.4 Effects of Cerebral Ischemia on Various Body Functions

In 1985, Chandler [103] used the gerbil as a model to study behavioral changes due to cerebral ischemia. Laidley et al. [104] studied the behavioral and histological variability in the Mongolian gerbil and concluded that "Thus, the gerbil model of forebrain ischemia, at least using Charles River animals, no longer produces consistent injury and behavioral alterations. Investigators are urged to consider adopting other models in future neuroprotection studies or ensure that their gerbil population lacks communicating arteries."

The effects of cerebral ischemia on the Cochlea and hearing loss were published in two reports [105, 106] from Japan. In 2013, Malek et al. [107] tested the effects of hyperbaric oxygen and hyperbaric air as a treatment after forebrain ischemia in the gerbil. They concluded that "the results presented here show for the first time that not only HBO but also HBA applied between 1 and 6 h after transient forebrain ischemia may prevent ischemia-induced neuronal damage in the CA1 region of the gerbil brain and that this protection is mostly likely due to a pressure-related inhibition of brain temperature increase which typically originates from ischemia. Although early applied NBO also resulted in neuronal protection, this treatment was less effective than delayed HBO or HBA therapy."

In 1955, Wahrman and Zahavi [108] described the cytological contributions to the Phylogeny and Classification of the rodent Genus *Gerbillus*. Later on in 1970, Weiss et al. described the karyotype of the Mongolian gerbil, *M. unguiculatus* [109].

The pathophysiological phenomenon of cortical spreading depression (CSD) was discussed in two publications. In 1987, Guedes et al. [110] tested the involvement of opioid mechanisms in the slow potential change and neuronal refractoriness during the CSD event. Later on in 2000, Hashimoto et al. [111] measured the dynamic changes of NADH fluorescence images and NADH content during CSD in the cerebral cortex of gerbils.

## References


1. Milne-Edwards A (1867) Observations. Sur Quelques Mammiferes Du Nord De La Chine. Ann Sci Nat (Zool) 7:375–377
2. Levine S, Payan H (1966) Effects of ischemia and other procedures on the brain and retina of the gerbil (*Meriones unguiculatus*). Exp Neurol 16:255–262
3. Schwentker V (1963) The gerbil, a new laboratory animal. Ill Vet 3:5
4. Walters GC, Pearl J, Rogers JV (1963) The gerbil as a subject in behavioral research. Psychol Rep 12:315–318

5. Thiessen DD, Friend HC, Lindzey G (1968) Androgen control of territorial marking in the Mongolian gerbil. Science 160:432–434
6. Gulotta EF (1971) Meriones unguiculatus. Mamm Species 3:1–5
7. Oldfield T (1908) The Duke of Bedford's zoological exploration in eastern Asia. In: IX. List of mammals from the Mongolian plateau. Proceedings of the Zoological Society, London, pp 104–110
8. Leiper RT (1933) Helminthology: a chapter in comparative medicine: (section of comparative medicine). Proc R Soc Med 27:127–134
9. Van Der Hoeden J (1954) The pathogenicity of leptospiras to field rodents in Israel; a new test animal for use in Leptospira research. J Infect Dis 95:213–219
10. Robinson PF (1959) Metabolism of the gerbil, *Meriones unguiculatus*. Science 130:502–503
11. Herrington LP (1940) The heat regulation of small laboratory animals at various environmental temperatures. Am J Physiol 129:123–139
12. Dawson WR (1955) The relation of oxygen consumption to temperature in desert rodents. J Mammal 36:543–553
13. Stockberger L (1977) The Momgolian gerbil. Iowa State Univ Vet 2:62–64
14. Fisher MF, Llewellyn GC (1978) The Mongolian gerbil: natural history, care, and maintenance. Am Biol Teach 40:557–560
15. Robinson DG (1975) Gerbil classification and nomenclature. In: The gerbil digest. Tumblebrook Farms, West Brookfield
16. Sontheimer H, Hoffmann KP (1987) Horizontal optokinetic reflex in light reared and dark reared Israelian gerbils (*Meriones tristrami*). Exp Brain Res 66:440–444
17. Breuer Z (1988) Detenination of brain vasculарure in the Mongolian gerbil and the biochemical correlated responses under ischemia. In: Life sciences. Bar-Ilan, Ramat Gan
18. Robinson DG (1980) The gerbil stroke model. Gerbil Digest 7:2
19. Robinson DG (1985) The gerbil stroke model. An update. Gerbil Digest 11:1
20. Robinson DG (1988) The gerbil stroke model – update II. Gerbil Digest 14:2
21. Levine S, Sohn D (1969) Cerebral ischemia in infant and adult gerbils. Relation to incomplete circle of Willis. Arch Pathol 87:315–317
22. Kahn KJ, Lawrence EP, Pranzarone GF (1971) The use of dextran-40 in modifying the course of experimental intraciion in the gerbil. Neurology 21:400
23. Wexler BC (1972) Pathophysiological responses to acute cerebral ischemia in the gerbil. Stroke 3:71–78
24. Kahn K (1972) The natural course of experimental cerebral infarction in the gerbil. Neurology 22:510–515
25. Hartmann JF, Becker RA (1973) Disaggregation of polyribosomes in intact gerbils following ischemia. An ultrastructural study. Stroke 4:964–968
26. Klatzo I, Ito U, Go KG, Spatz M (1974) Observations on experimental cerebral ischemia in Mongolian gerbils. In: Cervos-Navarro J (ed) Pathology of cerebral microcirculation. Walter de Gruyter, New York/Berlin, p 338
27. McGraw CP, Fleming DF Jr, Spruil JH (1974) Effect of methylprednisolone on experimental cerebral infarction in the Mongolian gerbil. Stroke 5:444–446
28. Spatz M, Go KG, Klatzo I (1974) The effect of ischemia on the brain uptake $^{14}C$ glucose analogues and $^{14}C$ glucose. In: Cervos-Navarre J (ed) Pathology of cerebral microcirculation. Walter de Gruyter, New York/Berlin, p 361
29. Wurtman RJ, Zervas NT (1974) Monoamine neurotransmitters and the pathophysiology of stroke and central nervous system trauma. J Neurosurg 40:34–36
30. Harrison MJ, Brownbill D, Lewis PD, Russell RW (1973) Cerebral edema following carotid artery ligation in the gerbil. Arch Neurol 28:389–391
31. Levy DE, Brierley JB (1974) Communications between vertebro-basilar and carotid arterial circulations in the gerbil. Exp Neurol 45:503–508
32. Berry K, Wisniewski HM, Svarzbein L, Baez S (1975) On the relationship of brain vasculature to production of neurological deficit and morphological changes following acute unilateral common carotid artery ligation in gerbils. J Neurol Sci 25:75–92

33. Molinari GF, Laurent JP (1976) A classification of experimental models of brain ischemia. Stroke 7:14
34. Lust WD, Mrsulja BB, Mrsulja BJ, Passonneau JV, Klatzo I (1975) Putative neurotransmitters and cyclic nucleotides in prolonged ischemia of the cerebral cortex. Brain Res 98:394–399
35. Levy DE, Brierley JB, Plum F (1975) Ischaemic brain damage in the gerbil in the absence of "no-reflow". J Neurol Neurosurg Psychiatry 38:1197–1205
36. Kelly RA, Halsey JH (1976) Comparison of local blood flow and oxygen availability at the same locus in the ischemic gerbil brain. Stroke 7:274
37. Mrsulja BB, Lust WD, Mrsulja JY, Passonneau IV, Klatzo I (1976) Post-ischemic changes in certain metabolites following prolonged ischemia in the gerbil cerebral cortex. J Neurochem 26:1099
38. Kobayashi M, Lust WD, Passonneau JV (1977) Concentrations of energy metabolites and cyclic nucleotides during and after bilateral ischemia in the gerbil cerebral cortex. J Neurochem 29:53–59
39. Levy DE, Duffy TE (1977) Cerebral energy metabolism during transient ischemia and recovery in the gerbil. J Neurochem 28:63–70
40. Ito U, Ohno K, Yamaguchi T, Tomita H, Inaba Y, Kashima M (1980) Transient appearance of "no-reflow" phenomenon in Mongolian gerbils. Stroke 11:517–521
41. Crockard A, Iannotti F, Hunstock AT, Smith RD, Harris RJ, Symon L (1980) Cerebral blood flow and edema following carotid occlusion in the gerbil. Stroke 11:494–498
42. Silberstein BR, Mayevsky A, Chance B (1978) In: Dutton PL, Leigh J, Scarpa A (eds) Metabolic responses of the gerbil brain cortex to anoxia, spreading depression, carotid occlusion and stroke. Academic Press, New York, pp 1477–1485
43. Ginsberg MD, Mela L, Wrobel-Kuhl K, Reivich M (1977) Mitochondrial metabolism following bilateral cerebral ischemia in the gerbil. Ann Neurol 1:519–527
44. Mayevsky A (1978) Pyridine nucleotide oxidation reduction state of the cerebral cortex in the awake gerbil. J Neurosci Res 3:369–374
45. Mayevsky A, Bar-Sagie D (1978) The interrelation between CBF, energy metabolism and ECoG in a new awake brain model. Adv Exp Med Biol 92:761–768
46. Haselgrove J, Barlow CH, Chance B (1980) The 3D distribution of metabolic states in the gerbil brain during the course of spreading depression. In: Passoneau JV, Hawkins RA, Lust WD, Welsh FA (eds) Cerebral metabolism and neural function. Williams & Wilkins, Baltimore/London, pp 72–76
47. Crowe W, Mayevsky A, Mela L (1979) The dynamics of $K^+$ leakage and recovery in cerebral ischemia. In: Leter A, Saba T, Mela L (eds) Advances in shock research. Alan R. Liss, New York, pp 221–232
48. Chance B, Barlow C, Nakase Y, Takeda H, Mayevsky A, Fischetti R, Graham N, Sorge J (1978) Heterogeneity of oxygen delivery in normoxic and hypoxic states: a fluorometer study. Am J Phys 235:H809–H820
49. Van Uitert RL, Levy DE (1978) Regional brain blood flow in the conscious gerbil. Stroke 9:67–72
50. Payan HM, Conrad JR (1977) Carotid ligation in gerbils. Influence of age, sex, and gonads. Stroke 8:194–196
51. Tuo QZ, Zou JJ, Lei P (2021) Rodent models of vascular cognitive impairment. J Mol Neurosci 71:1–12
52. Jarrott DM, Domer FR (1980) A gerbil model of cerebral ischemia suitable for drug evaluation. Stroke 11:203–209
53. Martinez NS, Machado JM, Perez-Saad H, Coro-Antich RM, Berlanga-Acosta JA, Salgueiro SR, Illera GG, Alba JS, Del Barco DG (2012) Global brain ischemia in Mongolian gerbils: assessing the level of anastomosis in the cerebral circle of Willis. Acta Neurobiol Exp (Wars) 72:377–384
54. Baskaya MK, Dogan A, Dempsey RJ (1999) Application of endovascular suture occlusion of middle cerebral artery in gerbils to obtain consistent infarction. Neurol Res 21:574–578

55. Matsumoto M, Hatakeyama T, Akai F, Brengman JM, Yanagihara T (1988) Prediction of stroke before and after unilateral occlusion of the common carotid artery in gerbils. Stroke 19:490–497
56. Delbarre G, Delbarre B, Barrau Y (1988) A suitable method to select gerbils with incomplete circle of Willis. Stroke 19:126–126
57. Kitagawa K, Matsumoto M, Handa N, Fukunaga R, Ueda H, Isaka Y, Kimura K, Kamada T (1989) Prediction of stroke-prone gerbils and their cerebral circulation. Brain Res 479:263–269
58. Goldblatt D, Konow A, Shotjlson I, Macmath T (1971) Effect of anticonvulsants on seizures in gerbils. Paper presented at the meeting of the American Academy of Neurology, New York
59. Robinson DCJ (1986) Animals suited to epileptic research. Sci News 93:16–18
60. Kaplan H, Miezejeski C (1972) Development of seizures in the mongolian gerbil (Meriones unguiculatus). J Comp Physiol Psychol 81:267–273
61. Loskota WJ, Lomax P, Rich ST (1974) The gerbil as a model for the study of the epilepsies. Seizure patterns and ontogenesis. Epilepsia 15:109–119
62. Kaplan H (1975) What triggers seizures in the gerbil, Meriones unguiculatus? Life Sci 17:693–698
63. Loskota WJ, Lomax P (1975) The Mongolian gerbil (*Meriones unguiculatus*) as a model for the study of the epilepsies: EEG records of seizures. Electroencephalogr Clin Neurophysiol 38:597–604
64. McCarty R (1975) Letter: magnesium deprivation and seizures in Mongolian gerbils. J Gen Psychol 92:3–4
65. Schonfeld AR, Glick SD (1979) Cerebrovascular abnormalities associated with seizure susceptibility in the Mongolian gerbil. Brain Res 173:147–151
66. Kaplan H (1981) Effects of fostering on seizure activity in the Mongolian gerbil. Dev Psychobiol 14:565–570
67. Donadio MF, Kozlowski PB, Kaplan H, Wisniewski HM, Majkowski J (1982) Brain vasculature and induced ischemia in seizure-prone and non-seizure-prone gerbils. Brain Res 234:263–273
68. Majkowski J, Kaplan H (1983) Value of Mongolian gerbils in antiepileptic drug evaluation. Epilepsia 24:609–615
69. Bertorelli R, Adami M, Ongini E (1995) The Mongolian gerbil in experimental epilepsy. Ital J Neurol Sci 16:101–106
70. Buckmaster PS (2006) Inherited epilepsy in Mongolian gerbils. In: Models of seizures and epilepsy. Elsevier/Academic Press, Amsterdam, pp 273–294
71. Fujisawa N, Maeda Y, Yamamoto Y, Sato NL, Niimura S (2003) Newly established low seizure susceptible and seizure-prone inbred strains of Mongolian gerbil. Exp Anim 52:169–172
72. Brown AW, Levy DE, Kublik M, Harrow J, Plum F, Brierley JB (1979) Selective chromatolysis of neurons in the gerbil brain: a possible consequence of "epileptic" activity produced by common carotid artery occlusion. Ann Neurol 5:127–138
73. Batchelder M, Keller LS, Ball Sauer M, West WL (2011) Gerbils. In: The laboratory rabbit, guinea pig, hamster, and other rodents. National Library of Medicine. Elsevier
74. Radtke-Schuller S, Schuller G, Angenstein F, Grosser OS, Goldschmidt J, Budinger E (2016) Brain atlas of the Mongolian gerbil (*Meriones unguiculatus*) in CT/MRI-aided stereotaxic coordinates. Brain Struct Funct 221(Suppl 1):1–272
75. Norris ML, Adams CE (2009) The growth of the Mongolian gerbil, *Meriones unguiculatus*, from birth to maturity. J Zool 166:277–282
76. Marston JH, Chang MC (1965) The breeding, management and reproductive physiology of the Mongolian gerbil (Meriones unguiculatus). Lab Anim Care 15:34–48
77. Kramer AW Jr (1964) Body and organ weights and linear measurements of the adult Mongolian gerbil. Anat Rec 150:343–348
78. Cheal ML (1986) The gerbil: a unique model for research on aging. Exp Aging Res 12:3–21

79. Institute of Laboratory Animal Resources (U.S.). Committee on Animal Models for Research on Aging (1981) Mammalian models for research on aging. National Academy Press, Washington, DC
80. Hara H, Onodera H, Kato H, Kogure K (1992) Effects of aging on signal transmission and transduction systems in the gerbil brain: morphological and autoradiographic study. Neuroscience 46:475–488
81. Bertrand N, Ishii H, Spatz M (1996) Cerebral ischemia in young and adult gerbils: effects on cholinergic metabolism. Neurochem Int 28:293–297
82. Guedes RC, Amorim LF, Teodosio NR (1996) Effect of aging on cortical spreading depression. Braz J Med Biol Res 29:1407–1412
83. Kuchinka J, Nowak E, Szczurkowski A, Kuder T (2008) Arteries supplying the base of the brain in the Mongolian gerbil (*Meriones unguiculatus*). Pol J Vet Sci 11:295–299
84. Kuchinka J (2018) The stapedial artery in the Mongolian gerbil (*Meriones unguiculatus*). Anat Rec (Hoboken) 301:1131–1137
85. Yoshimine T, Yanagihara T (1983) Regional cerebral ischemia by occlusion of the posterior communicating artery and the middle cerebral artery in gerbils. J Neurosurg 58:362–367
86. Hata R, Matsumoto M, Hatakeyama T, Ohtsuki T, Handa N, Niinobe M, Mikoshiba K, Sakaki S, Nishimura T, Yanagihara T et al (1993) Differential vulnerability in the hindbrain neurons and local cerebral blood flow during bilateral vertebral occlusion in gerbils. Neuroscience 56:423–439
87. Ito U, Hakamata Y, Yamaguchi T, Ohno K (2013) Cerebral ischemia model using mongolian gerbils-comparison between unilateral and bilateral carotid occlusion models. Acta Neurochir Suppl 118:17–21
88. Du XY, Zhu XD, Dong G, Lu J, Wang Y, Zeng L, Zhao TY, Ye HH, Li RS, Bai JY, Chen ZW (2011) Characteristics of circle of Willis variations in the mongolian gerbil and a newly established ischemia-prone gerbil group. ILAR J 52:E1–E7
89. Du X, Wang D, Li Y, Huo X, Li C, Lu J, Wang Y, Guo M, Chen Z (2018) Newly breeding an inbred strain of ischemia-prone Mongolian gerbils and its reproduction and genetic characteristics. Exp Anim 67:83–90
90. Du XY, Li CL, Guo M, Wang Y, Guo HG, Dai FW, Sa XY, Chen ZW (2018) Evaluation of an ischemic model in ischemia prone and general Mongolian gerbils by neurological symptom, injury, and sex difference. Anim Model Exp Med 1:62–67
91. Abe Y, Toyama K, Shinohara A, Nagura-Kato GA, Ikai Y, Koshimoto C, Spin JM, Hato N (2023) Message to researchers: the characteristic absence of a posterior communicating artery is easily lost in the gerbil. Anat Sci Int 98:426–433
92. Hara H, Kato H, Kogure K (1990) Protective effect of alpha-tocopherol on ischemic neuronal damage in the gerbil hippocampus. Brain Res 510:335–338
93. Kirino T (1982) Delayed neuronal death in the gerbil hippocampus following ischemia. Brain Res 239:57–69
94. Wang Y, Zhao P, Song Z, Du X, Huo X, Lu J, Liu X, Lv J, Li C, Guo M, Chen Z (2020) Generation of gene-knockout Mongolian gerbils via CRISPR/Cas9 system. Front Bioeng Biotechnol 8:780
95. Sommer C, Roth SU, Kuhn R, Kiessling M (2000) Metabotropic glutamate receptor subtypes are differentially expressed after transient cerebral ischemia without, during and after tolerance induction in the gerbil hippocampus. Brain Res 872:172–180
96. Goda W, Satoh K, Nakashima M, Hara A, Niwa M (2012) PBN fails to suppress in delayed neuronal death of hippocampal CA1 injury following transient forebrain ischemia in gerbils. Neurosci Lett 517:47–51
97. Nitatori T, Sato N, Waguri S, Karasawa Y, Araki H, Shibanai K, Kominami E, Uchiyama Y (1995) Delayed neuronal death in the CA1 pyramidal cell layer of the gerbil hippocampus following transient ischemia is apoptosis. J Neurosci 15:1001–1011
98. Levy DE, Duffy TE (1975) Effect of ischemia on energy metabolism in the gerbil cerebral cortex. J Neurochem 24:1287–1289

99. Martin RM, Halsey JH Jr, Reneau DD (1982) A critical evaluation of oxygen disappearance during stop flow in the gerbil brain. Neurol Res 4:21–34
100. Aldrich EF, Eisenberg HM, Saydjari C, Foulkes MA, Jane JA, Marshall LF, Young H, Marmarou A (1992) Predictors of mortality in severely head-injured patients with civilian gunshot wounds: a report from the NIH Traumatic Coma Data Bank. Surg Neurol 38:418–423
101. Alternative Therapy Evaluation Committee for the Insurance Corporation of British Columbia (2003) A review of the scientific evidence on the treatment of traumatic brain injuries and strokes with hyperbaric oxygen. Brain Inj 17:225–236
102. Colbourne F, Nurse SM, Corbett D (1993) Temperature changes associated with forebrain ischemia in the gerbil. Brain Res 602:264–267
103. Chandler MJ, DeLeo J, Carney JM (1985) An unanesthetized-gerbil model of cerebral ischemia-induced behavioral changes. J Pharmacol Methods 14:137–146
104. Laidley DT, Colbourne F, Corbett D (2005) Increased behavioral and histological variability arising from changes in cerebrovascular anatomy of the Mongolian gerbil. Curr Neurovasc Res 2:401–407
105. Watanabe F, Hakuba N, Gyo K (2009) Measurement of DPOAE after ischemia/reperfusion injury of the cochlea in gerbils. Neurosci Lett 467:135–138
106. Hakuba N, Koga K, Shudou M, Watanabe F, Mitani A, Gyo K (2000) Hearing loss and glutamate efflux in the perilymph following transient hindbrain ischemia in gerbils. J Comp Neurol 418:217–226
107. Malek M, Duszczyk M, Zyszkowski M, Ziembowicz A, Salinska E (2013) Hyperbaric oxygen and hyperbaric air treatment result in comparable neuronal death reduction and improved behavioral outcome after transient forebrain ischemia in the gerbil. Exp Brain Res 224:1–14
108. Wahrman J, Zahavi A (1955) Cytological contributions to the phylogeny and classification of the rodent genus Gerbillus. Nature 175:600
109. Weiss L, Mayeda K, Dully M (1970) The karyotype of the Mongolian gerbil, *Meriones unguiculatus*. Cytologia 35:102–106
110. Guedes RC, de Azeredo FA, Hicks TP, Clarke RJ, Tashiro T (1987) Opioid mechanisms involved in the slow potential change and neuronal refractoriness during cortical spreading depression. Exp Brain Res 69:113–118
111. Hashimoto M, Takeda Y, Sato T, Kawahara H, Nagano O, Hirakawa M (2000) Dynamic changes of NADH fluorescence images and NADH content during spreading depression in the cerebral cortex of gerbils. Brain Res 872:294–300

# Chapter 2
# Brain Energy Metabolism and Mitochondrial Function

**Abstract** This chapter introduces a new book that, for the first time, presents the history of the real-time monitoring of mitochondrial NADH redox state in the Mongolian gerbil brain. It also presents the history of the discovery of oxygen and mitochondria.

In addition, a general introduction to brain energy metabolism is presented. The following sections, after the general introduction, will be discussed:

1. Brain oxygen homeostasis
2. The discovery of oxygen
3. The discovery of mitochondria
4. The background to the monitoring of mitochondria in vivo

**Keywords** Oxygen metabolism · Mitochondrial redox state · Brain oxygen balance · Cerebral blood flow

## 2.1 Introduction

The level of oxygen in the atmosphere (21%) must be kept within a narrow range to avoid the development of pathological events in the human body, especially in the brain (hypoxia or hyperoxia).

Two historical discoveries made in the eighteenth and nineteenth centuries led to our current understanding of the processes involved in intracellular respiration, which provides the majority of ATP needed for cellular functions. The first event occurred about 250 years ago (1771–1775) when *Oxygen* was discovered by three scientists, namely, Carl Wilhelm Scheele, Joseph Priestley, and Antoine de Lavoisier [1]. The second event occurred around 110–120 years later, when Richards Altmann (1890) [2] and Carl Benda (1898) [3] discovered *Mitochondria*.



A. Mayevsky, *The Mongolian Gerbil Brain*,
https://doi.org/10.1007/978-3-031-69549-0_2

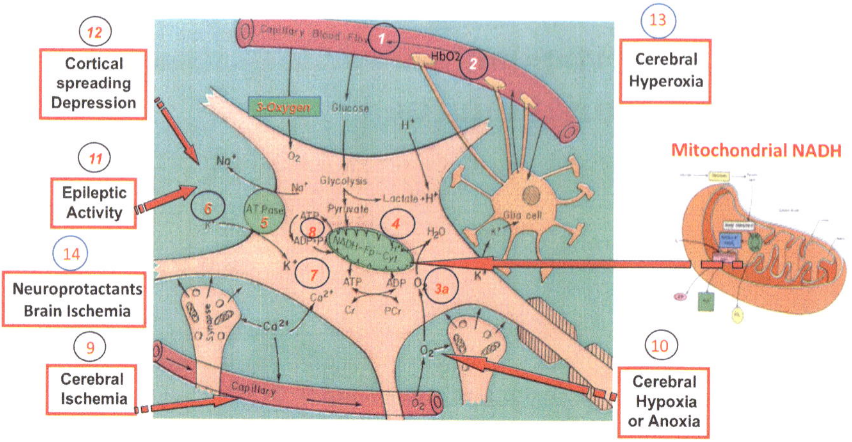


**Fig. 2.1** Schematic presentation of the "basic building stones" of a typical cerebral cortex tissue. The six pathological states (9–14) are related directly or indirectly to the studies made with the gerbil brain [4] (modified)

Figure 2.1 schematically illustrates the main elements that construct typical functional cerebral tissue. The cellular compartment contains typical neurons and glial cells as well as synapses. The small blood vessels—the capillaries—provide oxygen and other nutrients to the cells. As seen in Fig. 2.1, oxygen is supplied by the blood flow (1) containing the oxygenated hemoglobin (2) that releases oxygen to the extracellular (3) and intracellular (3a) compartments, finally reaching and being consumed by the mitochondria (4) and water is formed. The ATP produced is used for different physiological and biochemical activities. The majority of the ATP is consumed by ion pumps such as $Na^+K^+$ ATPase (5), which keeps the ionic homeostasis (6) and low intracellular calcium levels (7). The recharging of the ATP pool is done by oxidative phosphorylation in the mitochondria (8). There are four pathological events that develop in the brain of patients and experimental animals, as seen in Fig. 2.1. The major initial event is (9) cerebral ischemia (core area), caused by a decrease in cerebral blood flow (CBF) and oxygen supply to the brain tissue. The second event is cerebral hypoxia and anoxia, developed under a systemic decrease in blood oxygenation (10). The third event is epileptic activity (11), which may have developed under various pathologies. Cortical Spreading Depression (12) was initially developed by Leão in 1944 [5] by direct electrical stimulation and later was found in the penumbra zone around the ischemic core area or other pathological states. Another important event is cerebral hyperoxia (13), which may lead to oxygen toxicity. The very important treatment given after stroke is the use of neuroprotectant drugs against ischemia (14). Next, I will discuss the processes of oxygen uptake, transport by the blood, and utilization by the mitochondria in healthy and sick patients. These processes are the building stones of the basic concept named Tissue Oxygen Balance, which keeps the homeostasis of oxygen in the body. The concept of "Physiological Homeostasis," coined by Cannon in 1929, [6] was based on the general idea of Claude Bernard in 1878 that organisms keep the internal environment stable [7]. One of the critical elements in the body is the homeostasis of oxygen that should be kept stable. It means that tissue oxygen balance that represents the relationship between oxygen supply and oxygen demand will vary in a narrow range. Figure 2.2 demonstrates the homeostasis of oxygen in the brain. Oxygen demand (right side) is determined by the specific physiological processes that consume oxygen in the various physiological and biochemical processes in the brain. Actually, in all organs of the body, the main energy source is ATP synthesized in the mitochondria. The basic processes involved in oxygen supply and its regulation (left side) are very similar in all organs of the body, and the details will be discussed later. This concept of oxygen homeostasis in the brain was investigated at the end of the nineteenth century and was published by Roy and Sherrington in 1890 [10]. They studied the effect of various factors on the regulation of blood flow to the brain and reached the following conclusion:

> We conclude then, that the chemical products of cerebral metabolism contained in the lymph which bathes the walls of the arterioles of the brain can cause variations of the caliber of the cerebral vessels: that in this reaction the brain possesses an intrinsic mechanism by which its vascular supply can be varied locally in correspondence with local variations of functional activity.

The interpretation of their conclusion is that, in the brain, changes in oxygen consumption due to changes in functional activity are due to variations in the diameter of blood vessels. The next step in the development of the concept Oxygen Balance

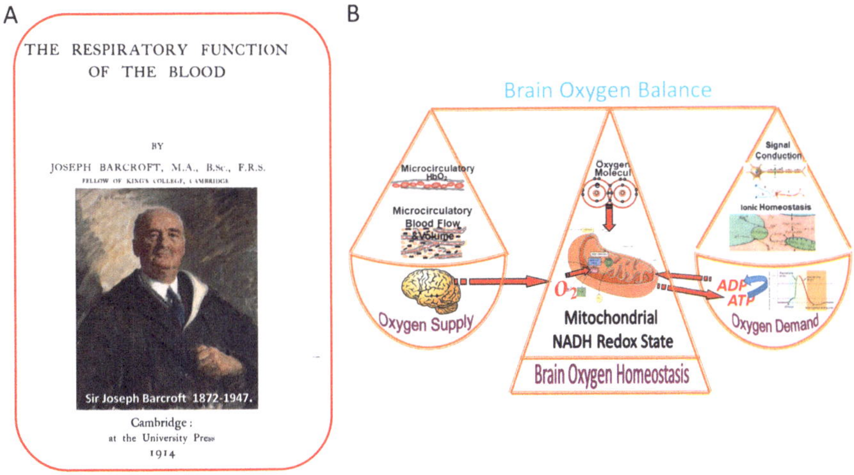


**Fig. 2.2** (**a**) The book published in 1914 by Barcroft regarding the call for oxygen by tissues at rest or during activation [8]. (**b**) Schematic presentation of brain oxygen balance evaluated by oxygen supply and demand. Oxygen supply could be evaluated by monitoring brain blood flow (CBF), brain blood volume (CBV), hemoglobin oxygenation ($HbO_2$), and brain tissue oxygen partial pressure. Oxygen demand includes ionic homeostasis, signal conduction, and biochemical synthesis. Mitochondrial NADH redox state served as an indicator for brain oxygen balance or brain oxygen homeostasis [9]

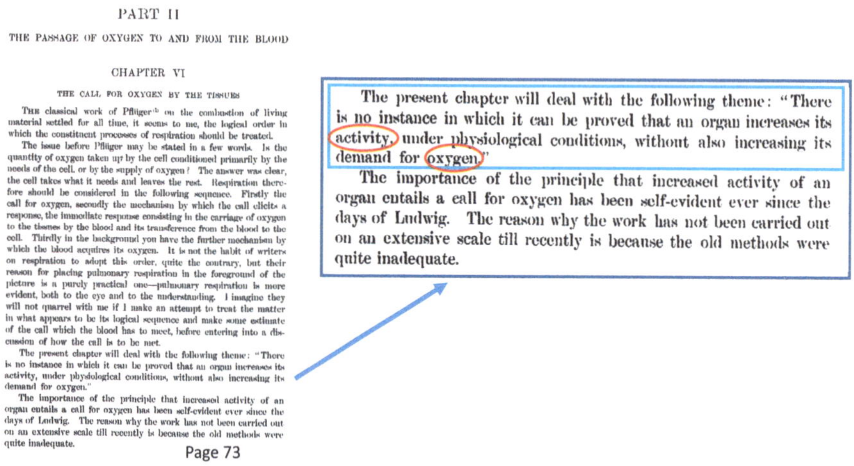

PART II

THE PASSAGE OF OXYGEN TO AND FROM THE BLOOD

CHAPTER VI

THE CALL FOR OXYGEN BY THE TISSUES

The classical work of Pflüger[1] on the combustion of living material settled for all time, it seems to me, the logical order in which the constituent processes of respiration should be treated.

The issue before Pflüger may be stated in a few words. Is the quantity of oxygen taken up by the cell conditioned primarily by the needs of the cell, or by the supply of oxygen? The answer was clear, the cell takes what it needs and leaves the rest. Respiration therefore should be considered in the following sequence. Firstly the call for oxygen, secondly the mechanism by which the call elicits a response, the immediate response consisting in the carriage of oxygen to the tissues by the blood and its transference from the blood to the cell. Thirdly in the background you have the further mechanism by which the blood acquires its oxygen. It is not the habit of writers on respiration to adopt this order, quite the contrary, but their reason for placing pulmonary respiration in the foreground of the picture is a purely practical one—pulmonary respiration is more evident, both to the eye and to the understanding. I imagine they will not quarrel with me if I make an attempt to treat the matter in what appears to be its logical sequence and make some estimate of the call which the blood has to meet, before entering into a discussion of how the call is to be met.

The present chapter will deal with the following theme: "There is no instance in which it can be proved that an organ increases its activity, under physiological conditions, without also increasing its demand for oxygen."

The importance of the principle that increased activity of an organ entails a call for oxygen has been self-evident ever since the days of Ludwig. The reason why the work has not been carried out on an extensive scale till recently is because the old methods were quite inadequate.

Page 73

The present chapter will deal with the following theme: "There is no instance in which it can be proved that an organ increases its activity, under physiological conditions, without also increasing its demand for oxygen."

The importance of the principle that increased activity of an organ entails a call for oxygen has been self-evident ever since the days of Ludwig. The reason why the work has not been carried out on an extensive scale till recently is because the old methods were quite inadequate.

**Fig. 2.3** Citations from the book published in 1914 by Barcroft regarding the call for oxygen by tissues at rest or during activation [8]

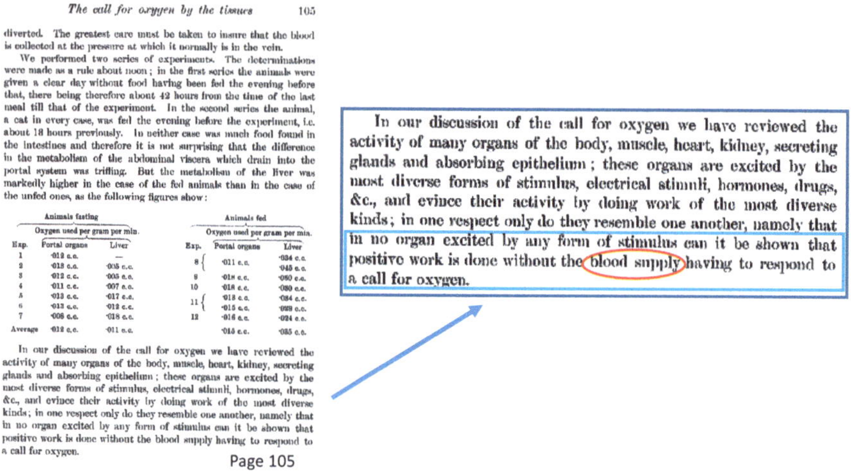

*The call for oxygen by the tissues* 105

diverted. The greatest care must be taken to insure that the blood is collected at the pressure at which it normally is in the vein.

We performed two series of experiments. The determinations were made as a rule about noon; in the first series the animals were given a clear day without food having been fed the evening before that, there being therefore about 42 hours from the time of the last meal till that of the experiment. In the second series the animal, a cat in every case, was fed the evening before the experiment, i.e. about 18 hours previously. In neither case was much food found in the intestines and therefore it is not surprising that the difference in the metabolism of the abdominal viscera which drain into the portal system was trifling. But the metabolism of the liver was markedly higher in the case of the fed animals than in the case of the unfed ones, as the following figures show:

Animals fasting

| Exp. | Oxygen used per gram per min. Portal organs | Liver |
|---|---|---|
| 1 | ·012 c.c. | — |
| 2 | ·013 c.c. | ·005 c.c. |
| 3 | ·012 c.c. | ·005 c.c. |
| 4 | ·011 c.c. | ·007 c.c. |
| 5 | ·013 c.c. | ·017 c.c. |
| 6 | ·013 c.c. | ·012 c.c. |
| 7 | ·008 c.c. | ·018 c.c. |
| Average | ·012 c.c. | ·011 c.c. |

Animals fed

| Exp. | Oxygen used per gram per min. Portal organs | Liver |
|---|---|---|
| 8 | ·011 c.c. | ·034 c.c. |
| | | ·045 c.c. |
| 9 | ·018 c.c. | ·050 c.c. |
| 10 | ·018 c.c. | ·030 c.c. |
| 11 | ·018 c.c. | ·034 c.c. |
| | ·015 c.c. | ·029 c.c. |
| 12 | ·016 c.c. | ·024 c.c. |
| | ·015 c.c. | ·035 c.c. |

In our discussion of the call for oxygen we have reviewed the activity of many organs of the body, muscle, heart, kidney, secreting glands and absorbing epithelium; these organs are excited by the most diverse forms of stimulus, electrical stimuli, hormones, drugs, &c., and evince their activity by doing work of the most diverse kinds; in one respect only do they resemble one another, namely that in no organ excited by any form of stimulus can it be shown that positive work is done without the blood supply having to respond to a call for oxygen.

Page 105

In our discussion of the call for oxygen we have reviewed the activity of many organs of the body, muscle, heart, kidney, secreting glands and absorbing epithelium; these organs are excited by the most diverse forms of stimulus, electrical stimuli, hormones, drugs, &c., and evince their activity by doing work of the most diverse kinds; in one respect only do they resemble one another, namely that in no organ excited by any form of stimulus can it be shown that positive work is done without the blood supply having to respond to a call for oxygen.

**Fig. 2.4** Citations from the book published in 1914 by Barcroft regarding the call for oxygen by tissues at rest or during activation [8]

Homeostasis was made by the leading English physiologist Joseph Barcroft who published the book "The Respiratory Function of Blood" in 1914 [8], as seen in Fig. 2.2. In Chapter VI of the book, seen in Fig. 2.3, he discussed the subject of "The call for Oxygen by the tissues". He proved that "There is no instance in which it can be proved that an organ increases its activity, under physiological conditions, without also increasing its demand for oxygen." Barcroft came to the second conclusion (seen in Fig. 2.4) that "in no organ excited by any form of stimulus can it be shown that positive work is done without the blood supply having to respond to a call for oxygen". In 2016, Longo [11] published a review article titled "Sir Joseph Barcroft: one Victorian physiologist's contributions to a half century of discovery". Longo discussed that "Barcroft's scientific career may be divided into four major phases of inquiry: that of oxygen ($O_2$) consumption in tissues; the definition of the oxyhemoglobin saturation [$HbO_2$] curve; establishing the limits of human tolerance to high altitude; and pioneering the exploration of various aspects of oxygenation of the fetus in utero".

It is important to remember that oxygen is a drug that may need to be optimized in the level given to patients thus avoiding under as well as overdosing [12, 13].

As seen on the right side of Fig. 2.2, the mitochondrial NADH redox state represents the tissue oxygen balance.

## 2.2 The Discovery of Oxygen

The final steps of oxygen discovery in the atmosphere happened about 250 years ago (1771–1775). There are two aspects that are related to the discovery of oxygen. Until about 25 years ago, the story of oxygen discovery was described and discussed in terms of oxygen as a chemical element and the most important discovery of chemistry and science [14–18]. In many published papers and Internet sites, the credit of oxygen discovery was given either to the American Joseph Priestley [19–21]. or to the German-Swedish Carl Wilhelm Scheele [19, 22, 23]. The available information indicates that Scheele was the first one to discover oxygen in 1771–1772, but he published it officially only a few years later in 1777. He claimed that in 1774 he sent a letter to the famous French chemist Antoine de Lavoisier, regarding his discovery, but Lavoisier claimed that he never received the letter. Priestley discovered oxygen, and in 1774, he notified his findings to Antoine de Lavoisier who studied oxygen chemistry for several years. Priestley actually published his discovery in 1775. In 2004, Wilkinson published a paper [1] describing the significant contribution of Lavoisier, Scheele, and Priestley, seen in Fig. 2.5, to the understanding of respiratory physiology in the eighteenth century. Here are his conclusions:

> There appear to be a great desire in history to define who was the first to do something. In reality it makes no great difference to today's work who was first to define oxygen in the eighteenth century. What can be seen from the above is the ability of these three great men to set new paradigms and work to prove their theories. Some were more accurate than others but it is the ability to take that lateral step that sets these characters apart from their peers. It could be considered that if Scheele first isolated oxygen then Priestley defined its properties and Lavoisier showed it was an element but in fact they all added a significant set of pieces to the physiological jigsaw puzzle of which oxygen was just a part.

In 1996, John Severinghaus published his first paper on this subject [25]. In his introduction, Severinghaus wrote that:

> AMERICANS are taught that the English Unitarian minister Joseph Priestley discovered oxygen in 1774. Scandinavians are taught that the Swedish apothecary Carl Wilhelm

| | Discovery year |
|---|---|
| 1. Ibn al Nafis (1213-1288). | 1250 |
| 2. Leonardo da Vinci (1452-1519). | 1515 |
| 3. Michael Servetus (1511-1553). | 1552 |
| 4. Michael Sendivogius (1566- 1636). | 1603 |
| 5. John Mayow (1641-1679). | 1668 |
| 6. Henry Cavendish (1731-1810). | 1783 |
| 7. Carl Wilhelm Scheele (1742-1786). | 1771 |
| 8. Antoine Laurent Lavoisier (1743-1794). | 1775 |
| 9. Joseph Priestley (1733-1804). | 1774 |

**Fig. 2.5** Nine scientists who were involved in the discovery of oxygen according to Severinghaus [24]

> Scheele generated oxygen in Uppsala in 1771–2, several years before Priestley. Scheele claimed that he wrote Lavoisier, describing the experiments in September, 1774. However, Lavoisier denied seeing or receiving his letter. Among Nordic historians, Lavoisier has never been forgiven for this rebuff. Some historians have doubted whether such a letter was ever sent or received, because Scheele first published in 1777. Priestley brought the news of his discovery to Lavoisier in October 1774. It was undoubtedly Priestley who triggered Lavoisier to realize over the next few years that air contained a new element, oxygen, which combined with combustible materials in fire. Lavoisier spent most of a decade in experiments to clarify the chemistry of oxygen and in publications overturning the conventional Phlogiston Theory, causing a revolution in chemistry.

One year later, Severinghaus published an extended paper on this topic [26] and concluded that Mary Ann Lavoisier received the original letter of Scheele and hid it until it was found 219 years later. Later, Severinghaus combined the discovery of Oxygen as an element in the eighteenth century to the description of the pulmonary circulation in the thirteenth century [24, 27].

## 2.3 The Discovery of Mitochondria

The discovery of the intracellular organelle, mitochondria, happened toward the end of the nineteenth century by Altmann in 1890 [2] and Benda in 1898 [3]. Its involvement in representing the tissue oxygen balance was formulated only close to 100 years later [28, 29]. At this stage, we were able to monitor microcirculatory blood flow, blood volume, hemoglobin oxygenation, and mitochondrial NADH redox state simultaneously.

As mentioned, Richards Altmann (1890) [2] and Carl Benda (1898) [3] discovered *Mitochondria*. It took another 60–70 years of intensive research to conclude that 90–95% of the consumed molecular oxygen by the organism is utilized by the cytochrome oxidase acting at the final step in the oxidative phosphorylation process performed by the respiratory chain located in the mitochondria [30–34]. The rest of the consumed $O_2$ is utilized by other oxidases such as the tyrosine hydroxylase converting the tyrosine to DOPA in the synthesis of Dopamine in the brain [35].

As we know today, the best-monitored parameter that represents mitochondrial function in vivo is the redox state of NADH. This section describes the historical background of its monitoring under in vivo conditions.

Chance et al. [36] reviewed the historical studies that made the connection between the mitochondrial activity and oxygen utilization as follows: "The accumulation of evidence since the pioneer work of Otto Warburg an 'Atmungsferment' [37] and David Keilin on cytochromes [38], as the keystones of cellular oxygen utilization led us to the study of the redox states of electron carriers in isolated mitochondria as a function of oxygen concentration and to develop techniques to measure the states of anoxia and normoxia in living tissues".

During the early 1950s, the study of mitochondrial activity in vitro and after 10 years in vivo was accelerated by the development of new optical monitoring devices. The UV-A light was applied to measure NADH absorption [39–42] and later on the measurement of NADH fluorescence was started [43, 44]. During those years, a large number of articles on the monitoring of NADH were published from the laboratory of Prof. B. Chance, and he became the leader of in vitro and in vivo monitoring of NADH fluorescence.

The detailed descriptions of the respiratory chain and oxidative phosphorylation in the mitochondria, published in 1955 by Chance and Williams, established our basic knowledge of the mitochondrial function [45–49]. In those five papers, Chance and Williams defined, for the first time, the metabolic states of isolated mitochondria in vitro, depending on the substrate, oxygen, and ADP levels. In addition, they correlated those metabolic states to the oxidation-reduction levels of the respiratory enzymes as seen in Fig. 2.6a. The physiological significance of those metabolic states was discussed in 1955–1956 by Chance and Williams [47, 51].

The "resting" state of mitochondria in vitro was defined as "State 4," where NADH is 99% in the reduced form, and ADP was the rate-limiting substance. If ADP is added to a suspension of mitochondria, ATP synthesis will be stimulated, oxygen consumption will increase, and the rate limit will be determined by the activity of the respiratory chain. During this State 3, or the "active state," the NADH redox state will decrease or become more oxidized (about 50%). When the "resting" mitochondria are deprived of oxygen, mitochondrial activity will stop and NADH will reach its maximum redox state (State 5).

A definitive description of the mitochondrial metabolic state has never been provided for in vivo conditions. Therefore, we described the in vivo mitochondrial conditions as recorded by NADH fluorescence in a representative tissue or organ, for example, the brain. While the range between minimal NADH (~0) and its maximal level was determined in vitro, it is almost impossible to determine this range in

A

TABLE 2 Metabolic States of Mitochondria and Associated Oxidation-Reduction Levels of Respiratory Enzymes

| Characteristics | | | | | | Steady state percentage reduction of components | | | | |
|---|---|---|---|---|---|---|---|---|---|---|
| State | $[O_2]$ | ADP level | Substrate level | Respiration rate | Rate-limiting substance | a | c | b | Flavo-protein[a] | $NADH_1$ |
| 1 | >0 | Low | Low | Slow | ADP | 0 | 7 | 17 | 21 | ~90 |
| 2 | >0 | High | ~0 | Slow | Substrate | 0 | 0 | 0 | 0 | 0 |
| 3 | >0 | High | High | Fast | Respiratory chain | 4 | 6 | 16 | 20 | 53 |
| 4 | >0 | Low | High | Slow | ADP | 0 | 14 | 35 | 40 | 99 |
| 5 | 0 | High | High | 0 | Oxygen | ~100 | ~100 | ~100 | ~100 | ~100 |

[a] These values are based on the amount of flavoprotein of the respiratory chain. Other fractions, reduced by anoxia, are present but disregarded here.
Source: After Chance and Williams [13].

B

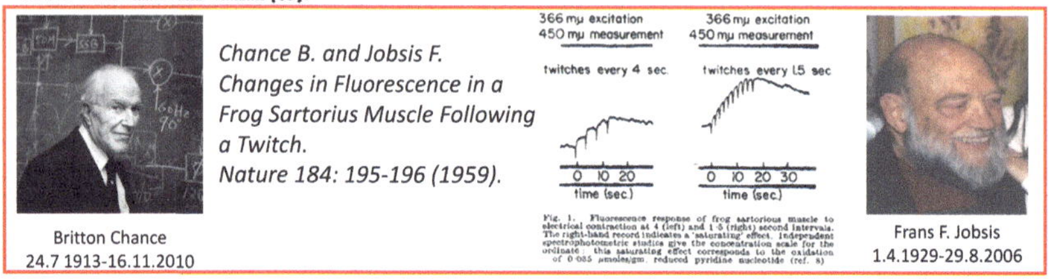


**Fig. 2.6** (**a**) Mitochondrial metabolic state, defined in vitro, by Chance and Williams, and opened up a new era in measurements of respiratory chain enzyme's redox state in vitro as well as in vivo [47]. (**b**) The two scientists significantly affected our knowledge on the mitochondrial function and NADH monitoring in vitro and in vivo [50]

the intact brain or other organs in vivo. For example, State 2, with a substrate-free medium, cannot be achieved in vivo because the tissue would die. On the other hand, the maximal level of NADH (State 5) can be monitored in vivo under complete deprivation of oxygen by anoxia or complete ischemia.

An intensive use of the real-time in vivo NADH monitoring approach started in 1962. The "classical" paper on in vivo monitoring of NADH was published in 1962 by Chance et al. [52]. They simultaneously monitored the brain and kidney of anesthetized rats using two microfluorometers. Chance and his collaborators elaborated on this type of in vivo monitoring and applied it to other rat organs [53–55].

Figure 2.6b presents the pictures of the two cited scientists who significantly contributed to the development of the theoretical and experimental technology for monitoring of mitochondrial NADH function in vitro and in vivo. In one of their initial publications, they showed the responses of intact skeletal muscle NADH to twitches measured in real time [56].

## 2.4 Monitoring of Brain Mitochondrial Function In Vivo

The discovery of the optical properties of reduced Nicotineamide Adenine Dinucleotide—NADH (earlier names: DPNH—diphosphopyridine nucleotide, or PN—pyridine nucleotide) has led to very intensive research since the early 1950s. The reduced form of this molecule, NADH (Fig. 2.7a), absorbs light at 320–380 nm (Fig. 2.7b) and emits fluorescent light in the range of 420–480 nm (Fig. 2.7c).

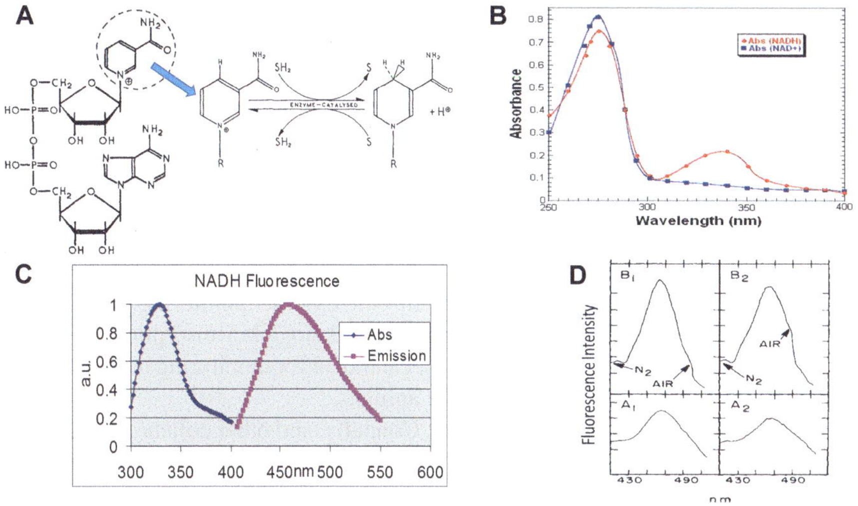


**Fig. 2.7** (**a**) Structures of NAD⁺. The nicotineamide group (broken ring) is the functional part of both molecules, that is, the portion of molecules where oxidation and reduction take place (see arrow). (**b**) Absorption spectra of $NAD^+$ and NADH in solution. (**c**) Absorption and emission spectra of NADH illuminated by UV light [57, 58]. (**d**) The emission spectra of rat cerebral cortex under aerobic conditions (lower trace—$A_1$ and $A_2$) and under anoxic conditions (upper trace—$B_1$ and $B_2$) [4, 58]

As the oxidized form $NAD^+$ (Fig. 2.7b, blue line) does not absorb light in this range, it is possible to evaluate the redox state of the mitochondria by monitoring the UV absorbance or blue fluorescence of NADH. Figure 2.7d presents the fluorescence spectra of the NADH measured from the exposed cortex of the rat brain. Under oxygen depletion (anoxia), there is a significant increase in fluorescence intensity, as seen in parts $B_1$ and $B_2$ of Fig. 2.7d.

The possibility of monitoring mitochondrial function in vivo, in both experimental animals and patients, is of great importance and could significantly contribute to the understanding of various pathological processes. Most of the information on mitochondrial function has been accumulated from in vitro studies, with a relatively small portion of published papers dealing with the monitoring of mitochondrial function in vivo and in real-time. During the past 50 years, we have published over 150 papers in this very significant area, including the largest number of studies using NADH redox state monitoring in patients as well.

The need for an intracellular $pO_2$ indicator, as a physiological and biochemical parameter of living tissue, has emerged 60 years ago [59]. Mitochondria are the intracellular organelles that consume most of the oxygen. Therefore, the redox state of electron carriers in isolated mitochondria in vitro as well as in vivo as a function of oxygen concentration has been extensively studied. Chance et al. concluded that "For a system at equilibrium, NADH is at the extreme low potential end of the chain, and this may be the oxygen indicator of choice in mitochondria and tissue as well" [36] Lubbers, in 1995, concluded that "the most important intrinsic luminescence indicator is NADH, an enzyme of which the reaction is connected with tissue respiration and energy metabolism" [60].

One of the main questions is why monitoring of NADH fluorescence is the leading technology in monitoring mitochondrial function in vivo and in real time. It is clear that other electron carriers in the respiratory chain may have an advantage since they are localized only in the mitochondria, whereas NADH appears also in the cytosol in addition to mitochondria. We have tried many options to monitor flavoproteins (FP) fluorescence together with NADH. It was found that FP measurement is dramatically affected and sensitive to hemodynamic artifacts, particularly changes in hemoglobin oxygenation. Cytochrome oxidase monitoring suffers from the same problems due to its overlapping with the hemoglobin absorption spectrum. So, in the near future, the NADH is the best candidate for monitoring mitochondrial function in both patients and experimental animals.

The pioneer work of Chance, Williams, Connelly, and other collaborators, in the early 1950s, opened up new possibilities for studying the mitochondrial function in vitro, and later on, in vivo monitoring became a reality [47, 51, 61–63]. The shift from studying mitochondrial NADH redox state in isolated mitochondria to higher cellular and tissue organization levels started in the late 1950s [56]. The main breakthrough occurred in 1962, a year that at least six articles described the in vivo monitoring of various organs in the anesthetized rat [52–55, 59, 64]. The crucial issue is whether we can extrapolate results obtained from in vitro monitoring of mitochondrial NADH to in vivo conditions Mitochondrial function and its metabolic states under in vivo conditions are completely different from the definition made for isolated mitochondria by Chance and Williams in 1955 and 1956 [45, 51]. The main differences between the two situations are as follows:

1. The isolated mitochondria are not interconnected to other intracellular organelles and components.
2. The amount of oxygen available in the mitochondrial medium is not regulated by changes in blood flow or hemoglobin oxygenation under in vitro conditions.
3. One of the largest oxygen or energy consumers, that is, various ions pumps, is missing from the isolated mitochondria preparation.
4. When ADP is added to State 4 mitochondria, ATP will be synthesized but will not be consumed in parallel as occurred under in vivo conditions.
5. The isolated mitochondria preparation is not exposed to systemic changes in hemodynamic and other physiological parameters occurring in the organism.

Can we compare and define the resting state (4) and active state (3) under in vivo situations to the in vitro definitions? As seen in Fig. 2.8, the maximal NADH level is achieved under complete $O_2$ deprivation, which can be induced under both in vitro ($A_1$) and in vivo ($B_1$) conditions. These results were obtained using the devices ($A_2$ and $B_2$) developed in 1951 by Theorell and Nyggard in 1954 [66] and Mayevsky and Chance in 1973 [67]. This signifies that this definitive point can be used to determine State 5 in vivo as well. As can be seen in Fig. 2.8, the steady-state NADH level at State 4 is 99%, namely that most of the NAD is in the reduced form NADH. Another possible explanation is that the NADH almost reaches its maximum allowed in the

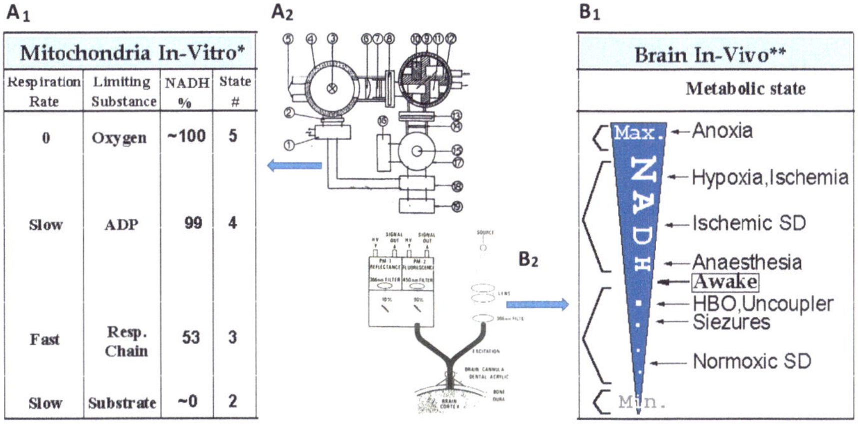


**Fig. 2.8** Comparison between mitochondrial metabolic state (in vitro), defined by Chance and Williams in 1955 ($A_1$ and $A_2$) and responses of the in vivo brain to changes in $O_2$ supply and demand induced by brain activation described by Mayevsky ($B_1$ and $B_2$) [65]

system, but this option of explanation is not in agreement with the technology used. When the mitochondria are shifted to State 3, by adding ADP, the NADH becomes more oxidized and the NADH level is decreased to 53% of the reduced form. Moving from State 4 to State 5 and eliminating the oxygen from the media will have a very small effect on the NADH level, namely, changing from 99% to 100%. It is important to note that other members of the respiratory chain behave somewhat similar to NADH. When the mitochondria are shifted to State 2, the NADH level is 0 since NAD is the dominant state in the absence of a substrate.

The problem is to determine the "resting" metabolic state of a tissue under an in vivo situation. If we adopt the in vitro value of a resting state (State 4), this would signify that the increase in NADH during State 5, induced by anoxia (0% $O_2$), would be only 1%, namely, from 99% to 100%. According to all in vivo studies, this is not the case, and during anoxia, the increase in NADH is much larger than the decrease under the State 4 to 3 transition.

When we started monitoring the NADH redox state using fiber optic–based fluorometry/reflectometry in 1972, we used the brain in vivo exposed to various conditions in real time [67]. We found that the NADH level could either increase or decrease from the baseline level calibrated in the normoxic animal [4]. The initial results were plotted against the perturbation used, as shown in the right side of Fig. 2.8. The relative changes in brain NADH under various perturbations as compared to the Chance and Williams definitions (Fig. 2.8 $A_1$) are presented. Although the maximal level of NADH could be achieved in vivo by exposing the animal to anoxia (100% nitrogen), the minimal level that could be monitored in vivo is very hard to achieve. We tried to build the scale between maximal and minimal levels of NADH by exposing the brain to many types of perturbations. In the table, the horizontal lines represent the relative change in NADH recorded during the perturbation. The values are not accurate but rather trying to map the scale of NADH changes. The question is how to locate the in vivo scale relative to the in vitro definition shown on the left side. According to the changes recorded, it seems that the so-called resting brain is not at State 4 (99% NADH) but rather around States 3.5 and 3.6. As seen, NADH will increase by various factors such as anesthesia, hypoxia, and partial ischemia. Under complete ischemia or anoxia, the NADH will reach its maximal level—State 5. Exposing the animal to hyperoxia or uncoupler led to a decrease in NADH toward State 3. Activation of the brain by inducing seizures or cortical spreading depression led to a very large decrease in NADH. The main issue in the scaling or the responses is how to "push" the brain in vivo to minimal NADH levels. As of now, it seems that this question remains open and unsolved.

Activation of the brain (CSD) under conditions of restricted amounts of oxygen, such as partial ischemia, will lead to an increase in NADH (reduction) instead of a decrease (oxidation) as found in the normoxic brain. Those results found in various experimental conditions led to a dilemma. According to Chance and Williams, ATP production related to the State 4 to State 3 transition is always correlated with a decrease or oxidation of NADH from 99% to 53%. The question is how ATP turnover could increase under NADH increase (reduction) and not decrease (oxidation). We want to make sure that the experimental results do not represent a technical artifact related to the monitoring itself. We assumed that using appropriate modeling methods, the dilemma could be clarified or resolved [68].

During the years 1972–2014, we were able to monitor the brain's NADH redox state in many pathophysiological conditions in experimental animals as well as in neurosurgical patients in the OR and ICU [57, 69]. As seen in Fig. 2.9, the experimental conditions included changes in oxygen supply (hypoxia, ischemia, hemorrhage, and hyperoxia). Brain activation effects were tested in models of epileptic activity as well as CSD. The effects of various pharmacological agents were tested as well [70].

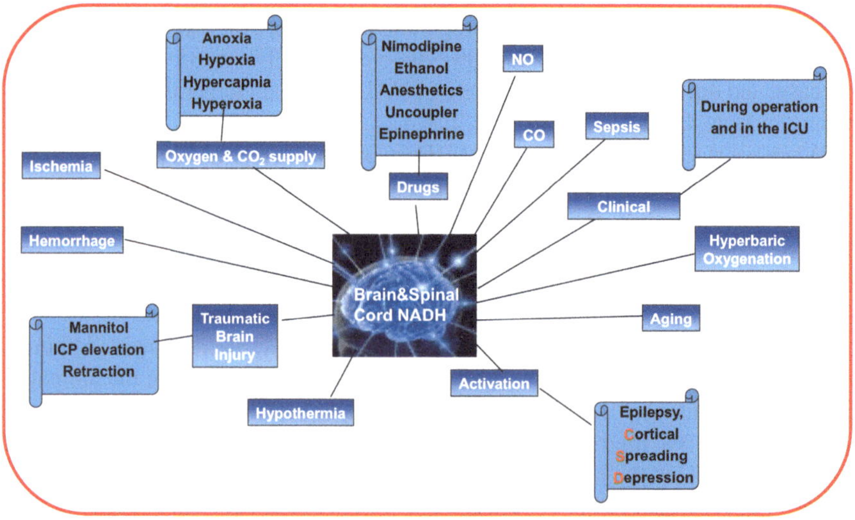


**Fig. 2.9** Presentation of NADH redox state monitored in various organs and perturbations used by Mayevsky et al. in experimental animals as well as in patients [65] (modified)

# References


1. Wilkinson DJ (2004) Resuscitation great. The contributions of Lavoisier, Scheele and Priestley to the early understanding of respiratory physiology in the eighteenth century. Resuscitation 61:249–255
2. Altmann R (1890) Die Elementarorganismen Und Ihre Beziehungen Zu Den Zellen. Veit & Comp, Leipzig
3. Benda C (1898) Ueber die Spermatogenese der Vertebraten und höherer Evertebraten, II. Theil: Die Histiogenese der Spermien. Arch Anat Physiol 73:393–398
4. Mayevsky A (1984) Brain NADH redox state monitored *in vivo* by fiber optic surface fluorometry. Brain Res Rev 7:49–68
5. Leão AAP (1944) Spreading depression of activity in cerebral cortex. J Neurophysiol 7:359–390
6. Cannon WB (1929) Organization for physiological homeostasis. Physiol Rev IX:399–430
7. Bernard C (1878) Leçons sur les phénomènes de la vie communs aux animaux et aux végétaux. Fisher – University of Toronto
8. Barcroft J (1914) The respiratory function of blood. Cambridge University Press, Cambridge
9. Mayevsky A, Meirovithz E, Livnat A, Tolmasov M, Kutai-Asis H, Mandelbaum M (2020) Urethral tissue metabolic score (TMS) as a surrogate marker of brain oxygen balance in stroke, ARDS and critical care. EC Emergency Medicine and Critical Care. SI.01, 05-40
10. Roy CS, Sherrington CS (1890) On the regulation of the blood-supply of the brain. J Physiol 11:85–158.17
11. Longo LD (2016) Sir Joseph Barcroft: one victorian physiologist's contributions to a half century of discovery. J Physiol 594:1113–1125
12. Branson RD, Robinson BR (2011) Oxygen: when is more the enemy of good? Intensive Care Med 37:1–3
13. Martin DS, Grocott MP (2013) III. Oxygen therapy in anaesthesia: the yin and yang of $O_2$. Br J Anaesth 111:867–871
14. West JB (2014) Joseph Priestley, oxygen, and the enlightenment. Am J Physiol Lung Cell Mol Physiol 306:L111–L119
15. Cassebaum H, Schufle JA (1975) Scheele's priority for the discovery of oxygen. J Chem Educ 52:442
16. Smeaton WA (1992) Carl Wilhelm Scheele (1742-1786): provincial Swedish pharmacist and world-famous chemist. Endeavour 16:128–131
17. Marshall JL, Marshall VR (2005) Rediscovery of the elements: Carl Wilhelm Scheele, University of North Texas Libraries, UNT Digital Library. Spring 2005. Crediting UNT College of Arts and Sciences, Indianapolis
18. Palmer W (2006) Carl Wilhelm Scheele (1742–1786): the life of a great chemist. In: IPS-USA-2006 Internet conference on advances in the Internet, Processing, Systems and Interdisciplinary Research, Palo Alto, CA, USA
19. Partington JR (1962) The discovery of oxygen. J Chem Educ 39:123–125
20. Miller FA (1987) Joseph Priestley, preeminent amateur chemist. J Chem Educ 64:745
21. Williams KR (2003) The discovery of oxygen and other Priestley matters. J Chem Educ 80:1129–1131
22. Smeaton WA (1963) New light on Lavoisier: the research of the last ten years. Hist Sci 2:51–69
23. Taylor G (2019) The chemical works of Carl Wilhelm Scheele. Ambix 66:90–92
24. Severinghaus JW (2016) Eight sages over five centuries share oxygen's discovery. Adv Physiol Educ 40:370–376
25. Severinghaus JW (2002) Priestley, the furious free thinker of the enlightenment, and Scheele, the taciturn apothecary of Uppsala. Acta Anaesthesiol Scand 46:2–9
26. Severinghaus JW (2003) Fire-air and dephlogistication. Revisionisms of oxygen's discovery. Adv Exp Med Biol 543:7–19

27. Severinghaus JW (2016) The most important discovery of science. In: Elwell CE, Leung TS, Harrison DK (eds) Oxygen transport to tissue XXXVII. Advances in experimental medicine and biology. Springer, New York, pp 1–16
28. Mayevsky A (1992) Cerebral blood flow and brain mitochondrial redox state responses to various perturbations in gerbils. Adv Exp Med Biol 317:707–716
29. Rampil IJ, Litt L, Mayevsky A (1992) Correlated, simultaneous, multiple-wavelength optical monitoring in vivo of localized cerebrocortical NADH and brain microvessel hemoglobin oxygen saturation. J Clin Monit 8:216–225
30. Wikstrom M, Krab K, Sharma V (2018) Oxygen activation and energy conservation by cytochrome c oxidase. Chem Rev 118:2469–2490
31. Pereira MM, Santana M, Teixeira M (2001) A novel scenario for the evolution of haem-copper oxygen reductases. Biochim Biophys Acta 1505:185–208
32. Waltemath CL (1970) Oxygen, uptake, transport, and tissue utilization. Anesth Analg 49:184–203
33. Jobsis FF (1964) Basic processes in cellular respiration. In: Fenn WO, Rabn H (eds) Handbook of physiology – respiration. American Physiological Society, Washington, DC, pp 63–124
34. Mason HS (1957) Mechanisms of oxygen metabolism. Adv Enzymol Relat Subj Biochem 19:79–233
35. Mayevsky A, Sjoquist B, Fri CG, Samuel D, Sedvall G (1973) Incorporation of $^{18}O$ into homovanillic acid of rat brain during exposure to oxygen 18 containing atmospheres. Biochem Biophys Res Commun 51:746–755
36. Chance B, Oshino N, Sugano T, Mayevsky A (1973) Basic principles of tissue oxygen determination from mitochondrial signals. In: Oxygen transport to tissue, Advances in experimental medicine and biology, vol 37A. Springer, Boston, pp 277–292
37. Warburg O (1949) Heavy metal prosthetic groups and enzyme action (trans: Lawson TA). Clarendon Press, Oxford
38. Keilin D (1925) On cytochrome, a respiratory pigment, common to animals, yeast, and higher plants. Proc R Soc Lond B Biol Sci 98:312–339
39. Theorell H, Bonnichsen R (1951) Studies on liver alcohol dehydrogenase I. Equilibria and initial reaction velocities. Acta Chem Scand 5:1105–1126
40. Theorell H, Chance B (1951) Studies on liver alcohol dehydrogenase II. The kinetics of the compound of horse liver alcohol dehydrogenase and reduced diphosphopyridine nucleotide. Acta Chem Scand 5:1127–1144
41. Chance B (1952) Spectra and reaction kinetics of respiratory pigments of homogenized and intact cells. Nature 169:215–221
42. Chance B (1953) Dynamics of respiratory pigments of ascites tumor cells. Trans NY Acad Sci 16:74–75
43. Duysens LNM, Amesz J (1957) Fluorescence spectrophotometry of reduced phosphopyridine nucleotide in intact cells in the near-ultraviolet and visible region. Biochim Biophys Acta 24:19–26
44. Chance B, Baltscheffsky H (1958) Respiratory enzymes in oxidative phosphorylation (VII – Binding of intramitochondrial reduced pyridine nucleotide). J Biol Chem 233:736–739
45. Chance B, Williams GR (1955) Respiratory enzymes in oxidative phosphorylation (IV – The respiratory chain). J Biol Chem 217:429–438
46. Chance B, Williams GR, Holmes WF, Higgins J (1955) Respiratory enzymes in oxidative phosphorylation (V – A mechanism for oxidative phosphorylation). J Biol Chem 217:439–451
47. Chance B, Williams GR (1955) Respiratory enzymes in oxidative phosphorylation (III – The steady state). J Biol Chem 217:409–427
48. Chance B, Williams GR (1955) Respiratory enzymes in oxidative phosphorylation (I – Kinetics of oxygen utilization). J Biol Chem 217:383–393
49. Chance B, Williams GR (1955) Respiratory enzymes in oxidative phosphorylation (II – Difference spectra). J Biol Chem 217:395–407
50. Mayevsky A (2015) Mitochondrial function *in vivo* evaluated by NADH fluorescence. Springer, Cham

51. Chance B, Williams GR (1956) The respiratory chain and oxidative phosphorylation. In: Nord FF (ed) Advances in enzymology. Interscience Publisher, Inc., New York, pp 65–134
52. Chance B, Cohen P, Jobsis F, Schoener B (1962) Intracellular oxidation-reduction states *in vivo*. Science 137:499–508
53. Chance B, Schoener B, Fergusson JJ (1962) *In vivo* induced oxidation by adrenocorticotrophic hormone of reduced pyridine nucleotide in the adrenal cortex of hypophysectomized rats. Nature 195:776–778
54. Chance B, Schoener B (1962) Correlation of oxidation-reduction changes of intracellular reduced pyridine nucleotide and changes in electro-encephalogram of the rat in anoxia. Nature 195:956–958
55. Chance B, Legallias V, Schoener B (1962) Metabolically linked changes in fluorescence emission spectra of cortex of rat brain, kidney and adrenal gland. Nature 195:1073–1075
56. Chance B, Jobsis F (1959) Changes in fluorescence in a frog sartorius muscle following a twitch. Nature 184:195–196
57. Mayevsky A, Barbiro-Michaely E (2013) Shedding light on mitochondrial function by real time monitoring of NADH fluorescence: I. Basic methodology and animal studies. J Clin Monit Comput 27:1–34
58. Mayevsky A, Meirovithz E (2020) Oxygen homeostasis and mitochondrial function in COVID-19 and ARDS patients: the ultimate vital signs. J Biosci Biomed Eng 1:1–24
59. Chance B, Cohen P, Jobsis F, Schoener B (1962) Localized fluorometry of oxidation-reduction states of intracellular pyridine nucleotide in brain and kidney cortex of the anesthetized rat. Science 136:325
60. Lubbers DW (1995) Optical sensors for clinical monitoring. Acta Anaesthesiol Scand Suppl 39:37–54
61. Chance B, Williams GR (1954) Steady-state of reduced pyridine nucleotides in phosphorylating rat liver mitochondria. Am Soc Biol Chem 13:Abstract 633
62. Chance B, Williams GR (1955) A method for the localization of sites for oxidative phosphorylation. Nature 176:250–254
63. Connelly CM, Chance B (1954) Kinetics of reduced pyridine nucleotides in stimulated frog muscle and nerve. Am Physiol Soc 13:29
64. Chance B, Schoener B (1963) Control of oxidation-reduction state of NADH in the liver of anesthetized rats. In: Symp Regul Enzyme Act Synth Norm Neoplast Tissues Proc, pp 169–185
65. Mayevsky A, Kutai-Asis H, Tolmasov M (2020) Mitochondrial function and brain metabolic score (BMS) in ischemic stroke: evaluation of "neuroprotectants" safety and efficacy. Mitochondrion 50:170–194
66. Theorell H, Nygaard AP (1954) Kinetics and equilibria in flavoprotein systems I. A fluorescence recorder and its application to a study of the dissociation of the old yellow enzyme and its resynthesis from riboflavin phosphate and protein. Acta Chem Scand 8:877–888
67. Mayevsky A, Chance B (1973) A new long-term method for the measurement of NADH fluorescence in intact rat brain with implanted cannula. In: Bicher HI, Bruley DF (eds) Oxygen transport to tissue. Plenum Press, New York, pp 239–244
68. Vatov L, Kizner Z, Ruppin E, Meilin S, Manor T, Mayevsky A (2006) Modeling brain energy metabolism and function: a multiparametric monitoring approach. Bull Math Biol 68:275–291
69. Mayevsky A, Barbiro-Michaely E (2013) Shedding light on mitochondrial function by real time monitoring of NADH fluorescence: II. Human studies. J Clin Monit Comput 27:125–145
70. Mayevsky A, Rogatsky G (2007) Mitochondrial function *in vivo* evaluated by NADH fluorescence: from animal models to human studies. Am J Physiol Cell Physiol 292:C615–C640

# Chapter 3
# Brain Real-Time Monitoring Techniques Used in Mongolian Gerbils

**Abstract** The basic principles and technology of NADH monitoring were discussed in detail in my book published in 2015 (Mayevsky A, Mitochondrial function in vivo evaluated by NADH fluorescence. Springer, Cham, 2015). However, this chapter presents specific details regarding monitoring of NADH and other functions in the Mongolian gerbil brain. In addition to monitoring of the mitochondrial NADH, this chapter will present the development of multiparametric monitoring systems that provide real-time information on the cerebral hemodynamic, ionic, and electrical activities during various perturbations. The various combinations of multiparametric assemblies will be discussed in detail.

**Keywords** NADH fluorometry · IONIC homeostasis monitoring · Brain oxygen level monitoring · Multisite monitoring of NADH · Electrical activity of the brain

## 3.1 Introduction

The ability to measure NADH using the blue fluorescence spectrum emitted under UV illumination was presented in Chap. 2. In the early stages, NADH monitoring was based on the difference in the absorption of NADH and $NAD^+$. At 320–380 nm, only the reduced form, NADH, absorbs light, while $NAD^+$ does not (Fig. 2.7b). Therefore, when a mixture of NADH and $NAD^+$ is illuminated in a cuvette at 320–380 nm, only NADH will affect the absorption spectrum peak at 340 nm. Since the early 1950s, fluorescence spectrophotometry of NADH has been employed in various in vitro and in vivo models. The emission of NADH fluorescence, under illumination at 320–380 nm, has a very broad spectrum (420–480 nm) with a peak at 450–460 nm (Fig. 2.7c). NADH fluorescence has been identified by Chance and


A. Mayevsky, *The Mongolian Gerbil Brain*,
https://doi.org/10.1007/978-3-031-69549-0_3

his collaborators as a good indicator of the intramitochondrial oxidation-reduction state (Fig. 2.7d) [2].

The basic features of NADH fluorometers consist of the following elements:

1. A light source unit (including appropriate filters).
2. An optical path to the preparation unit and back to the detection unit.
3. Detection and signal processing units.
4. Signal recording and storage units.

In our earlier reviews published in 1984 [3] and in 2007 [4] and in the book published in 2015 [1], we extensively specified the light-guide-based fluorometry used in our studies [3].

In this chapter, I will provide the reader with extensive reference information, gathered over the years, on the construction of fluorometers. Priority will be given to articles mainly dealing with NADH monitoring in vivo, though in vitro fluorometer types will be also considered. Duysens and Amez [5] schematized the first fluorescence spectrophotometer used for intact cells. They utilized the mercury arc—providing a very strong band at 366 nm, even though not at the maximal NADH absorption peak of NADH (340 nm).

In 1959, Chance and Legallais [6] described a differential fluorometer that heralded a new era in monitoring NADH fluorescence in vivo as an indicator of mitochondrial function. They used a microscope, serving as the fluorometer basis, with two light sources: tungsten and mercury lamps with appropriate filters. In 1959, Chance and Jobsis [7] and Chance [8] proved that mechanical muscle activity is associated with NADH oxidation measured in excised muscle.

The first in vivo NADH monitoring device was presented in the late 1950s and early 1960s. At that stage, the effects of scattered light and tissue absorption due to blood were not taken into consideration when monitoring NADH fluorescence. The first detailed results of in vivo NADH fluorescence measurements were published in 1962 [9, 10].

These "classical" papers described two microfluorometers that were modifications of previous designs [5, 11]. The same instrumentation was used in other in vivo studies, including those of Chance's group [12–18], Dora and Kovach's group [19, 20], Rievich's group [21, 22], Jobsis and collaborators [23–28], Schuette et al. [29], Gosalvez et al. [30, 31], and Anderson and Sundt [32–34]. A new model of time-sharing spectrofluorometer and fluorometer was developed at the Johnson Research Foundation in 1971 [35].

The effect of blood on NADH fluorescence was discussed earlier by Chance et al. The monitoring of a second parameter in tissue fluorometry in vivo was reported by Chance and Legallais [12]. They showed that "changes due to the deoxygenation of oxyhemoglobin do not interfere with measurement of the time course of fluorescence changes in the tissue studies".

The addition of a second monitoring signal, namely, tissue reflectance at the excitation wavelength, was reported in 1968 by Jobsis and Stansby [36]. It was based on a previous model described by Jobsis et al [37]. In two other papers by

Jobsis and collaborators [25, 26], the measurement of 366-nm reflectance was used for correction of the NADH fluorescence signal from the brain [38].

## 3.2 Monitoring of NADH in the Cerebral Cortex of Gerbils

### *3.2.1 Introduction*

The first attempt to monitor NADH fluorescence in the gerbil brain was made in my laboratory and reported in 1978 [39]. Later on, other monitoring systems were developed and used in gerbils as seen in Table 3.1, which shows the 12 monitoring systems that were developed in my laboratory.

In order to enable the monitoring of brain NADH fluorescence in anesthetized or unanesthetized animals, a flexible means was needed to connect the fluorometer with the tested brain. This was achieved in 1972, when UV-transmitting quartz

**Table 3.1** Milestones in the development and use of brain monitoring of NADH fluorescence and other physiological parameters in vivo in the Gerbil Brain by *Mayevsky et al.*

| Protocol | Year | Discovery/activity | Reference |
|---|---|---|---|
| 3.2 | 1978 | The first fiber-optic-based time-sharing fluorometer-reflectometer used in the brain of an anesthetized Mongolian gerbil for the monitoring of NADH redox state and electrocortical activity (ECoG) | [120] |
| 3.3 | 1980 | Simultaneous monitoring of NADH in vivo together with extracellular $K^+$, DC steady potential, and ECoG | [46] |
| 3.4 | 1980 | Monitoring of brain NADH together with tissue $pO_2$ and ECoG | [78] |
| 3.5 | 1981 | Bilateral monitoring of NADH and ECoG under ischemia | [114, 121] |
| 3.6 | 1982 | The first multiparametric assembly for NADH, extracellular $K^+$, $H^+$, DC steady potential, and ECoG | [75] |
| 3.7 | 1983 | Multisite monitoring of NADH in gerbils under various pathological conditions | [114] |
| 3.8 | 1983 | Responses to ischemia in vivo and freeze-trapped redox scanning of the gerbil brain | [115] |
| 3.9 | 1984 | Simultaneous real-time monitoring of brain NADH, $pO_2$, ECoG, DC potential, and extracellular $K^+$ | [116] |
| 3.10 | 1990 | Simultaneous real-time monitoring of brain NADH, $HbO_2$, ECoG, DC potential, and extracellular $K^+$ and $Ca^{2+}$ | [117, 122] |
| 3.11 | 1990 | Simultaneous real-time monitoring of brain NADH, CBF, ECoG, DC potential, and extracellular $K^+$ and $Ca^{2+}$ | [100] |
| 3.12 | 1993 | Simultaneous monitoring of brain NADH, CBF, ECoG, DC potential, and extracellular $K^+$, $Na^+$, $Ca^{2+}$, and $H^+$ | [123] |
| 3.12 | 1997 | Multiparametric monitoring of NADH, CBF, ECoG, extracellular $K^+$, $Ca^{2+}$, and DC potential inside NMR magnet measuring ATP, PCr, and Pi | [124] |

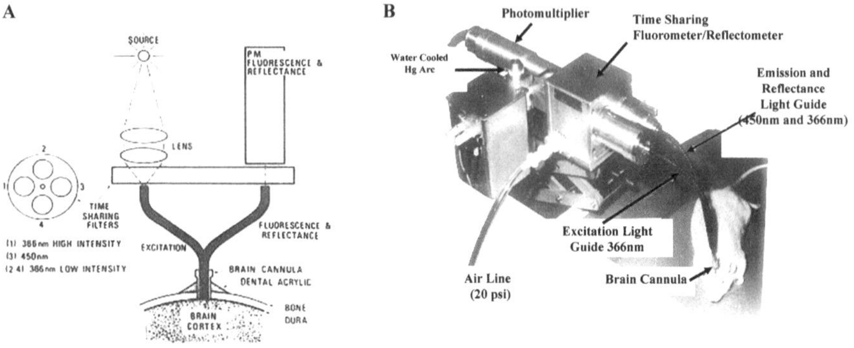


**Fig. 3.1** The structure of the first type of fiber-optic-based fluorometer developed. In Part **a** [3], the time-sharing device is presented and in Part **b** [42], the picture of the device is presented. In the picture seen in Part **b**, the fiber-optic probe is connected to the skull of a small animal (a rat or a gerbil)

fibers became available (Schott Jena Glass, Germany). We have used the light-guide-based fluorometer for in vivo monitoring of the brain [40, 41] subjected to anoxia or cortical spreading depression (CSD). The historical development of light-guide-based fluorometry-reflectometry is shown in Fig. 3.1. The original device functioned on the time-sharing principle (Fig. 3.1a), where four filters were placed in front of a 2-arms light guide. Filters 1 and 3 enabled the measurement of NADH fluorescence, while filters 2 and 4 were used to measure tissue reflectance at the excitation wavelength. The reflectance trace was used to correct the NADH signal for hemodynamic artifacts, and to indicate relative changes in the blood volume of the sampled tissue. In this original system, only one photomultiplier tube was used for the detection of the two signals. Figure 3.1b presents one of the first in vivo brain monitoring time-sharing setups connected to the brain of an anesthetized rat [42] using a special light-guide holder developed.

### 3.2.2 *Methods*

In this section, I present the typical results obtained in 1 of a group of 10 gerbils (*Meriones tristami*) used in our experiments. The gerbil was anesthetized by intraperitoneal injection of Equithesin (0.3 ml/100 g) and mounted on a head holder. The skull was exposed, and a 5-mm hole was drilled in the parietal bone. A plexiglas cannula (light-guide holder) was attached epidurally with screws. Two pairs of stainless steel screws were implanted and served as electrocorticography (ECoG) electrodes. A push-pull stainless steel cannula was implanted epidurally near the light-guide holder for epidural application of KCl solution in order to elicit cortical spreading depression. The cannulae and the screws were cemented to the skull by dental acrylic cement (for more details, see Reference [43]). The common carotid arteries were dissected in the neck, and loops of ligature were placed around the vessels. The oxidation-reduction state of NADH was measured using a time-sharing fluorometer/reflectometer, as described previously. The animal was connected to the fluorometer by a Y-shaped light guide (Schott, Mainz, Germany), and ECoG was monitored bilaterally. The calibration and correction for hemodynamic artifacts were done as described previously [43, 44].

### 3.2.3 Results and Discussion

Figure 3.2 shows the response of a slightly anesthetized gerbil to anoxia ($N_2$ exposure). The upper trace shows the response of the reflected light, which decreased during the anoxia, probably due to increase in blood volume in the tissue that was measured. The second and third traces from the top show the response of NADH fluorescence to $N_2$. The uncorrected trace shows a 40% increase in fluorescence, but after correction, the change is almost 70% in this particular gerbil. The recovery of the NADH to its baseline level had two phases. In the first phase, a rapid decrease in NADH was recorded, and in the second phase, there was a slow decrease of NADH to the normal oxygen level. The ECoG response to anoxia is shown in the two traces at the bottom of Fig. 3.2. The ECoG became isoelectric when NADH reached 80% of its maximal increase [43]. It appears that the secondary slow recovery of NADH is due to slow response of the blood vessels after anoxia. In the rat

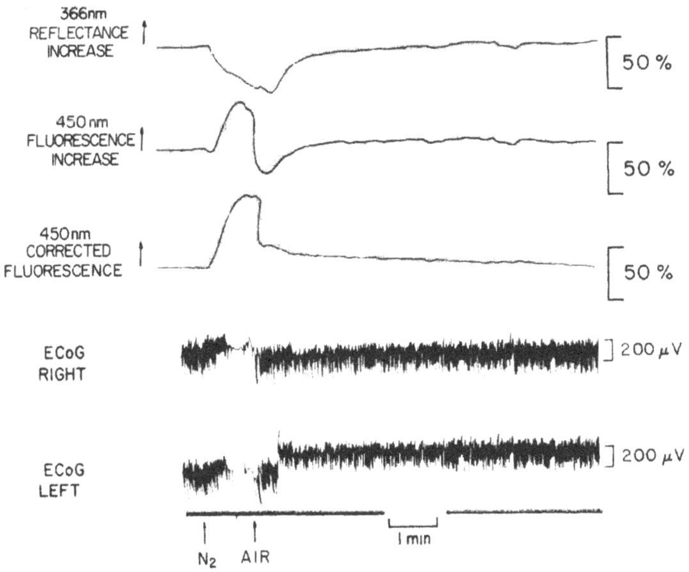


**Fig. 3.2** The effect of anoxia ($N_2$ breathing) on the oxidation-reduction state of NADH, reflected light, and the electrical activity of the brain of the gerbil [39]

brain model, we never saw this type of response [43, 44]. During anoxia, the blood vessels in the brain dilate, and owing to an increase in blood volume in the tissue, the reflectance decreases. When oxygen is available, the blood vessels return to their normal tone and within 10 min, the reflectance and the uncorrected traces return to their baseline level, having the same slope [43]. In the gerbil, the response of the vessels is probably different from that which occurs in the rat, and slow recovery is observed.

In order to elicit cortical spreading depression, 0.5 M KCl solution was washed through the push-pull cannula, and the results (Fig. 3.3) are very similar to those obtained from the brain of an awake rat [43, 45, 46]. The oxidation-reduction cycles were stopped by washing the brain with NaCl solution (W in Fig. 3.3). The oxidation-reduction cycles can be seen in the uncorrected as well as the corrected traces. Owing to the biphasic response of the reflectance, the corrected trace shows cycles which are more symmetrical and have a larger amplitude of oxidation.

The effect of ischemia was tested by temporal ligation of one or both common carotid arteries. It seems that in a few animals, unilateral carotid ligation did not result in a large increase in NADH in the ipsilateral hemisphere, but after bilateral ligation NADH increased, as in $N_2$ exposure. These results show that not all gerbils can be used as good models for ischemia, as was found by others [47–50]. In another group of gerbils, unilateral occlusion diminished blood supply to the ipsilateral hemisphere. Figure 3.4 shows the results obtained from a slightly anesthetized gerbil. The occlusion of the left carotid artery caused a transient increase in

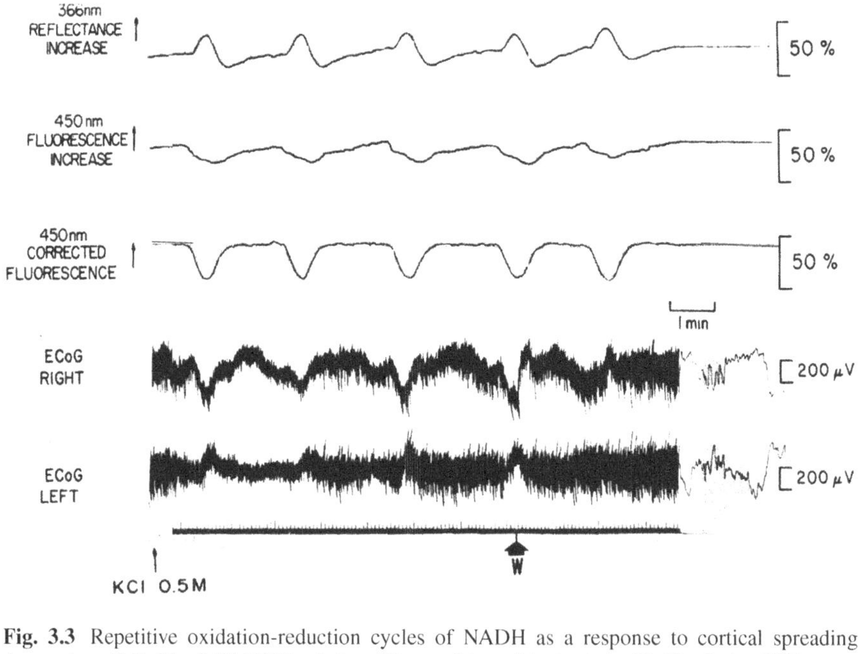


**Fig. 3.3** Repetitive oxidation-reduction cycles of NADH as a response to cortical spreading depression elicited by 0.5 M KCl solution. W—washing of the brain by NaCl solution [39]

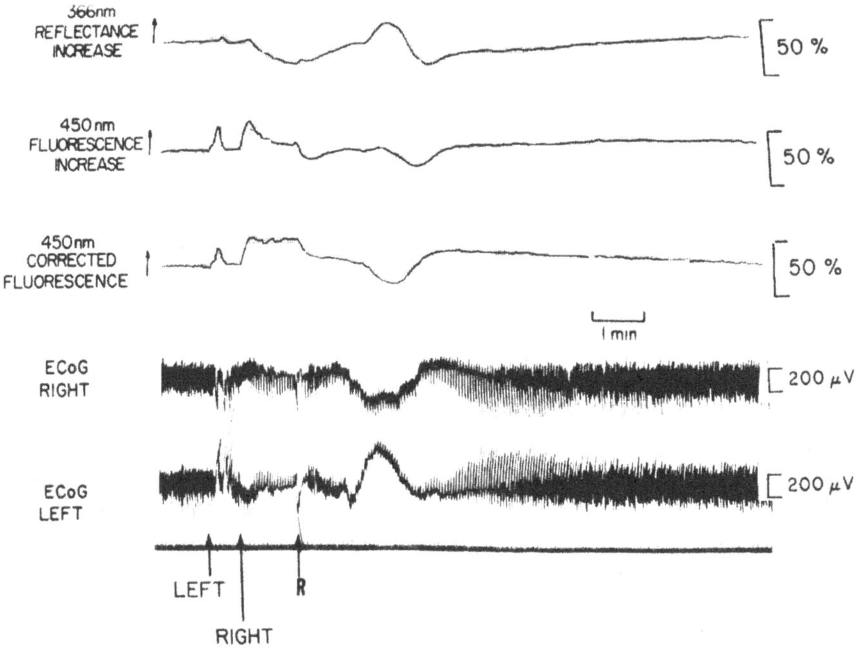


**Fig. 3.4** The effect of ischemia (carotid artery occlusion) on brain NADH oxidation-reduction state and ECoG of an awake gerbil [39]

fluorescence in the contralateral hemisphere, which recovered to normal while the artery was occluded. The occlusion of the right artery (when the left was occluded) led to an increase of 25% in NADH, and this high level remained until the reopening of both carotid arteries (R, Fig. 3.4). The recovery of the brain from the ischemic condition was accompanied by an oxidation-reduction cycle of NADH, and simultaneously the ECoG was depressed. The transient increase in NADH, when the left artery was occluded, was probably due to a sudden change in blood pressure, which affects the distribution of blood between the two hemispheres. When only the right artery was occluded, the same increase in NADH was detected (not shown in Fig. 3.4). From these results, we concluded that in the gerbil (experimental results of which are shown in Fig. 3.4), the blood vessel connections between the left and right hemispheres were not minimal.

The spreading depression-like response, which follows the ischemia episode, seems to be identical to the connection between the anoxia episode and oxidation-reduction cycle we described in the rat [43]. We can assume that $K^+$ ions are accumulated in the extracellular space during ischemia, and when blood containing oxygen is restored, the $K^+$ ions are pumped back into the cells by an active process that consumes energy [43, 45, 46]. The gerbil has the advantage of having a peculiar blood supply to the brain; ischemia can be investigated in more detail in a small awake animal.

## 3.3 Simultaneous Monitoring of NADH In Vivo Together with Extracellular $K^+$, DC Steady Potential, and ECoG

In order to correlate the responses of NADH and ECoG to changes in extracellular $K^+$, we adopted two approaches. Initially we used a double-barreled microelectrode [51, 52], seen in Fig. 3.5a, inserted 1 mm into the cerebral cortex of a rat [46]. This technology was based on the development of a potassium-sensitive ion-exchanger [53] filled in one barrel, while the other barrel was filled with 2 M KCl for the measurement of DC steady potential. Typical recording of the response to CSD presented in Fig. 3.5b was done with a double-barreled microelectrode in the cortical tissue. The waves recorded with the microelectrode were more sharply defined, with fast rise times, fast signal-decay times, and a short duration.

The measurement of extracellular $K^+$ was combined with that of NADH and ECoG, and the responses are seen in Fig. 3.5c [54]. An spreading depression wave was initiated by passing a solution of 0.6 M KCI epidurally through the cannula. We repeated the same experiment in five different animals, and the results were almost the same. Three SD cycles were developed and then stopped by washing the dura with 0.85% NaCl. The first cycle was recorded 2 min after the application of KCI. The DC potential showed a negative shift followed by a plateau and a slower

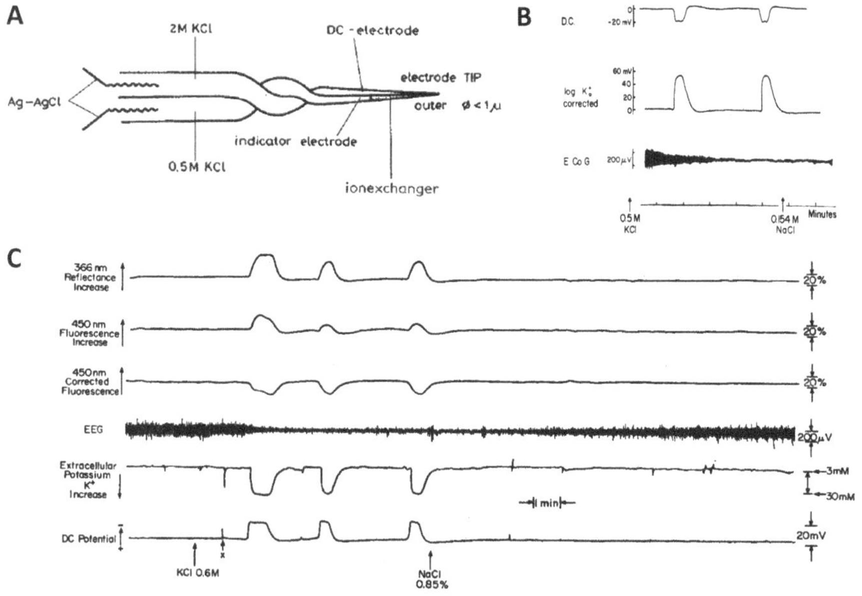


**Fig. 3.5** (**a**) Schematic presentation of the microelectrodes used in order to measure brain extracellular $K^+$ and DC steady potential in a rat model [112]. (**b**) Microelectrode recording from brain in spreading depression using a two-barrel microelectrode and separate ECoG electrodes [55]. (**c**) The effect of cortical spreading depression (elicited by 0.6M KCI) on the extracellular potassium, DC potential, ECoG, and NADH fluorescence of the exposed rat brain cortex [46]

recovery phase, which had a small positive wave. The extracellular potassium concentration increased very quickly and, after a plateau phase, decreased slowly to the original level. The ECoG showed a depression response and recovered to the normal pattern 10–15 min after washing the dura with NaCl. The 366-nm reflectance, NADH fluorescence, and the corrected fluorescence (CF) show the response to CSD as described previously [45]. The corrected fluorescence trace shows a decrease of 25% in NADH in comparison to the normoxic level. This decrease in NADH, which represents an oxidation (25%), is identical in the three cycles. The first cycle was longer, but the level of oxidation was the same. In our present experiments, a larger increase in reflectance was observed in most of the animals tested, but the subsequent decrease in the reflectance response described previously was either smaller or obscure. Since the size of the light guide (0.3 mm) and the microelectrode (~1 μm) is not the same, and the fields they sample are not identical, it is difficult to correlate the onset of changes in NADH and $K^+$. However, the duration of the $K^+$ cycle and NADH oxidation cycle is the same. In each set of cycles elicited by KCl, the first cycle was longer than the following cycles, which were identical one to the other. The main differences between the results obtained with the two light guides are the size of the reflectance response and the duration of the CSD cycle, which is twice as long when the larger light guide is used. In one animal, we cemented the microelectrode to the skull, and the animal was released from the head holder and allowed to become awake. The duration of the $K^+$ cycle in the awake animal was shorter than when the same animal was anesthetized.

In order to avoid damage to the brain tissue by inserting the microelectrode into the cortex, we decided to develop a $K^+$-sensitive surface electrode based on an impregnated polymer membrane. All technical details have been given in our published paper [55]. In the experiments described here, a plexiglas cannula containing the combination electrode and having a pair of ECoG leads incorporated in the walls (see Fig. 3.6a, b) was fitted in place to make a tight fit with the skull and

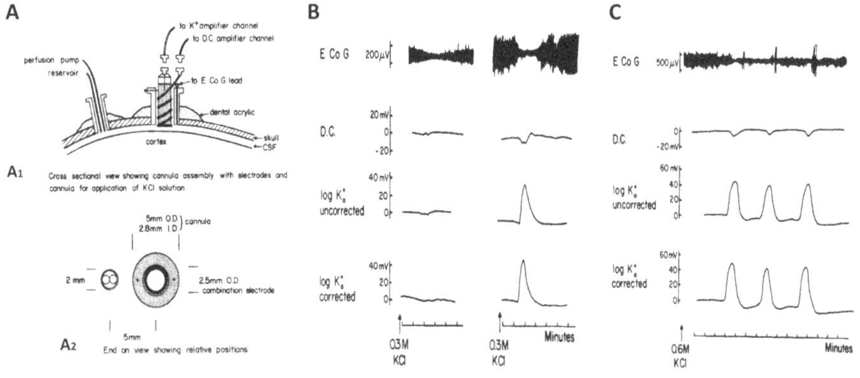


**Fig. 3.6** (**a**) Representation of combination electrode ECoG leads and cannulae for four-channel recording from surface of brain. (**b**) Recording from brain using surface combination electrode before (left side) and after (right side) removal of dura mater. (**c**) Recording made with a combination surface electrode during cortical spreading depression [55]

communicate with the cortex through a layer of cerebrospinal fluid. The cannula was sealed in place using dental acrylic cement. A second skull cannula (2 mm OD), containing two stainless steel tubes, was implanted in the skull above the dura mater at a distance of 3 mm from the electrode (Fig. 3.6a*1*, a*2*). This cannula was connected through one line of polyethylene tubing to a reservoir filled with 0.3 M KCl solution and through the other line to a syringe or a Harvard infusion/withdrawal pump, allowing KCl to be drawn over the intact dura mater above the cortex to initiate spreading depression in the cortex [56]. In some cases, a higher concentration of KCl solution was needed to elicit cortical spreading depression.

In three recordings, measurements were attempted with the dura mater intact to determine whether $K^+$ ions freely diffuse through this membrane. There would be a marked improvement in animal preparation if the dura could be left in place. However, as Fig. 3.6b (left side) indicates, the $K^+$ signal is severely attenuated by the dura. Figure 3.6b (right side) shows the results immediately after removing the dura and measuring the same area of the brain. Typical four-channel recording is shown in Fig. 3.6c. The top line shows the ECoG, as recorded from the two stainless steel leads in the wall of the support cannula. Depression of this trace is one of the indications that cortical spreading depression (CSD) has been elicited. The ECoG leads are positioned to be on a line roughly parallel to the advancing wave front of the CSD wave. The second and third traces give the outputs from the combination electrode. The DC electrode shows the characteristic surface negativity as the CSD wave passes.

A combination surface electrode was used to provide continuous measurements of potassium activity and slow-wave surface potential. The active elements of the electrode were a valinomycin-impregnated polymer membrane for measuring $K^+$ and an Ag/AgCl wire with salt bridge for measuring slow wave potential. An electronic subtraction was done to provide a measure of $K^+$ activity corrected for slow-wave electrical changes.

Potassium electrodes were calibrated in isotonic solutions containing varying proportions of NaCl and KCl so that $K^+_e$ values could be interpreted directly in units of concentration. Crowe et al. in 1977 [55] have discussed in detail the preparation and use of these electrodes.

Electrode signals were measured using voltage amplifiers (gain set at 1) with input impedance greater that $10^{13}$ ohms. Recordings were made on a Beckman Offner dynagraph eight-channel recorder. ECoG, recorded differentially from bipolar leads (stainless steel, 0.5-mm diameter), was amplified with a Beckman EEG preamplifier module and amplifier module.

The energy state of the brain was monitored by measuring the oxidation-reduction state of NADH by surface fluorometry. The excitation and emitted light from the fluorometer to the brain and back was transmitted via a Y-shaped flexible light guide.

The fluorometer used in our experiments was a "DC type"; this instrument has been described in detail by Mayevsky [2]. A schematic presentation of the special cannulae used in our experiments is shown in Fig. 3.7a. The localization of the various cannulae implanted above the brain is shown in Fig. 3.7b. In most gerbils used

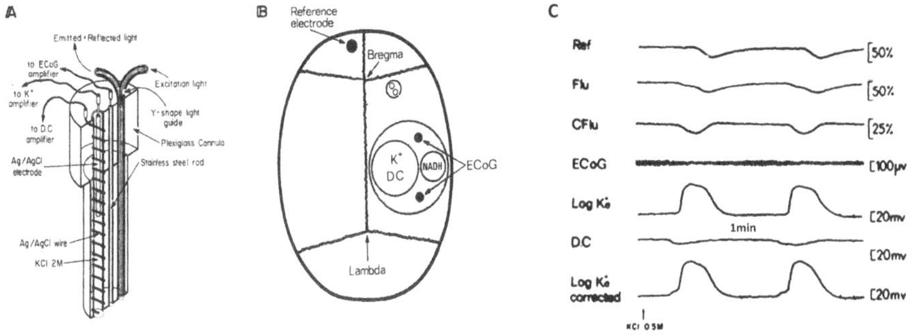


**Fig. 3.7** (**a**) Schematic presentation of the longitudinal section of the combined cannula and (**b**) a topical view of the skull of a gerbil implanted with cannulae and electrodes. (**c**) The effects of spreading depression on the metabolic, electrical, and extracellular $K^+_e$ activity in the gerbil [113]

in this study, one combination electrode for $K^+$/DC and one light guide were implanted on the right hemisphere.

The anesthetic used was Equithesin (Jensen-Salsberg Laboratories, U.S.A.). Each animal was given 0.3 ml/100 g by intraperitoneal injections to provide sufficient anesthesia for the implantation of the cannulae. Holes were drilled in the calvarium to accommodate the cannulae, and the dura mater was removed at the site of $K^+$ measurement. The combination electrode/light pipe was located in place with micromanipulator and cemented to the calvarium together with the other cannula with dental acrylic cement. During the experiment, each gerbil was kept minimally anesthetized by injection of small amounts (0.03–0.05 ml) of Equithesin every 30 min.

Figure 3.7c shows the typical response of a normal brain to spreading depression elicited by 0.5 M KCl solution applied epidurally with a push-pull cannula. The response of the DC potential was very small, so that $K^+_e$ uncorrected and $K^+_e$ corrected traces were quite similar. $K^+_e$ increased quickly, reaching its maximum level within 10–20 s. The recovery of $K^+_e$ took about 2 min, and simultaneously the NADH trace (CFlu) showed a typical oxidation cycle. In this animal, the reflected light trace showed a decrease, which indicates an increase of blood volume in the tissue. The ECoG trace showed a typical depression during CSD.

## 3.4 Monitoring of Brain NADH Together with Tissue $pO_2$ and ECoG

When $O_2$ consumption was evaluated, Leao and Morrison [57] found a decreased $O_2$ concentration in blood samples taken from the sagittal sinus during CSD. Van Harreveld and Stamm [58] measured a decreased $pO_2$ in the depressed region. The same fall in $pO_2$ during CSD was also described by Marshall [59] and Lukyanova and Bures [60]. The described hemodynamic and $pO_2$ responses to CSD are indicators of increased metabolic activity during the recovery phase of CSD.

In this study, we correlated for the first time the changes in NADH fluorescence, changes in the reflected light, and the changes in $pO_2$ measured with a surface oxygen electrode. The interrelation between the three parameters measured was tested under various conditions, such as anoxia, hypoxia, cortical spreading depression (CSD), anesthesia, and ischemia (Table 3.2).

Male gerbils (60–70 g) were used in our study. The animal was anesthetized by intraperitoneal injection of 0.3 ml/100 g of Equithesin (a mixture of pentobarbital, chloral hydrate, magnesium sulfate). The skull was exposed by a midline incision, and a 5–6-mm-diameter hole was drilled in the parietal bone to fit a combined NADH-$pO_2$ cannula (Fig. 3.8a). A 2-mm-diameter hole was drilled in the frontal bone to fit a push-pull cannula for epidural application of KCl in order to elicit cortical CSD. Two stainless steel screws were implanted on the left parietal bone for holding the cement to the skull. The cannula and the screws were cemented to the skull using dental acrylic cement. Oxidation-reduction state of NADH was monitored using a standard DC fluorometer/reflectometer with a 1.5-mm-diameter, Y-shaped light guide. The $pO_2$ was measured by an open-type oxygen electrode 2.5 mm in diameter. It consisted of a platinum wire 25 microns in diameter and a silver wire 250 microns in diameter embedded in epoxy resin and ground so that their surfaces were flush with the glass capillary tube. Polarization potential was 0.80 V. The electrode was dipped in 5% cellulose diacetate to decrease sensitivity to stirring under calibration conditions. ECoG was measured by two stainless steel rods (0.5 mm in diameter) inserted through the wall of the combined cannula. Recordings were made on a Beckman multichannel dynograph.

**Table 3.2** Detailed parameters measured in the various protocols

| Parameters / Protocol | NADH | ECoG | $K^+$ | DC St. Pot | $pO_2$ | $H^+$ | $Ca^{2+}$ | $Na^+$ | CBF | $HbO_2$ | NMR | Number in the References List |
|---|---|---|---|---|---|---|---|---|---|---|---|---|
| 3.2 | + | + | | | | | | | | | | [39] |
| 3.3 | + | + | + | + | | | | | | | | [63,114] |
| 3.4 | + | + | | | + | | | | | | | [78] |
| 3.5 | ++ | ++ | | | | | | | | | | [114,121,125] |
| 3.6 | + | + | + | + | | + | + | | | | | [3,75,79,98,126,127] |
| 3.7 | ‡‡ | | | | | | | | | | | [66] |
| 3.8 | ++ | | | | | | | | | | | [65,85,103,115,128] |
| 3.9 | + | + | + | + | + | | | | | | | [76,80,101,116,129,130] |
| 3.10 | + | + | + | + | | | + | | | + | | [117,122] |
| 3.11 | + | + | + | + | | | + | | + | | | [89,100,108,118,131,132] [133] |
| 3.12 | + | + | + | + | | + | + | + | + | | | [134] |
| 3.13 | + | + | + | + | | | + | | + | | NMR* | [119] |

*NMR – $P_i$, PCr, ATP, PCr/$P_i$

*NMR—$P_i$, PCr, ATP, PCr/$P_i$

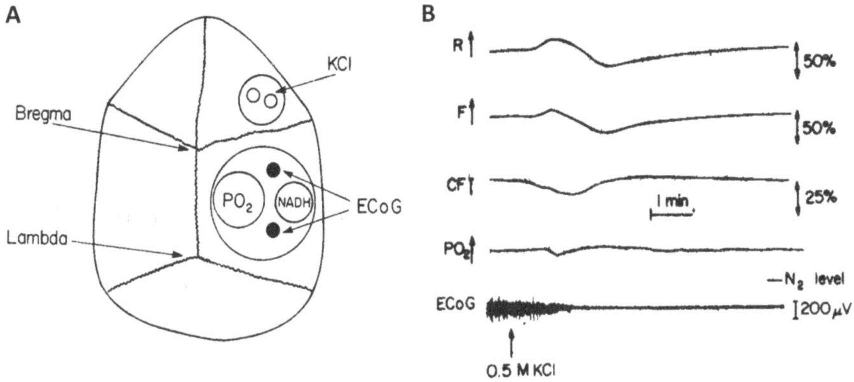


**Fig. 3.8** Schematic presentation of the location of probes above the gerbil brain. (**a**) The large cannula contained the light guide for NADH measurement, the $pO_2$, and ECoG electrodes. The cannula for KCl application was located 2–3 mm anterior to the large cannula [78]. (**b**) The effects of cortical spreading depression on NADH, $pO_2$, and ECoG in the awake gerbil brain [78]

In Part B of Fig. 3.8, the responses to CSD after elicitation by topical application of KCl solution are shown. The reflectance trace (R) shows a typical biphasic response, and the corrected fluorescence (CF) shows the oxidation-reduction cycle. The $pO_2$ trace presents a decreased level of oxygen during the CSD cycles. The ECoG was depressed during the cycles, although recovery to the original amplitude was very slow and is not shown in the figure.

## 3.5 Bilateral Monitoring of NADH and ECoG Under Ischemia

The measurement of intramitochondrial NADH oxidation-reduction state from the brain of an awake animal is a well-established technique and has been in use for the last few years. We had used this animal model in studying the effects of hypoxia, spreading depression, epileptiform activity, [43] or ischemia [61] on the metabolic activity. In all of the studies carried out by our group or other groups, only one site or side of the brain was monitored fluorometrically. In the ischemia studies, we found that under certain experimental conditions it is very important to measure metabolic activity from the "experimental" site as compared to the "control" site or side simultaneously. In our awake model, we used a topical application technique to study the effects of drugs on the cortical tissue without any effects on the rest of the brain. This model also requires a bilateral measurement technique. We present here a new type of fluorometer/reflectometer by which NADH fluorescence, as well as reflected light, can be measured simultaneously from two points of an awake brain model. This model was exposed to anoxia, spreading depression, or ischemia, and the results are presented.

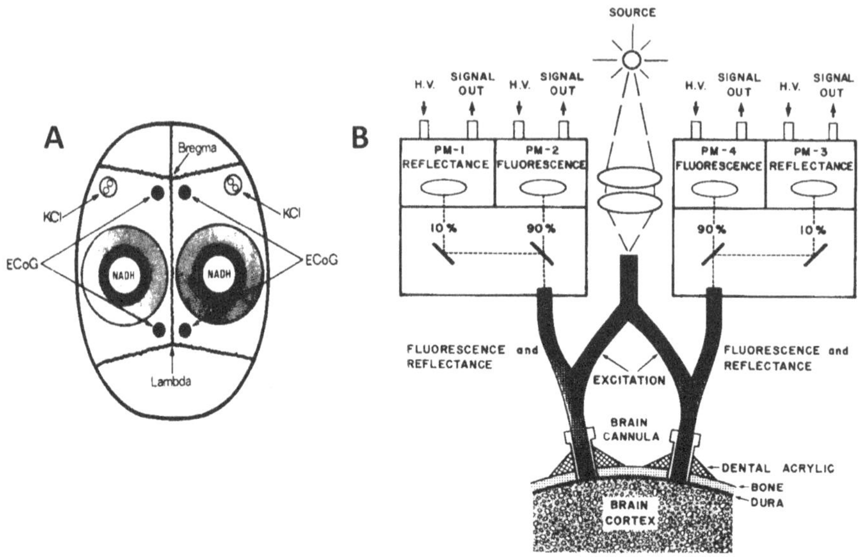


**Fig. 3.9** (**a**) Schematic presentation of the location of various cannulae and electrodes on the skull of a gerbil [63]. (**b**) Two-channel DC fluorometer-reflectometer for simultaneous measurement of NADH from two points of an anesthetized or awake gerbil [62]

Figure 3.9 presents the two-channel fluorometer/reflectometer used in the study of the gerbil brain. The technical details were published in 1978 by Mayevsky and Bar-Sagie [62] and are presented in Fig. 3.9.

Figure 3.10 illustrates the response of the gerbil brain to anoxia. In the two hemispheres, the changes in NADH were very similar due to the lack of oxygen. In addition, a spontaneous cortical spreading wave was developed 2–3 min after recovery from anoxia to the normoxic state. The same type of response has been described previously [43, 63].

## 3.6 The First Multiparametric Assembly for Recording of NADH, Extracellular $K^+$, $H^+$, DC Steady Potential, and ECoG

Surface probes became available for the monitoring of parameters reflecting normal metabolic and electrical activity of living tissues as well as pathological disturbances in their homeostatic mechanisms. Such probes include ion-selective electrodes for extracellular $K^+$ activity [55], flat glass pH electrodes, surface $pO_2$ electrodes, and light guides for monitoring changes in the oxidation-reduction (redox) state of fluorescent components of the respiratory chain. Our aim was to

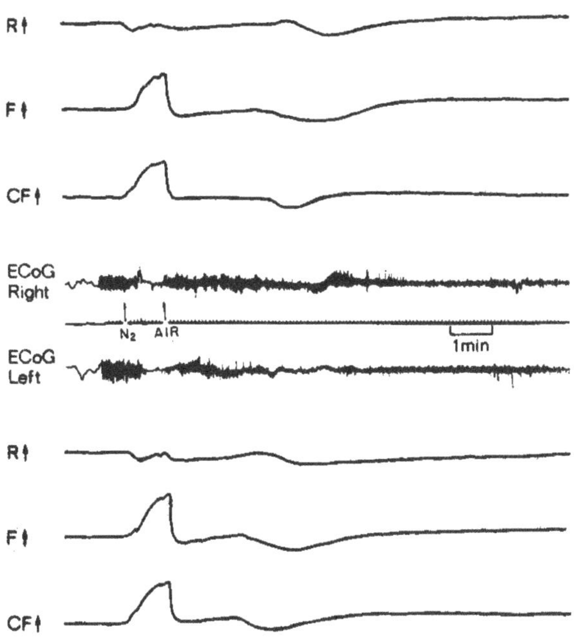


**Fig. 3.10** Bilateral intramitochondrial metabolic responses to anoxia in the gerbil brain [114]

construct a new holding system allowing an easy implantation of surface probes in various combinations on the brain of a small animal model.

Ideally, the multiprobe assembly (MPA) designed for our project had to be rugged enough to withstand routine use by nonspecialists and yet be miniaturized enough to fit the limited surface available for implantation on rat and gerbil skulls (about 6–7 mm in diameter on each hemisphere). For ischemic studies, interest was focused on monitoring mean values over large areas, suggesting the use of larger electrodes. In contrast, following propagated events such as spreading depression would have required the tip of the multiprobe assembly to be concentrated in a tiny space seeing the same phase of the wave unless each an auxiliary signal. The second option was retained by recording the wave of the DC potential concentrically to each sensor. In addition, the complete assembly had to provide adequate protection and shielding of the high-impedance ion-sensitive electrodes ($K^+$ and pH) and a stable enough reference junction for use in unanesthetized animals. With $K^+$ surface electrodes, concentric DC potential has been proposed as the best approximation for the DC bipotential component to be subtracted from the sensor signal [55]. In this study, we also attempted to check the accuracy of the method by recording DC potential differentially between a central barrel and the peripheral slit and to reevaluate its usefulness in correcting surface $K^+$ and pH measurements under various conditions.

### 3.6.1 Electrode Assembly

The solution adopted for the electrode holder is basically a modification of the Lucite cannula described by Mayevsky et al. [63] for the light guide and the potassium-sensitive electrodes. To offer space for more probes, the new cannula was shaped as a truncated cone instead of a cylinder (Fig. 3.11a). The holes accommodating short electrode probes ($K^+$, pH) are made convergent toward the lower surface to occupy less space on the brain and divergent at the top to facilitate handling and sealing of the probes. An additional hole is drilled obliquely from the upper surface to merge with each sensor channel at about mid-distance of the lower surface. This hole accommodates an AgCl wire used to record the local DC potential concentrically to the sensor.

The long and rigid steel stem of the light guide (L) used in this study occupies a straight vertical hole in the cannula and serves as an axis to hold the cannula (c) and the cable holder (h) at a convenient distance of each other (4 cm). Steel rods (preferentially threaded) can be used as additional or replacement pillars to fix the cannula to the cable holder. The arrangement leaves optimal access to the electrodes

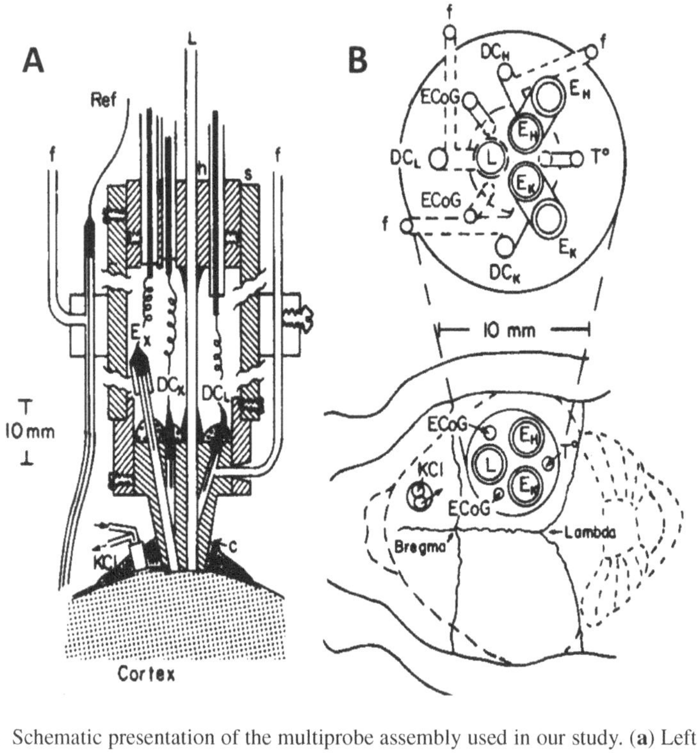


**Fig. 3.11** Schematic presentation of the multiprobe assembly used in our study. (**a**) Left longitudinal section of cannula and assembly located above cortex. A small push-pull cannula was placed near the assembly to initiate spreading depression. (**b**) Lower right: topical view of gerbil skull together with location of various electrodes. Top right: cross section of wide end of cannula [75]

and electrical connections for assembling and replacement. The complete assembly is protected and shielded by a sliver-pointed Lucite sleeve sliding over the cable holder. If a stronger construction is desired, the sleeve can also be permanently screwed into the cannula and cable holder, with a half-cylinder piece cut out as a removable cover.

When the holder is assembled, the fixed steel pillars are screwed and/or glued in first. To protect the solder point, chloritized silver wires are glued with two-component epoxy glue onto short pieces of Teflon that fit over the sensor electrode shaft and into the DC channel in the cannula. The Teflon-sleeved probes (Yellow Springs Instrument temperature probe 511) with paraffin melted by a thermocautery tip. To avoid tension on the electrode wires, they are connected to the input cable through a flexible coil of 36-guage isolated copper wire.

Once the electrodes were in place, the DC channel and the lower part of the sensor channel were filled with fluid from a lateral hold connected to a reservoir syringe. Thus, it can be ensured that each sensor and DC electrode is tightly sealed to the top of the cannula and will not let saline reflow back into the cable housing compartment. The system also allows the extracellular fluid and blood to be cleaned from the slit around each probe after the cannula is removed from an implantation site. In the final assembly (Fig. 3.11a, b), a protective sleeve is used as an anchoring point for the refill tubes to the DC and sensor channels (f) and for the reference liquid junction (REF in Fig. 3.11a).

### 3.6.2 Correction for Local DC Biopotential

In the measurement of electrode transmembrane potential, an additional emf may appear in series in the circuit: the tissue biopotential existing between the ion-selective membrane and the reference junction. With microelectrodes, this potential can be picked up within a few microns of the sensor tip by an adjacent DC glass microelectrode and subtracted from the potential recorded through the sensor. Applying the same correction principle with surface electrodes may introduce an error as the ion-sensing and DC-sensing areas are not superimposed. In 1977, Crowe et al. [55] proposed using the DC potential recorded from a concentric slit around $K^+$-sensitive electrodes as the best approximation from the DC component to subtract from the sensor potential. Although the peripheral slit sees cortical spreading depression (CSD) waves somewhat sooner and later than the central electrode, Crowe et al. [55] consider that the correction will not be seriously compromised, because the width of the SD wave is as large or larger than the width of the electrode assembly. To reevaluate this assumption, we measured DC potential changes simultaneously from central electrodes and from their concentric slits, replacing the central ion-sensitive electrodes by a glass or polyethylene barrel of the same diameter filled with agar-saline.

### 3.6.3 NADH Fluorescence Measurement

In this study we used a one-channel DC fluorometer-reflectometer [2, 64]. The source for the 366-nm excitation light was a 100-W water-cooled mercury arc passing via the excitation fiber-optic bundle to the brain. The emitted light from the brain transmitted through another fiber-optic bundle was split into 90:10 ratios in order to measure the fluorescence light (F; 90%) and the reflected light (R; 10%). The calibration of the signals as well as other details was published previously [61–63].

The effect of complete ischemia was also tested in a gerbil in which the carotids were occluded for more than 4 min (Fig. 3.12). As a result, NADH reached its maximal level (CF) for about 2 min, and then a transient apparent oxidation was recorded on the corrected trace. This "oxidation" may be an artifact due to the changes in light absorption by the tissue and blood existing in it and was not corrected by the 1:1 correction factor used. One can see a large increase in R as well as in the uncorrected F during this episode, which is a general depolarization occurring during complete ischemia. NADH (CF) showed a fast recovery after reopening of the carotids. The extracellular $K^+$ showed a two-step response to complete ischemia. Immediately after occlusion, a small and slow increase in extracellular $K^+$ was

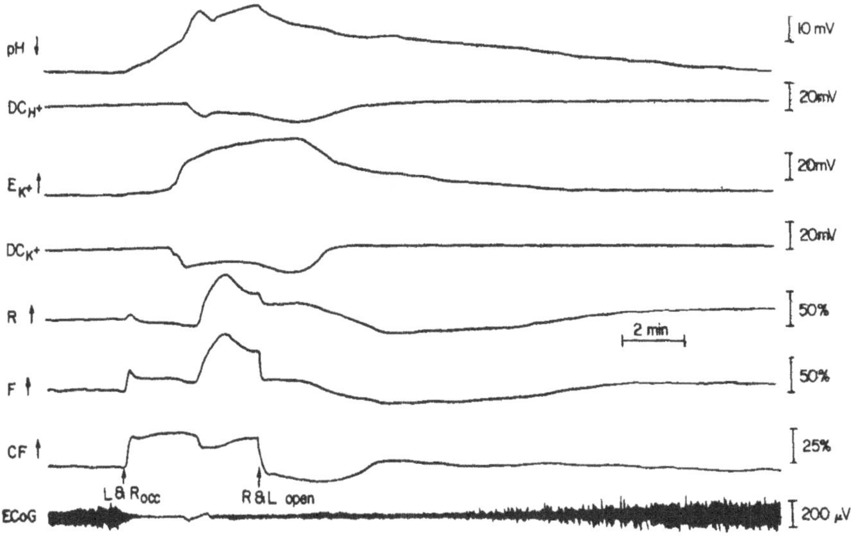


**Fig. 3.12** Responses of gerbil brain to about 4 min of complete ischemia and recovery after reopening of both carotid arteries. Arrows near signal label on the left represent an increase in the level of parameter. For abbreviations, see text [75]

recorded, and after 1.5 min a sharp and large increase occurred simultaneously with a negative shift in the DC potential surrounding the $K^+$ electrode. The pH showed a monotonic acidification response until reopening, but 1.5–2 min after occlusion a large DC negative shift was recorded near the pH electrode, as can also be seen in the corrected pH trace. In this particular gerbil, recovery of the pH started immediately after reopening of the blood vessels, while the extracellular $K^+$ showed a delay of about 2 min before it started to decrease. The time needed for complete recovery of the pH and $K^+$ was much longer than the NADH recovery time. The ECoG showed a silent period during occlusion and gradually recovered after reperfusion of the brain.

In conclusion, the new multiprobe approach described in this study opens up new possibilities for studying mechanisms of brain pathology in a very detailed manner. We now use this approach in our laboratory to study hyperbaric oxygen toxicity, long-term ischemia, and recovery.

## 3.7 Multisite Monitoring of NADH in Gerbils Under Various Conditions

The use of the four-channel DC fluorometer-reflectometer to monitor four different locations on the same brain was the next step in our technological development. In this fluorometer we used one light source to simultaneously illuminate the four monitored areas in the same brain. This arrangement avoids the possible differences in the intensity of excitation light delivered to the four points and therefore decreases variation in the result measured from the four sites when exposed to the same treatment. Figure 3.13 shows the schematics of the monitoring system. The emitted light from the four monitoring sites is collected separately and treated in four identical detection and amplification units. The four light guide arms could be placed in the same hemisphere or in both hemispheres of the brain [65, 66].

Figure 3.13 presents the four-channel fluorometer/reflectometer used in the gerbil brain [65, 66].

Figure 3.14 shows the typical response of the gerbil brain to anoxia [66]. These responses were typical and were found in all normoxic gerbil brains tested. The animal was breathing air spontaneously, and when it was exposed to 100% nitrogen, a typical two-step decrease in reflectance was recorded together with a large increase in corrected NADH fluorescence, which reached a plateau shortly thereafter. When air breathing was restored, a fast oxidation of NADH was recorded in all four sites.

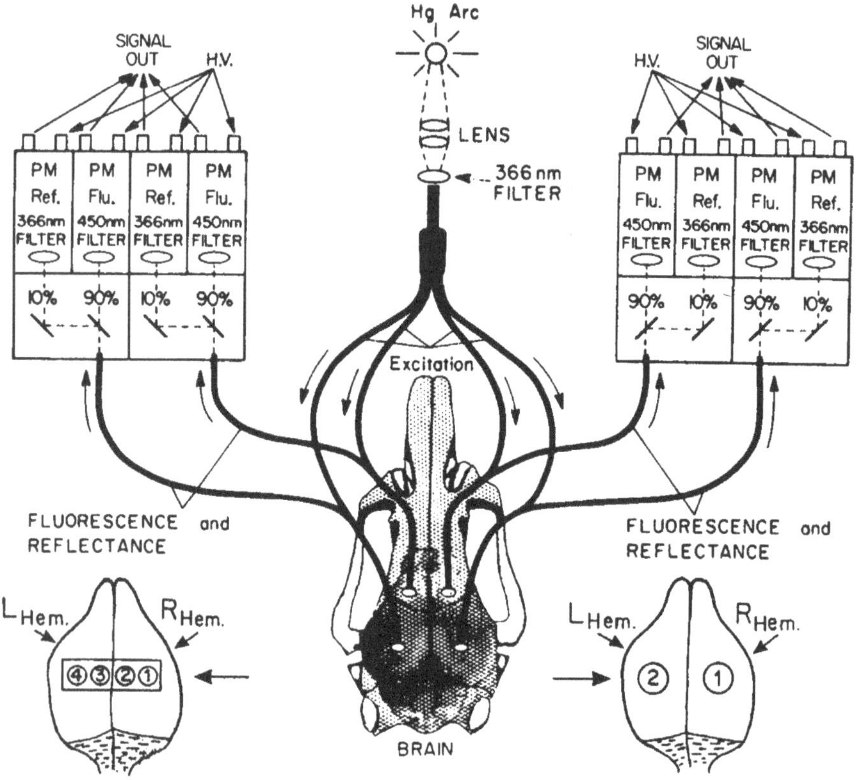


**Fig. 3.13** Schematic presentation of the four-channel DC fluorometer-reflectometer used in this study. Each channel contained a 366-nm as well as 450-nm measurement for reflected light and fluorescence, respectively. *PM* photomultiplier, *HV* high voltage [66]

## 3.8 Responses to Ischemia In Vivo and Freeze-Trapped Redox Scanning of the Gerbil Brain

### 3.8.1 In Vivo Surface Fluorometry/Reflectometry

In order to evaluate the redox state of the brain, a two-channel [62] DC fluorometer/reflectometer was used in various experiments. The light for excitation of NADH (a 100-W water-cooled Hg arc) passed through a bundle of UV-transmitting fiber optic toward the brain. The emitted light was split in 90:10 ratios for fluorescence and reflectance measurements, as described previously [62]. In most of the experiments, the common part of the fiber optic was 2 mm in diameter. The diameter of the common part of the fiber optic probe in the last experiment to be shown was 1 mm outer diameter (0.8 mm fiber diameter). The signals from the brain were amplified and calibrated as described previously [62, 66].

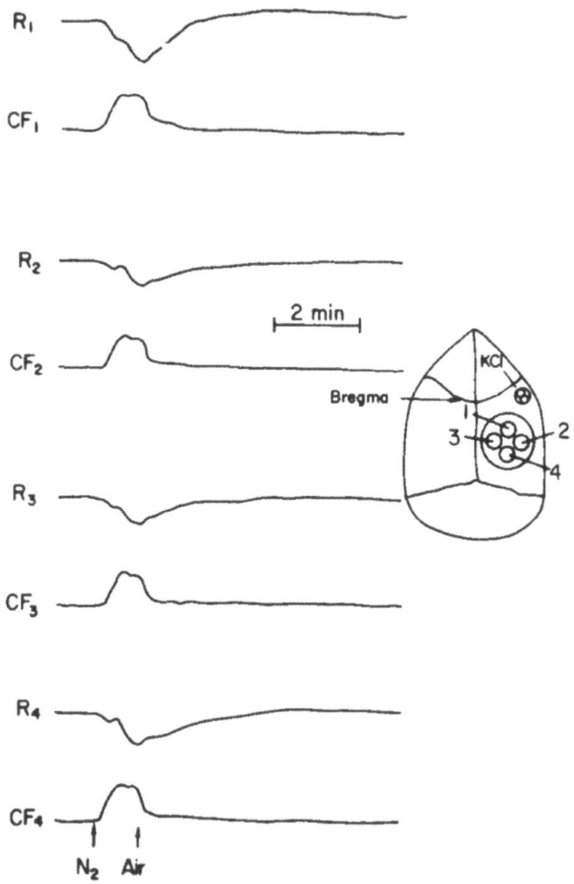


**Fig. 3.14** The effects of anoxia (A) and cortical spreading depression (B) on the reflectance (R) and corrected fluorescence (CF) measured from four sites in the gerbil brain (1–4 in the center schematic) [66]

## *3.8.2 Frozen Brain Redox State Evaluation*

The metabolic state of the frozen brain was studied by recording the fluorescence signals of the mitochondrial NADH and flavoproteins (Fp) [67]. The excitation wavelengths were 366 and 460 nm, and the emitted light was measured at 440 and 540 nm for reduced NADH and oxidized flavoproteins (Fp) correspondingly. The study of fluorescent intensities of the enzymes will give a metabolic picture of the distributions of redox states, as has recently been described by Chance et al. [68]. The ratio between the two signals measured gave the corrected redox state not affected by changes in mitochondrial concentration or hemoglobin level in the tissue analyzed. The distribution of the fluorescence signals in three dimensions throughout the brain was recorded with an online scanner as described in detail [69].

### 3.8.3 Gerbil Preparation

The gerbil (50–70 g) was anesthetized by intraperitoneal injection of Equithesin (0.3 ml/100 g) and mounted onto a head holder for in vivo studies. Two or four 4-mm-diameter holes were drilled in appropriate locations on the skull, and the dura mater remained intact. Special light-guide holders were placed above the brain and cemented to the skull with dental acrylic. In gerbils used for the frozen brain studies, the parietal bone was removed bilaterally without removing the dura mater. The animal remained in the head holder till the end of the freezing procedure. The light guide was held by a special micromanipulator above the cortex for the in vivo monitoring and was removed before freezing the brain.

### 3.8.4 Funnel Freezing of the Brain

Owing to the small size of the gerbil brain, the barrel of a 30-ml syringe was used as a funnel for liquid $N_2$. The rim of the funnel was covered with Vaseline for better sealing during freezing. When cessation of respiration was observed, the gerbil was removed from the holder and immersed in a container of liquid $N_2$. The head was severed by a band saw, and a block of the head was transferred to the low-temperature scanner for redox state analysis (Fig. 3.15).

### 3.8.5 Typical Results in One Gerbil

Figure 3.16 presents the typical results obtained in one gerbil. The results obtained in the frozen brain studies confirmed and correlated with those found in the in vivo stage but showed a few interesting findings that can be summarized as follows.

1. The in vivo NADH increase during ischemia was correlated to the PN changes measured after freezing. High NADH levels are shown as bright areas in the 2-D scan.
2. In most brains, the Fp fluorescence was also correlated to the level of ischemia: the ischemic area is the darker one (less fluorescence of the oxidized Fp form).
3. Blood is supplied to the two olfactory bulbs through the same artery, as can be seen from the identical response to ischemia. The redox state of the olfactory bulbs was different from that of the rest of the cortex.
4. In many of the brains tested, the borderline of the ischemic area was not the midline, as expected from the anatomy of blood vessels to the cortex as reported [70–74]. A narrow area of ischemic or normoxic area was found in the contralateral hemisphere, as can be seen in Fig. 3.16.

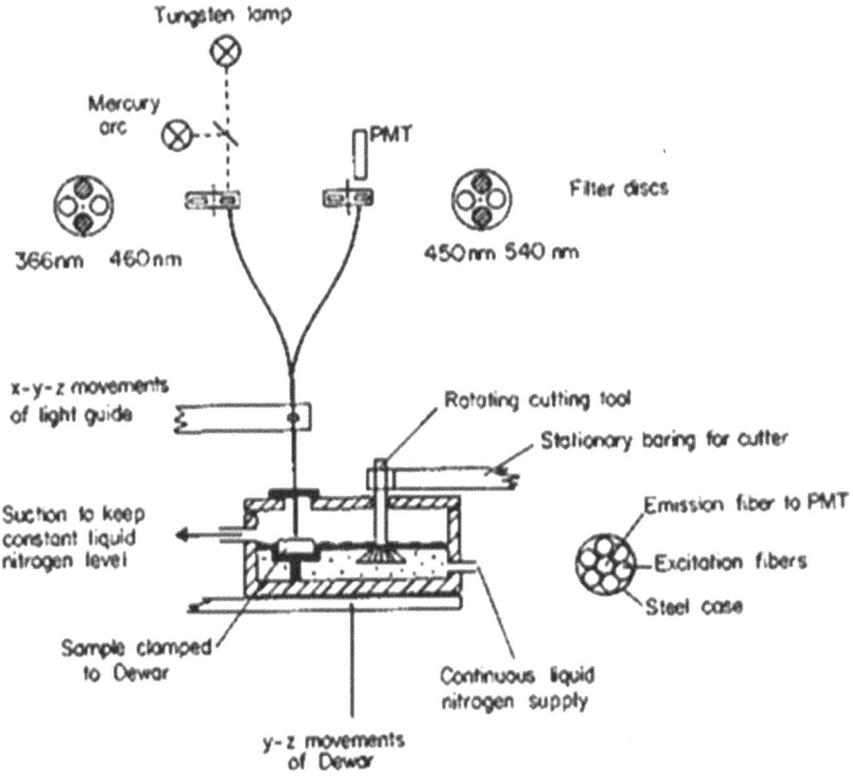


**Fig. 3.15** Illustrating the method of time-sharing Fp and PN signals, 2-D lightguide scanning of tissue section, and low-temperature grinding of tissue surface for scans of consecutive sections [67]

## 3.9 Simultaneous Real-Time Monitoring of Brain NADH, $pO_2$, ECoG, DC Potential, and Extracellular $K^+$

The multiparameter assembly can measure up to eight parameters representing the metabolic, ionic, and electrical activities in the cerebral cortex. Metabolic activity is evaluated by measuring tissue $O_2$ tension and by monitoring the intramitochondrial NADH oxidation-reduction state. The ionic state of the tissue is evaluated by monitoring extracellular potassium (KI) activity. Electrical activity was monitored by bipolar electrocorticography (ECoG) and DC steady potential. In addition, we monitored the surface temperature with a miniature thermistor. The multiprobe assembly has the following main features.

1. All parameters have the same type of noninvasive surface contact with the cortex.
2. The multiprobe assembly can be removed without any damage from the skull very easily at the end of the experiment, so that repetitive experiments can be performed in a short period of time.

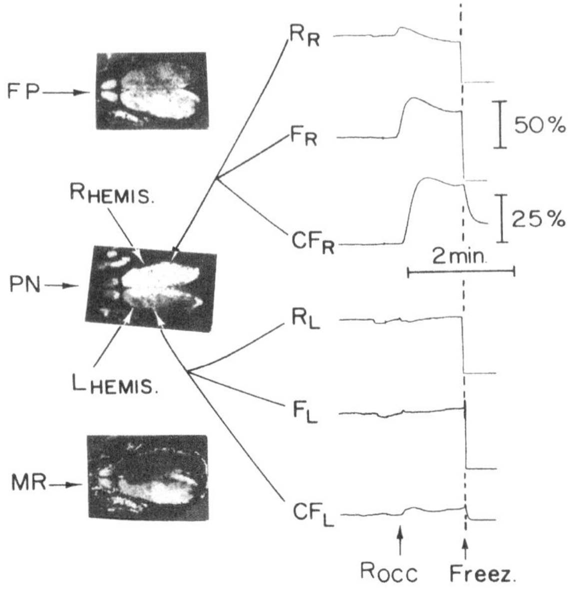


**Fig. 3.16** Correlation between in vivo and frozen-brain metabolic responses under right carotid artery occlusion in the nonseizure-prone (NSP) gerbil. An ischemic area was also detected in the midline region of the left hemisphere [115]

3. All signals monitored by the multiprobe assembly show a very small sensitivity to movements of the animal while exposed to hyperbaric pressure.

## *3.9.1 The Multiprobe Assembly*

Figure 3.17 presents, in a schematic way, the connection between the multiprobe assembly and the cerebral cortex of the rat or gerbil brain (A) and the location of the various probes in relation to the skull (B). The various electrodes were held in the multiprobe assembly as described in detail elsewhere [75, 76]. A general description follows. The probes were mounted on a Lucite holder (cannula) that can be cemented directly to the skull. The holes accommodating the specific sensing electrodes converge at the implantation site to occupy less space on the brain and diverge at the top of the cannula to leave maximal room for mounting and replacing the electrodes. Another hole was drilled obliquely from the top to intersect the first chamber approximately half way down. This hole houses an Ag-AgCl wire to record DC potential concentrically to the sensing electrodes. One more hole was drilled horizontally into the chamber of the DC wire for filling with saline and for cleaning extracellular fluid and blood from around the electrodes at the end of an experiment. A Lucite sleeve was placed over the entire assembly and was held in place with set screws. This sleeve holds the assembly together as one unit, protects the electrodes and wiring, and also provides shielding for the $K^+$ electrode's high impedance. The sensing electrodes, Ag-AgCl wires for DC measurement, stainless steel wires for ECoG measurement, and a polyethylene-sleeved temperature probe were held in

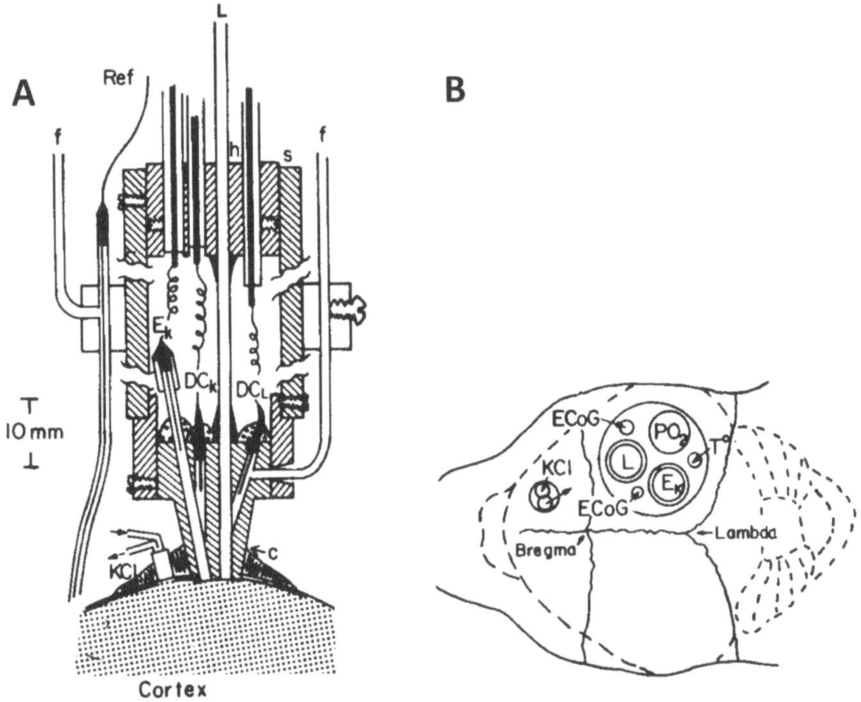


**Fig. 3.17** Schematic presentation of multiprobe assembly (MPA). (**a**) longitudinal section of MPA; (**b**) the relative location of various probes above the brain. Ref —reference Ag–AgCl electrode; f—refill tube for Ref or DC electrode; c—Lucite cannula; s—plexiglass sleeve; L—light guide; h—cable holder; Ek—$K^+$ electrode; $DC_K$, $DC_L$, Ag–AgCl electrode; $pO_2$—$O_2$ electrode; ECoG—electrocorticography electrode [94]

and sealed to the cannula with 5-min epoxy or by melting wax around them. Another block of Lucite was attached to the outer shield that holds the reference assembly and the tubes for filling the DC chambers.

## 3.9.2 *Preparation of Electrodes*

**Oxygen Electrode**

The electrodes were constructed inside 1.6-mm (OD) polyethylene (PE-160) tubing. A platinum wire (25 pm) was sealed in the glass by a flame, and an insulated lead wire was attached to the other end of the platinum wire. Two assemblies, along with a Teflon-coated 250-pm silver wire (for use as a reference), were pulled into the PE tubing so that the glass-sealed ends of the platinum were flush with the end of the PE tubing. The Teflon on the end of the silver wire was removed, and this bare section could extend beyond the edge of the PE tubing. The electrode's zero response was tested in saline bubbled with $N_2$, and its linearity was tested in saline bubbled with different mixtures of $O_2$ and $N_2$. A cellulose diacetate membrane was put on the tip by dipping the electrode in 5% cellulose diacetate solution. The calibration and calculation of POT in the brain are only a relative term, since the absolute value of POT is very sensitive to the electrode location on the brain.

**Extracellular $K^+$ Electrode**

A $K^+$-sensitive membrane made of valinomycin mixed in a silicone rubber compound was fixed to one end of a glass tube (OD = 1.5 mm) with a silicone rubber adhesive sealant. The electrode was filled with 3 mM KCl-150 mM NaCl solution and an Ag-AgCl wire was immersed in the solution through the open end. This end was then sealed with epoxy. The response of the electrode was tested and calibrated in solutions of $K^+$ in the physiological range (3, 7.5, 15, 30, and 60 mM).

**DC Electrodes**

DC was measured concentrically around the $K^+$ probe and the light guide. Both probes had a saline bath around their perimeter, and one Ag-AgCl wire was immersed in each bath.

**Reference Electrode**

An Ag-AgCl wire sealed in a plexiglas block, with the wire sticking out through a hole into which PE-100 tubing would fit snugly was used. The PE-100 was stuffed with a string that expanded when wet, which came into contact with the wire at one end, and stuck out 2–3 cm at the other end fastened under the skin in the neck area of the animal.

**Surface Fluorometry**

The intramitochondrial NADH redox state was monitored by a light-guide DC fluorometer-reflectometer [77]. The source of the 366-nm light was a 100-W mercury arc cooled by water. The emitted light was split in a 90:10 ratio for the measurement of NADH fluorescence and reflectance, respectively. The common part of the light guide was cemented inside the multiprobe assembly in the same way as the other electrodes. A short piece of the excitation and emission bundle of fibers was connected to the common part. The signals were calibrated as described previously [44, 77].

### 3.9.3 *Typical Results*

The typical response to partial ischemia is shown in Fig. 3.18. After occlusion of the right carotid artery, the $pO_2$ dropped simultaneously with increased level of intramitochondrial NADH. The occlusion time was 5 min, and there were two steps in the responses during this period. The occlusion led to partial ischemia due to the compensation of blood flow coming from the left carotid artery through the anterior part in the circle of Willis. One can see a small increase in $pO_2$ and oxidation of NADH 30 s after the occlusion. During this stage, the ECoG was somewhat depressed, as seen in the integrated ECoG also. About 2 min after occlusion the second phase started, and one could name it "ischemic depolarization (ID)", having the property to start in the most sensitive area of the cortex and spread through the entire hemisphere. This event is labeled as "1" in Fig. 3.18, and one can see that it started near the light guide and propagated to the $K^+$ electrode area. During the ischemic depolarization (ID), few events occurred. We believe that the increased level of $K^+$ due to decreased energy availability is the major factor leading to the development of ID. As a result, $K^+$ accumulated and reached the maximum level, which, in this gerbil, was more than a 10-fold increase as compared to the baseline level. Due to the ID phenomenon, which was also recorded as a negative shift in the DC traces and ECoG depression ($DC_F$, $DC_{K+}$), a large increase in the reflection signal was recorded, leading to an apparent oxidation of NADH, but we showed that this is an artifact of the correction factor used for the F signal [25]. In many gerbils, we observed a drop in $pO_2$ during the ID period, as can also be seen in Fig. 3.3, which suggested that a constriction of blood vessels has occurred.

During the recovery period (after reopening of the carotid artery), a clear correlation between the intensity of ischemic insult and reestablishment of normal cortical activities was noted. Simultaneously, with opening of the vessels, a decrease in the R signal was recorded together with an increase in $pO_2$ (2 in Fig. 3.18). A second decrease in the R signal was recorded 1 min later. The two step of the R changes are probably due to increased blood volume in the tissue and absorption of more light by the hemoglobin. The recovery of the R signal to the preocclusion level was very slow and was parallel to the decrease of the overshoot in $pO_2$ levels. The recovery of $K^+_e$ and DC traces occurred as soon as energy was available and, in this occlusion (partial ischemia only), took about 6–7 min. The ECoG also recovered very fast after this ischemic episode.

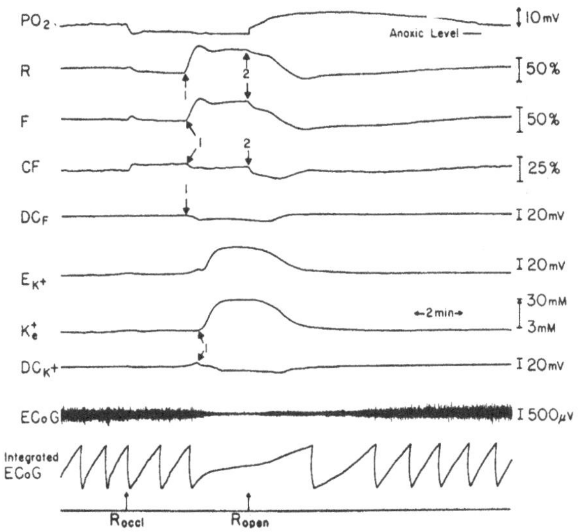


**Fig. 3.18** Metabolic ionic and electrical responses to partial ischemia induced by right carotid artery occlusion in the Mongolian gerbil. Arrow 1 represents the beginning of the ischemic depolarization (ID) and arrow 2 represents the recovery of blood flow after reopening of the carotid artery. $pO_2$ values are not an absolute one, and the anoxic level obtained under $N_2$ breathing is indicated [116]

## 3.10 Simultaneous Real-Time Monitoring of Brain NADH, $HbO_2$, ECoG, DC Potential, Extracellular $K^+$ and $Ca^{2+}$

### 3.10.1 Introduction

The online evaluation of brain physiological and biochemical functions in vivo is expected to contribute to a better understanding of the development of a pathophysiological state. The development and usage of a multiprobe assembly (MPA) for monitoring of the metabolic, ionic, and electrical activities from the cerebral cortex of experimental animals was the subject of our previous publications [75, 78–80]. In this study, we further extended the MPA to include a second fiber-optic bundle for monitoring of intracapillary hemoglobin oxygenation (oxy-deoxy Hb) using the EMPHO-I device [81–83].

As an animal model for the studies we adopted the Mongolian gerbil (*Meriones unguiculatus*), which provides an easy way to produce reversible partial or complete ischemia [39, 47, 71, 84, 85]. In the same animal we also tested the effects of graded hypoxia as well as cortical spreading depression (CSD) [59, 86–88]. The effect of CSD wave on brain oxygenation, as evaluated by oxy-deoxy Hb spectral analysis, together with NADH redox state, is described here for the first time.

### 3.10.2 Multiprobe Assembly

The basic concepts of the MPA, as previously described [75, 79, 80], were adopted and modified to accommodate more probes in the same surface area of the brain. As shown in Fig. 3.19, the center hole was used for the NADH light guide, which was developed to include fibers for relative cerebral blood flow (RCBF) measurement by laser Doppler flowmeter (LDF) [89] in addition to NADH fluorescence. In this study, two light guides were used, namely the center one (L), for NADH and cerebral blood flow (CBF) measurements, and a microlight guide used for $HbO_2$ monitoring [83]. In order to prevent interference between the measurements of $HbO_2$ and CBF, we used them alternatively with the NADH measurements. The longitudinal section of the MPA, shown on the left side of the figure, corresponds to the cross-sectional line A shown on the right side. In order to see the two light guides in the same longitudinal section, a different sectioning line would need to be used (B). Extracellular $K^+$ and $Ca^{2+}$ electrodes were made by WPI Inc. (Sarasota, FL, USA) using 1.2-mm diameter PVC tubing (Cole Flex, W, Babylon USA).

The electrodes were assembled in the holder by the same basic approach described previously for connection of the DC steady potential electrode [75]. For the DC electrode we used the electrode holder (containing Ag/AgCl pellet) provided by WPI Inc., and polyethylene tubing was connected to it and inserted into the holder. The center light guide (NADH) was covered by a rigid brass tube and was glued by a

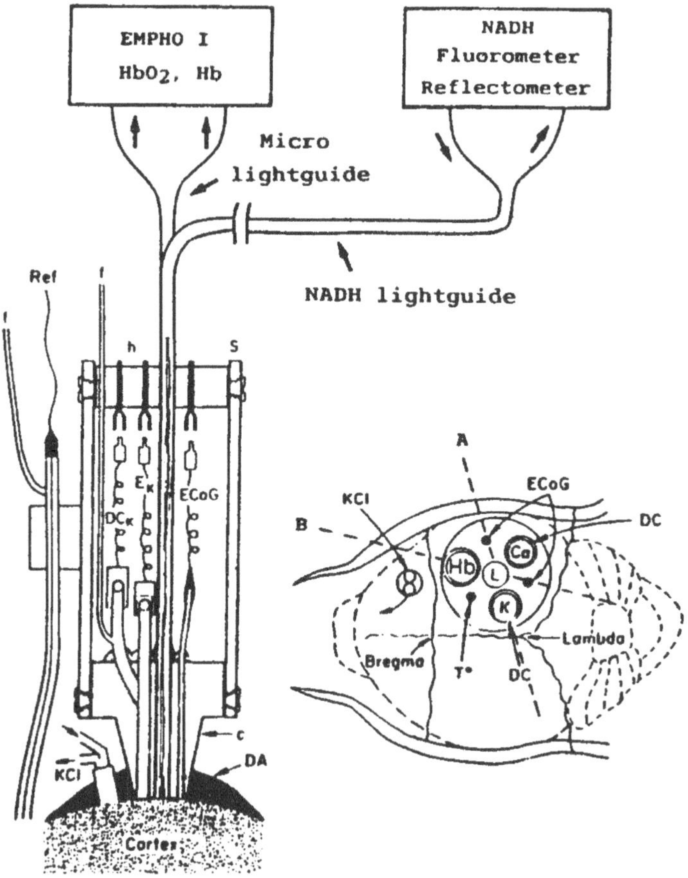


**Fig. 3.19** Schematic presentation of the multiprobe assembly (MPA) used in this study. The longitudinal section, at the left, shows the two light guides connected to EMPHO-I (microlight guide) and NADH fluorometer-reflectometer (NADH light guide). At the right side of the figure, the various probes are shown as located on the gerbil cerebral hemisphere. L—NADH light guide; Hb—EMPHO-light guide; K, Ca—extracellular potassium and calcium electrodes; DC—steady potential electrodes; T—temperature probe; ECoG—electro-corticography electrodes; KCl—push–pull cannula for topical application of KCl solution; Ref—reference electrode; DCk, Ek, ECoG—cables connecting the various electrodes to the cable holder; f—filling tubes connected to the reference and the DC electrodes; C—plexiglass cannula; DA—dental acrylic cement; h, s—plexiglass cable holder and sleeve [117]

5-min epoxy so that after cementation to the cannula (screwed and glued) it served as an axis to hold the cannula (c) and the connector holder (h) at a fixed and convenient distance.

### 3.10.3 NADH Fluorescence Measurement

In this study we used a single-channel DC fluorometer-reflectometer [3]. The source for the 366-nm excitation light was a 100-W air-cooled mercury arc passing via an excitation fiber-optic bundle to the brain (240-μ diameter quartz fibers). The emitted light from the brain transmitted through another fiber-optic bundle was split into 90:10 ratios in order to measure the fluorescent light (F: 90%) and the reflected light (R: 10%). The calibration of the signals as well as other details was published previously [3, 90].

### 3.10.4 Hemoglobin Spectrophotometry

In order to evaluate the level of hemoglobin oxygenation, we used the Erlangen microlight guide spectrophotometer EMPHO-I [83]. The light source was a water-cooled xenon high-pressure arc lamp (XBO 75W/2 Osram) operated by a starting unit and a power supply (Muller, Moosinning, Germany). The light, (400–1200 nm) passing through a heat absorption filter (KGI, Schott), located between two focusing lenses, entered a central 250-μ diameter micro-light-guide fiber, closely surrounded by a hexagon of six detecting fibers (70-μ diameter), and was transmitted to the brain surface. The emitted light from the brain was passed through a rotating interference band-pass filter disc (Anders, Nabburg), which works in the spectral domain at 502–650 nm. All other details regarding data collection, calibration, and calculations were published recently [83].

### 3.10.5 Typical Result

Figure 3.20 shows the response of the gerbil brain to ischemia induced by unilateral as well as bilateral carotid artery occlusion while the MPA was located on the right hemisphere. The 3-min occlusion of the right carotid artery led to partial ischemia, [85] as seen by deoxygenation of $HbO_2$ as well as increased level of reduced NADH (CF trace). The addition of left artery occlusion (while the right one was occluded) created a more severe level of ischemia, during which the ECoG became depressed, and a small increase in extracellular $K^+$ was noted. No change in extracellular $Ca^{2+}$

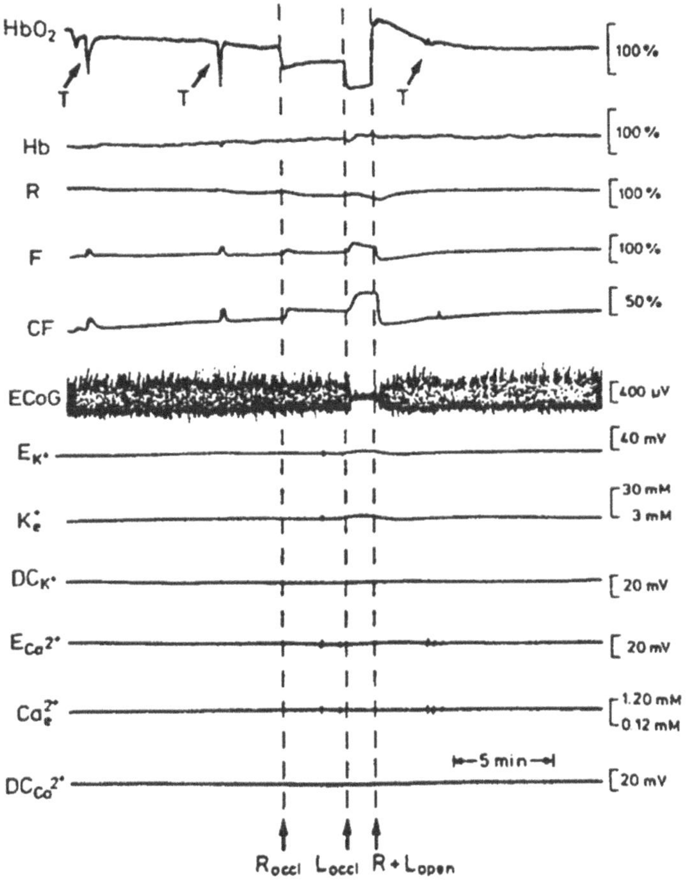


**Fig. 3.20** The effects of unilateral (Roccl) and bilateral (Loccl) carotid artery ligation on the various parameters monitored from the gerbil brain. $HbO_2$—hemoglobin oxygenation (the pre-ischemic value was taken as 100%); Hb—hemoglobin volume or content; R, F, CF—reflectance, fluorescence, and corrected fluorescence of NADH; EcoG—electrocoticogram; $E_{k+}$, $ECa_{2+}$ —uncorrected potassium and calcium potential; $K^{+}_{e}$, $Ca^{2+}_{e}$—corrected extracellular potassium and calcium levels; $DC_{k+}$, $DC_{Ca2+}$—DC steady potential measured around the potassium and calcium electrodes; T—oxygen consumption test [117]

level or DC steady potential was noted. When the two carotid arteries were opened, the brain became oxygenated within a few seconds, as seen by overshoot of the $HbO_2$ trace and recovery of the NADH redox state to the pre-ischemic level. The letter T in Fig. 3.20 represents the test for $O_2$ consumption performed before and after the ischemic episode, as will be discussed below.

## 3.11 Simultaneous Real-Time Monitoring of Brain NADH, CBF, ECoG, DC Potential, Extracellular $K^+$ and $Ca^{2+}$

The interrelation between cerebral blood flow and metabolism is an important factor in understanding brain functions under normal and pathological situations. Since few of the situations, such as spreading depression and epilepsy, are transient and are not in steady state it became necessary to monitor the various parameters under observation in real time and in a continuous mode. Until the late 1980s, CBF monitoring was limited to a single time point per animal using mapping technique or to a 10–15-min' interval between single measurements. The development of laser Doppler flowmeter (LDF) opened up the possibility of monitoring CBF in a real-time continuous mode [91]. As shown by a few groups, the CBF monitored by the LDF technique was significantly correlated to the flow measured by other quantitative techniques [92, 93]. Since CF is only one of the important parameters to be measured we incorporated a laser Doppler (LD) probe into the multiprobe assembly developed by us a few years ago [75, 94]. Since all the probes used in the MPA are surface probes we also monitored the CBF from the surface of the brain.

### *3.11.1 Multiprobe Assembly*

The multiprobe assembly (MPA) used in this study has the following features:

(a) All parameters have the same type of noninvasive surface contact with the cortex.
(b) The multiprobe assembly can be removed without any damage from the skull at the end of the experiment, so that repetitive experiments can be performed in a short period of time with minimal technical support.
(c) All signals monitored by the MPA show only very slight sensitivity to movements of the animal. All details regarding the preparation of electrodes were described previously [75, 94].

Figure 3.21 shows the multiprobe assembly used. For the present studies, a modified assembly which is under development was used and has the following differences:

1. Two light guides inserted into the MPA provide information on the NADH redox state (PN) as well as the relative CBF using a laser Doppler flow meter (LDF).
2. A $Ca^{2+}$ electrode was added to the MPA.
3. The DC steady potential is measured by a highly stable Ag/AgCl electrode located around the $K^+$ and the $Ca^{2+}$ electrodes.
4. The reference electrode is an Ag/AgCl electrode (via a salt bridge) located below the skin of the animal's neck.
5. The $Ca^{2+}$ electrode has the same construction as the $K^+$ electrode and is based on the calcium ionophore described by Dr. Simon's group [95].

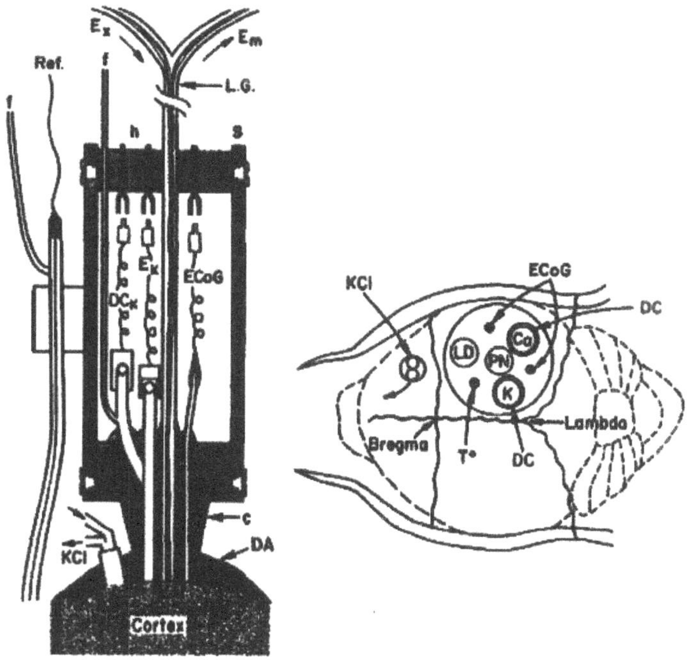


**Fig. 3.21** Schematic presentation of the multiprobe assembly (MPA) in a longitudinal section as well as its location on the gerbil brain. L.G.—light guide, c—cannula, f—filling tube for the DC and reference electrodes, h—connector holder, s—plexiglass sleeve, DA—dental acrylic, T°—temperature probe, Ex—excitation, EM—emission, KCL—push-pull cannula to elicit SD, ECoG—electrocortical electrodes, PN–NADH monitoring light guide, LD—laser Doppler flow meter light guide [98]

**NADH Redox State Fluorometry** The principle of NADH monitoring from the surface of the brain is that excitation light (366 nm) is passed from the fluorometer to the brain via a bundle of optical fibers made of quartz. The emitted light (450 nm), together with the reflected light at the excitation wavelength, is transferred to the fluorometer via another bundle. The changes in the reflected light are correlated to changes in tissue blood volume and also serve for correction of hemodynamic artifacts appearing in the NADH measurement. Calibration of the system was done as follows: the reflectance and fluorescence signals obtained from the photomultipliers (RCA931B) were calibrated to a standard signal (0.5 V) by variation of photomultiplier dynode voltage obtained from the high-voltage power supply. The standard 0.5 V signal was set to give a half-scale deflection on the recorder (2.5 cm) with pen resting at mid-scale. The gain was increased as required by a factor of two or four to give 50% or 25% of the full scale, respectively. The changes in fluorescence and reflectance signals were calculated relative to the calibrated signals under normoxic conditions. This type of calibration is not an absolute one but provides reliable and reproducible results from different animals and between different laboratories. More details on this technique have been published [3, 90].

In order to measure in real time, the CBF from the same cortical area of the MPA location, we used the laser Doppler flow meter (LDF) technique [91]. In the last year, this technique was calibrated against the $H_2$ clearance probe [92] or the $^{14}C$-iodoantipyrine in autoradiography, [93] two well-established ways to monitor CBF quantitatively. It seems that this LDF provides only relative changes, which correlated very well with relative changes in RCBF measured with the other two approaches. We used the 1.4-mm light guide (TSI probe model P-433) inside the MPA and correlated it to the NADH redox state as well as other parameters.

### 3.11.2 Animal Preparation

Gerbils (50–70 g) were used in this study. The animal was anesthetized by Equithesin (each ml contains: pentobarbital 9.72 mg, chloral hydrate 42.51 mg, magnesium sulfate 21.25 mg, propylene glycol 44.24% w/v, alcohol 11.5%, and water) injected IP (0.3 ml/100 g). We have been using this anesthetic for more than 50 years, and it never showed a significant effect on mitochondrial activity unless we gave an overdose. The skull was exposed, and a 6-mm hole was drilled in the parietal bone. The dura mater was removed gently, and the MPA was located on the brain so that any pressure on the brain was avoided. The MPA cannula and screws inserted in two to four different locations were cemented to the skull by dental acrylic. In experiments in which CSD was induced, a special push-pull cannula (2-mm diameter) was located epidurally 1–2 mm from the MPA. Additional injections of anesthetic were given to the gerbils every 30 min during the measurements (0.06 ml Equithesin/100 g body weight of gerbil).

### 3.11.3 Typical Results

During the ischemic episode (Fig. 3.22), a two-step decrease in CBF (LDF) was recorded, which correlated with an increase in NADH redox state (CF). During the ischemic depolarization, the large increase in the reflectance signal (R), which is interpreted as a decrease in blood volume, was correlated with the same pattern of change in the laser Doppler volume parameter. Also a very clear correlation between the R trace and the laser Doppler traces was recorded during the recovery phase. The hyperemic response (increase in flow and volume) was measured simultaneously by the reflectance trace.

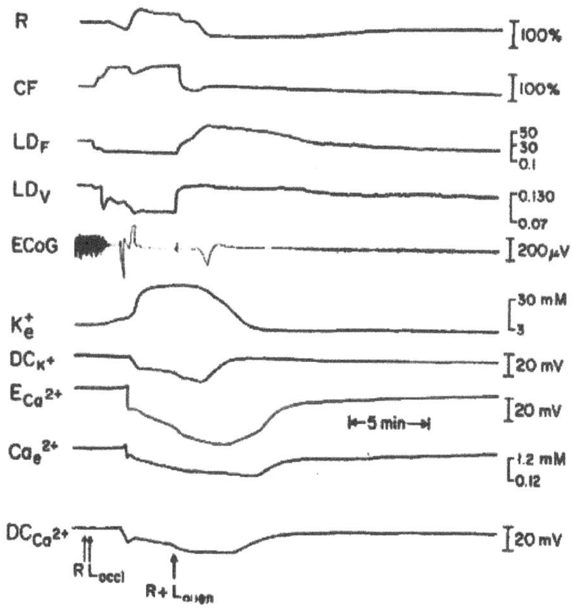


**Fig. 3.22** Responses of blood flow and volume as well as other metabolic and ionic parameters to bilateral carotid artery occlusion in the gerbil brain. The light guide of the laser Doppler was located inside the MPA, as seen in Fig. 5 [98]

# 3.12 Simultaneous Monitoring of Brain NADH, CBF, ECoG, DC Potential, Extracellular $K^+$, $Na^+$, $Ca^{+2}$, and $H^+$

## *3.12.1 Introduction*

Since the ionic shifts occurring under brain ischemia and cortical spreading depression (SD) are very similar [96], it is of significant importance to correlate those changes to the hemodynamic and metabolic responses in the same animal monitored. Due to variations in the speed of propagation (between animals), the real-time monitoring approach of all relevant parameters will provide a better understanding of the events occurring under those two pathological states. Using the multiparametric monitoring approach, we accumulated enough information which suggests that the difference between the two phenomena depends upon the oxygen balance in the tissue before, during, and after the event [58, 97, 98]. The multiprobe assembly (MPA) used in this study contained optical fibers and surface electrodes by which the hemodynamic, metabolic, ionic, and electrical activities were monitored simultaneously in real time. The $O_2$ balance was evaluated by monitoring the $O_2$ supply (relative CBF using laser Doppler flowmeter) as well as the intracellular $O_2$ levels (intramitochondrial redox state using NADH fluorometry). The ionic homeostasis was evaluated by monitoring the extracellular levels of various ions ($K^+$, $Ca^{2+}$. $Na^+$, $H^+$) using surface minielectrodes. The spontaneous electrical activity (ECoG) and DC steady potential were used as two electrical parameters.

## *3.12.2 Methods*

### 3.12.2.1 Multiprobe Assembly

This approach to monitor the brain in vivo in real time was continuously developed since the original description in 1982 [75, 94]. In this study, we are presenting results obtained using the typical MPA described in several publications [79, 98–101] as well as using a new assembly containing four minielectrodes ($K^+$, $Ca^{2+}$, $Na^+$, $H^+$) and one DC electrode as well as a new metaflow probe (MFP). The new probe is made up of a combined bundle of optical fibers to monitor relative CBP (flow) and NADH redox state (metabolism) from the same tissue volume (Fig. 3.23).

### 3.12.2.2 Relative Cerebral Blood Flow (RCBF)

In order to measure in real-time, the CBF from the same cortical area of the MPA location, we used the laser Doppler flowmeter (LDF) technique. In the last few years, this technique was calibrated against the $H_2$ clearance probe or the $^{14}C$-iodoantipyride in autoradiography, two well-established ways to quantitatively monitor CBF [102]. It seems that this LDF provides only relative changes which correlated very well with relative changes in RCBF measured with the other two approaches. We used the 1.4-mm light guide (TSI probe model P-433) inside the MPA and correlated it to the NADH redox state as well as the other parameters. In studies where the metaflow probe was in the monitoring system, the unit manufactured by Perimed Inc. (Model PF2B) was in use.

### 3.12.2.3 NADH Redox State Fluorometry

The principle of NADH monitoring from the surface of the brain is that excitation light (366 nm) is passed from the fluorometer to the brain via a bundle of optical quartz glass fibers. The emitted light (450 nm), together with the reflected light at the excitation wavelength, is transferred to the fluorometer via another bundle [10, 40]. The changes in the reflected light are correlated to changes in tissue blood volume and also serve for correction of hemodynamic artifacts appearing in NADH measurement. More details on this technique have been published [3, 41, 90].

### 3.12.2.4 Typical Result

Figure 3.24 shows the effects of bilateral carotid artery occlusion (4 min). In this particular gerbil, the connection between the two anterior cerebral arteries [85] was very good; therefore, the occlusion of one artery was not enough to create ischemia.

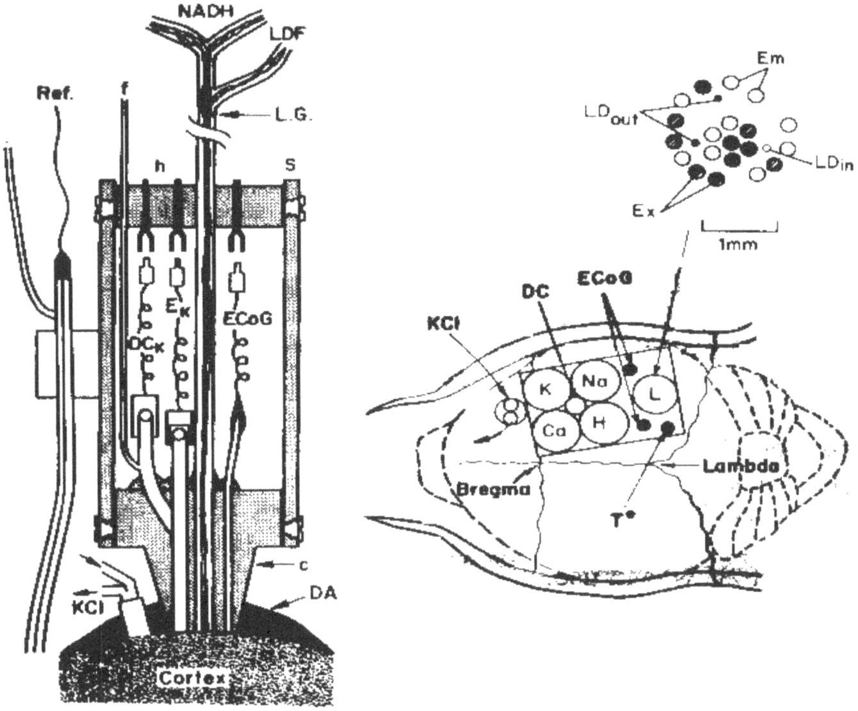


**Fig. 3.23** Schematic presentation of the developed multiprobe assembly (MPA) used in this study. LG—light guide monitoring the NADH redox state and relative CBF (LDF); ECoG—electrocorticography electrodes; K, Ca, Na, H—minielectrodes for extracellular potassium, calcium, sodium, and hydrogen monitoring; T°—temperature thermistor probe; DC—DC steady potential electrodes; DA—dental acrylic cement; Ref—reference electrode; KCl—push–pull cannula for KCl application (spreading depression); Ex, Em—excitation and emission fibers of the fluorometer; $LD_{in}$, $LD_{out}$—fibers connected to the laser Doppler flowmeter; C—plexiglass probe holder; h—connector holder; s—plexiglass sleeve; f—filling tube of reference or a DC electrode; $E_K$—cable connected to the $K^+$ minielectrode via Ag/AgCl holder; $DC_K$—cable connected to the DC steady potential electrode [118]

Under complete ischemia, the CBF reached very low values, as also indicated by the maximum level of the NADH redox state. An ischemic depolarization (ID) develops within 2–2.5 min, characterized by a massive vasoconstriction (large increase in R signal and decrease in blood volume (LDV)). The ionic responses to ID were typical as described before [46, 79].

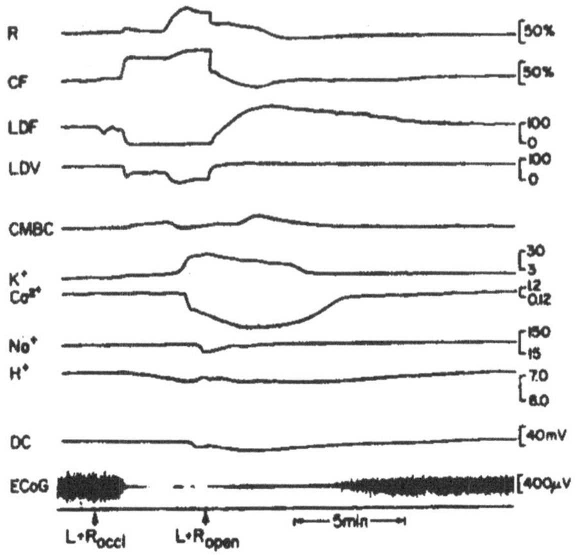


**Fig. 3.24** The effects of bilateral carotid occlusion on the various parameters monitored in the brain of the Mongolian gerbil [118]

## 3.13 Multiparametric Monitoring of NADH, CBF, ECoG, Extracellular $K^+$, $Ca^{2+}$, and DC Potential Inside NMR Magnet Measuring ATP, PCr, and Pi

### 3.13.1 Introduction

Various approaches have been used for studying the effects of ischemia on various parameters in the gerbil brain [103] (for review, see [85]). Several studies have used the Mongolian gerbil as a model for ischemia due to the incompleteness of its circle of Willis [47, 85]. Magnetic resonance spectroscopy (MRS) of $^{31}P$ has been used as a noninvasive tool for continuous monitoring of brain energetics under various conditions [104–106].

We developed a new multiprobe assembly, which can be used inside the magnet, such that the hemodynamic, metabolic, electrical ad ionic activities could be monitored continuously and simultaneously with 31P-MRS. Relative CBF was measured by a laser Doppler flowmeter using a fiber-optic probe [102]. Mitochondrial activity was evaluated by monitoring the NADH redox state using surface fluorometry [3, 10, 90]. Ionic homeostasis was evaluated by the extracellular levels of $K^+$ and $Ca^{2+}$ measured by surface nonselective minielectrodes [55, 107]. The electrical activities were assessed by the DC steady potential as well as the electrocorticogram (ECoG).

This study was carried out in order to determine whether it is possible to measure, simultaneously and in real time, parameters that are measurable by the multiprobe assembly with parameters that are measurable using NMR spectroscopy so as to more precisely determine injury during pathological interventions or conditions.

## 3.13.2 Methods

The details regarding monitoring of NADH fluorescence and cerebral blood flow have been presented in the previous sections of this chapter.

#### 3.13.2.1 Multiprobe Assembly

The $K^+$ and $Ca^{2+}$ electrodes (WPI Inc., USA), DC electrodes, ECoG electrodes as well as the light guide were held in a multiprobe assembly as described previously in detail [75, 94, 108]. The probes were mounted on a Lucite holder (cannula), which can be cemented directly to the skull (Fig. 3.25) and can be removed very easily from the skull at the end of the experiment without any damage, so that the same electrodes can be used for performing repetitive experiments within a short period of time.

#### 3.13.2.2 $^{31}$P MRS Data Acquisition

An Otsuka Electronics (USA) lnc., magnetic resonance spectrometer was used with a 2.1 Tesla superconducting 30-cm bore Oxford magnet. A 35.8-MHz radio frequency pulse with a 20–30-microsecond pulse width was transmitted through a 2.5-cm two-turn double-tuned copper surface coil placed directly on the skull to give approximately a 90° flip angle. This pulse was preceded by a long, low-power pulse to saturate a very broad bone peak [109]. Each free induction decay (FID) was collected for 80 msec. The pulse interval was 4 s, so that data was accumulated at a. rate of 15 FID/min, with a total of 60 FID for each 4-min data collection period. For the final analysis, a 10-Hz line broadening was applied, and baseline correction was achieved. by a 600-Hz convolution difference. These manipulations gave a flat baseline and a signal-to-noise level of 6 at the beta-ATP peak. The data were Fourier transformed, phased, and fitted. Details of this method have been published previously [110]. Quantitation of the metabolite values was determined from the areas of the peak obtained by MRS, as described previously [111].

#### 3.13.2.3 Animal Preparation

The gerbils were anesthetized by an IP injection (0.3 ml/100 g) of Equithesin (each ml contained 9.72 mg pentobarbital, 42.51 mg chloral hydrate, 21.25 mg magnesium sulfate, 44.34% w/v propylene glycol, 11.5% alcohol and water). Equithesin was also given 1 h after the initial dose in 30-min intervals (0.1 ml/100 g). The animals were placed in a head holder, and the skin above the dorsal area of the skull was removed A 6-mm hole was drilled in the parietal bone arc to accommodate the lower part of the MPA cannula. The dura mater was removed gently, and the MPA

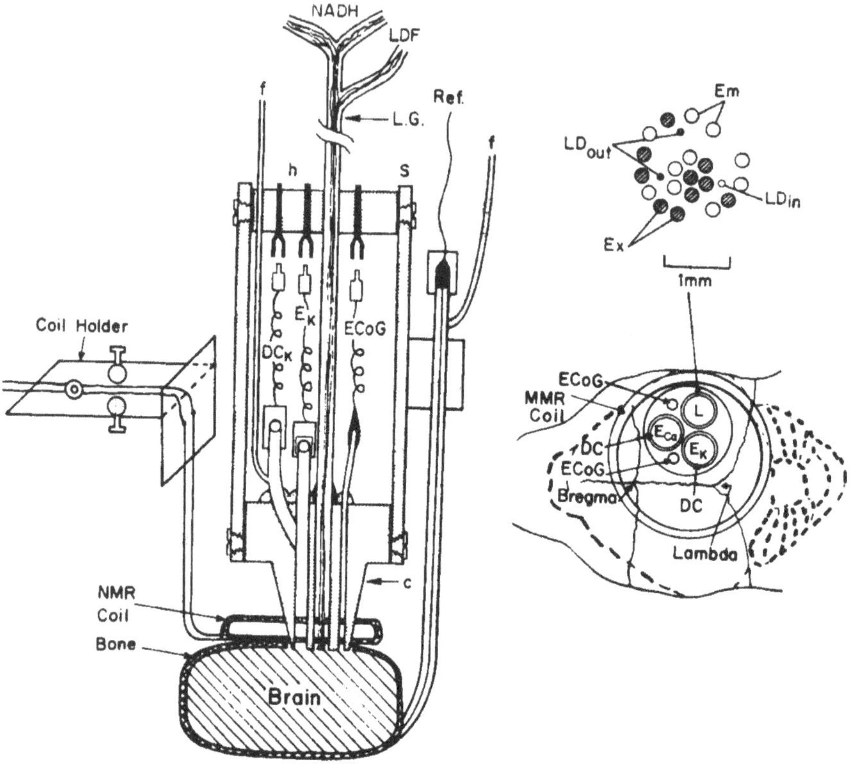


**Fig. 3.25** Schematic presentation of the experimental setup used in the study. A longitudinal section of the multiprobe assembly connected to the NMR coil and the gerbil brain is shown on the left side. The relative locations of the various probes on the brain are shown on the right side. ECoG—electrocorticographic electrodes; $E_k$, $E_{Ca}$ $DC_k$—potassium, calcium, and DC electrodes; c—plexiglass multiprobe holder; h—connection holder; s—plexiglass sleeve; Ref—reference electrode; F—filling tube of DC or reference electrode; L.G.—light guide; Ex, Em—excitation and emission fibers for NADH monitoring; $LD_{in}$, $LD_{out}$—fibers for CBF monitoring by laser Doppler flowmeter [119]

was placed on the brain using a micromanipulator attached to the operating table. During this procedure, the two-turn copper surface coil (15-mm diameter) was located on the exposed skull (Fig. 3.25). The MPA and the coil were cemented to the skull by dental acrylic and held with four plastic screws inserted into the skull. The two common carotid arteries were isolated and 4-0 silk sutures were placed around them for later occlusion. Body temperature was maintained by wrapping the gerbil in a water heating mantle. The animal was released from the head holder and transferred into the center of a 2.1-Tesla magnet together with the cemented MPA and MRS coil.

All the parameters were monitored via the long fibers and wires connected to the measuring devices located out of the range of the magnetic field. The various parameters were recorded on a Grass multichannel pen recorder.

#### 3.13.2.4 Typical Results

In order to compare the data accumulated by the MPA and the MRS, quantitative analysis of all measured parameters was performed at 5-min intervals. Figure 3.26 shows the typical effect of 20-min bilateral carotid artery occlusion on the various parameters as calculated. Part A exhibits the hemodynamic changes occurring in the three major parameters, that is, CBF, CBV (volume), and NADH redox state. After four control points (total 15 min) ischemia was induced and CBF, as well as CBV decreased to the minimal levels within 5 min. Concomitantly, the intramitochondrial NADH redox state reached maximal levels. The low CBF and high NADH remained unchanged until the reopening of the two carotid arteries. NADH returned to its pre-ischemic levels within 10 min, while CBF showed a large hyperemic response during the recovery phase. The CBV showed a small overshoot response.

The energy state of the gerbil brain during ischemia is shown in Fig. 3.26b. PCr and ATP levels decreased gradually during the ischemic episode and reached minimal levels 15 min after the occlusion. Ten minutes lapsed following reopening of the vessels before the high-energy compounds returned to their pre-ischemic levels.

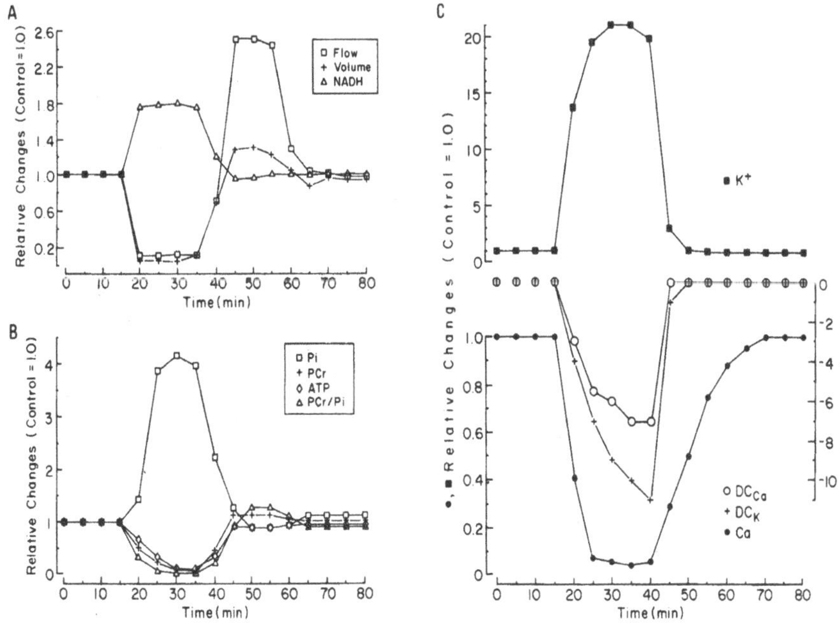


**Fig. 3.26** Typical changes of the various parameters monitored from the gerbil brain located in the NMR magnet and exposed to 20 min of ischemia. The original numbers measured were normalized in order to present a few of them on the same ordinate scale. Only the DC potential is presented in absolute units (mv). (**a**) cerebral blood flow and volume, NADH redox state; (**b**) Pi, PCr, ATP PCr/Pi; (**c**) $K^+$, $DCCa^{2+}$, $DCK^+$, $Ca^{2+}$ [119]

The typical disturbance in ionic homeostasis caused by the ischemic event is shown in Fig. 3.26c. The major changes in extracellular $K^+$ and $Ca^{2+}$ occurred during the first 10 min of ischemia followed by small changes during the remaining 10 min. In this particular gerbil, recovery of the two ions did not exhibit identical kinetics. Most of the $K^+$ recovered within 10 min (after a slow initial phase), while the recovery of $Ca^{2+}$ was much slower and was complete only after 30 min. The DC steady potential demonstrated a negative shift when measured concentric to the $K^+$ and $Ca^{2+}$ electrodes. Return of the DC potential to baseline levels occurred after 5–10 min of reperfusion whereas during the initial 5 min, changes were minimal.

## References


1. Mayevsky A (2015) Mitochondrial function *In Vivo* evaluated by NADH fluorescence. Springer, Cham
2. Mayevsky A (1978) Shedding light on the awake brain. In: Dutton PL, Leigh J, Scarpa A (eds) Frontiers in bioenrgetics: from elrctrons to tissues. Academic, New York, pp 1467–1476
3. Mayevsky A (1984) Brain NADH redox state monitored *in vivo* by fiber optic surface fluorometry. Brain Res Rev 7:49–68
4. Mayevsky A, Rogatsky G (2007) Mitochondrial function *in vivo* evaluated by NADH fluorescence: from animal models to human studies. Am J Phys Cell Phys 292:C615–C640
5. Duysens LNM, Amesz J (1957) Fluorescence spectrophotometry of reduced phosphopyridine nucleotide in intact cells in the near-ultraviolet and visible region. Biochim Biophys Acta 24:19–26
6. Chance B, Legallias V (1959) Differential microfluorimeter for the localization of reduced pyridine nucleotide in living cells. Rev Sci Instrum 30:732–735
7. Chance B, Jobsis F (1959) Changes in fluorescence in a frog sartorius muscle following a twitch. Nature 184:195–196
8. Chance B (1959) The response of mitochondria to muscular contraction. Ann NY Acad Sci 81:477–489
9. Chance B, Cohen P, Jobsis F, Schoener B (1962) Intracellular oxidation-reduction states *in vivo*. Science 137:499–508
10. Chance B, Cohen P, Jobsis F, Schoener B (1962) Localized fluorometry of oxidation-reduction states of intracellular pyridine nucleotide in brain and kidney cortex of the anesthetized rat. Science 136:325
11. Chance B, Thorell B (1959) Fluorescence measurements of mitochondrial pyridine nucleotide in aerobiosis and anaerobiosis. Nature 184:931–934
12. Chance B, Legallias V, Schoener B (1963) Combined fluorometer and double-beam spectrophotometer for reflectance measurements. Rev Sci Instrum 34:1307–1311
13. Chance B, Legallias V, Schoener B (1962) Metabolically linked changes in fluorescence emission spectra of cortex of rat brain, kidney and adrenal gland. Nature 195:1073–1075
14. Chance B, Mayer D, Rossini L (1970) A time-sharing instrument for direct readout of oxidation- reduction states in intracellular compartments of cardiac tissue. IEEE Trans Biomed Eng BME-17:118–121
15. Chance B, Graham N, Mayer D (1971) A time sharing fluorometer for the readout of intracellular oxidation-reduction states of NADH and flavoprotein. Rev Sci Instrum 42:951–957
16. Chance B (1966) The identification and control of metabolic states. Genootschap ter Bevordering van Natuur-, Genees-, en Heelkunde te Amsterdam:5–37
17. Chance B, Schoener B (1965) A correlation of absorption and fluorescence changes in ischemia of the rat liver, *in vivo*. Biochem Z 341:340–345

18. Chance B, Williamson JR, Jamieson D, Schoener B (1965) Properties and kinetics of reduced pyridine nucleotide fluorescence of the isolated and *in vivo* rat heart. Biochem Z 341:357–377
19. Dora E, Kovach AGB (1982) Effect of acute arterial hypo- and hypertension on cerebrocortical NAD/NADH redox state and vascular volume. J CBF Metab 2:209–219
20. Dora E (1984) A simple cranial window technique for optical monitoring of cerebrocortical microcirculation and NAD/NADH redox state. Effect of mitochondrial electron transport inhibitors and anoxic anoxia. J Neurochem 42:101–108
21. Ginsberg MD, Reivich M, Frinak S, Harbig K (1976) Pyridine nucleotide redox state and blood flow of the cerebral cortex following middle cerebral artery occlusion in the cat. Stroke 7:125–131
22. Harbig K, Chance B, Kovach AGB, Reivich M (1976) *In vivo* measurement of pyridine nucleotide fluorescence from cat brain cortex. J Appl Physiol 41:480–488
23. Jobsis FF (1964) Basic processes in cellular respiration. In: Fenn, Rabn (eds) Handbook of physiology – respiration, pp 63–124
24. Jobsis FF, Duffield JC (1967) Oxidative and glycolytic recovery metabolism in muscle. Fluorometric observations on their relative contributions. J Genet Physiol 50:1009–1047
25. Jobsis FF, O'Connor M, Vitale A, Vreman H (1971) Intracellular redox changes in functioning cerebral cortex. I. Metabolic effects of epileptiform activity. J Neurophysiol 3465:735–749
26. Jobsis FF, O'Connor MJ, Rosenthal M, Van Buren JM (1971) Fluorometric monitoring of metabolic activity in the intact cerebral cortex. In: Somjen GG (ed) Excerpta medica international congress series, NO. 253, Paris, pp 18–26
27. Rosenthal M, Martel D, LaManna JC, Jobsis FF (1976) *In situ* studies of oxidative energy metabolism during transient cortical ischemia in cats. Exp Neurol 50:477–494
28. Rosenthal M, Jobsis FF (1971) Intracellular redox changes in functioning cerebral cortex. II. Effects of direct cortical stimulation. J Neurophysiol 34:750–762
29. Schuette WH, Lewis DV, O'Connor M, Van Buren JM (1976) The design and operation of a dual-beal long-focal-length fluorometer for monitoring the oxidative metabolism *in vivo*. Med Biol Eng 14:235–238
30. Gosalvez M, Thurman RG, Chance B, Reinhold H (1972) Mammary tumours *in vivo* demonstrated by fluorescence of pyridine nucleotide. Br J Radiol 45:510–514
31. Gosalvez M, Thurman RG, Chance B, Reinhold HS (1972) Indication of hypoxic areas in tumours from *in vivo* NADH fluorescence. Eur J Cancer 8:267–269
32. Anderson RE (1975) Instrumentation for *in vivo* cerebral NADH studies in squirrel monkey. IEEE Trans Biomed Eng BME-22:220–224
33. Anderson RE (1978) Comparison of dark-field and bright-field incident illumination for *in vivo* measurements of reduced pyridine nucleotides. Anal Biochem 91:496–508
34. Sundt TM, Anderson RE (1975) Reduced nicotinamide adenine dinucleotide fluorescence and cortical blood flow in ischemic and nonischemic squirrel monkey cortex. I. Animal preparation, instrumentation, and validity of model. Stroke 6:270–278
35. Mayer D, Chance B, Legallias V (1971) Time-sharing in spectrophotomertry and fluorometry. Probes of structure and function of macromolecules and membranes. In: Probes and membranes function. Academic, New York, pp 527–534
36. Jobsis FF, Stainsby WN (1968) Oxidation of NADH during contractions of circulated mammalian skeletal muscle. Respir Physiol 4:292–300
37. Jobsis F, Legallias V, O'Connor M (1966) A regulated differential fluorometer for the assay of oxidative metabolism in intact tissues. IEEE Trans Biomed Eng BME-13:93–99
38. Gyulai L, Dora E, Kovach AGB (1982) NAD/NADH: redox state changes on cat brain cortex during stimulation and hypercapnia. Am J Phys 243:H619–H627
39. Mayevsky A (1978) Pyridine nucleotide oxidation reduction state of the cerebral cortex in the awake gerbil. J Neurosci Res 3:369–374
40. Chance B, Oshino N, Sugano T, Mayevsky A (1973) Basic principles of tissue oxygen determination from mitochondrial signals. Oxygen transport to tissue. Adv Exp Med Biol 37A:277–292

41. Mayevsky A, Chance B (1973) In: Bicher HI, Bruley DF (eds) A new long-term method for the measurement of NADH fluorescence in intact rat brain with implanted cannula. Plenum Press, New York, pp 239–244
42. Mayevsky A, Chance B (1976) The effect of decapitation on the oxidation-reduction state of NADH and ECoG in the brain of the awake rat. Adv Exp Med Biol 75:307–312
43. Mayevsky A, Chance B (1975) Metabolic responses of the awake cerebral cortex to anoxia hypoxia spreading depression and epileptiform activity. Brain Res 98:149–165
44. Mayevsky A (1976) Brain energy metabolism of the conscious rat exposed to various physiological and pathological situations. Brain Res 113:327–338
45. Mayevsky A, Chance B (1974) Repetitive patterns of metabolic changes during cortical spreading depression of the awake rat. Brain Res 65:529–533
46. Mayevsky A, Zeuthen T, Chance B (1974) Measurements of extracellular potassium, ECoG and pyridine nucleotide levels during cortical spreading depression in rats. Brain Res 76:347–349
47. Levine S, Payan H (1966) Effects of ischemia and other procedures on the brain and retina of the gerbil (*Meriones unguiculatus*). Exp Neurol 16:255–262
48. Levine S, Sohn D (1969) Cerebral ischemia in infant and adult gerbils. Relation to incomplete circle of Willis. Arch Pathol 87:315–317
49. Mrsulja BB, Lust WD, Mrsulja JY, Passonneau IV, Klatzo I (1976) Post-ischemic changes in certain metabolites following prolonged ischemia in the gerbil cerebral cortex. J Nevrochem 26:1099
50. Levy DE, Duffy TE (1975) Effect of ischemia on energy metabolism in the gerbil cerebral cortex. J Neurochem 24:1287–1289
51. Zeuthen T (1971) A method to fill glass microelectrodes by local heating. Acta Physiol Scand 81:141–142
52. Zeuthen T, Hiam RC, Silver IA (1974) Microelectrode recording of ion activity in brain. In: Berman HJ, Hebert NC (eds) Ion-selective microelectrodes. Advances in experimental medicine and biology. Springer, Boston, pp 145–156
53. Walker JA (1971) Ion specific liquid ion exchanger microelectrodes. Anal Chem 43:89A–93A
54. Kobayashi S, Kaede K, Nishiki K, Ogata E (1971) Microfluorometry of oxidation-reduction state of the rat kidney *in situ*. J Appl Physiol 31:693–696
55. Crowe W, Mayevsky A, Mela L (1977) Application of a solid membrane ion selective electrode to *in vivo* measurements. Am J Phys 233:C56–C60
56. Bures J, Buresova O, Krivanek J (1974) The mechanism and applications of Leão's spreading depression of electroencephalographic activity. Academic, New York
57. Leão AAP, Morison R (1945) Propagation of spreading cortical depression. J Neurophysiol 8:33–45
58. Van Harreveld A, Stamm JS (1952) Vascular concomitants of spreading cortical depression. J Neurophysiol 15:487–496
59. Marshall WH (1959) Spreading cortical depression of Leão. Physiol Rev 39:239–279
60. Lukyanov LD, Bures J (1967) Changes in $pO_2$ due to spreading depression in cortex and nucleus caudatus of rat. Physiol Bohemoslov 16:449–455
61. Mayevsky A (1978) Ischemia in the brain: the effects of carotid artery ligation and decapitation on the energy state of the awake and anesthetized rat. Brain Res 140:217–230
62. Mayevsky A, Bar-Sagie D (1978) The interrelation between CBF, energy metabolism and ECoG in a new awake brain model. Adv Exp Med Biol 92:761–768
63. Mayevsky A, Crowe W, Mela L (1980) The interrelation between brain oxidative metabolism and extracellular potassium in the unanesthetized gerbil. Neurol Res 1:213–226
64. Chance B, Legallias V, Sorge J, Graham N (1975) A versatile time-sharing multichannel spectrophotometer reflectometer and fluorometer. Anal Biochem 66:498–514
65. Mayevsky A, Breuer Z (1992) Brain vasculature and mitochondrial responses to ischemia in gerbils. I. Basic anatomical patterns and biochemical correlates. Brain Res 58:242–250

66. Mayevsky A, Chance B (1983) Multisite measurements of NADH redox state from cerebral cortex of the awake animal. Adv Exp Med Biol 159:143–155
67. Chance B, Barlow C, Haselgrove J, Nakase Y, Quistorff B, Matschinsky F, Mayevsky A (1978) Microheterogeneities of redox states of perfused and intact organs. In: Srere P (ed) Microenvironments and metabolic compartmentation. Academic, New York, pp 131–148
68. Chance B, Schoener B, Oshino R, Itshak F, Nakase Y (1979) Oxidation-reduction ratio studies of mitochondria in freeze- trapped samples. NADH and flavoprotein fluorescence signals. J Biol Chem 254:4764–4771
69. Haselgrove J, Barlow CH, Chance B (1980) The 3D distribution of metabolic states in the gerbil brain during the course of spreading depression. In: Passoneau JV, Hawkins RA, Lust WD, Welsh FA (eds) Cerebral metabolism and neural function. Williams & Wilkins, Baltimore/London, pp 72–76
70. Robinson DG (1980) The gerbil stroke model. Gerbil Digest 7
71. Donadio MF, Kozlowski PB, Kaplan H, Wisniewski HM, Majkowski J (1982) Brain vasculature and induced ischemia in seizure-prone and non-seizure-prone gerbils. Brain Res 234:263–273
72. Berry K, Wisniewski HM, Svarzbein L, Baez S (1975) On the relationship of brain vasculature to production of neurological deficit and morphological changes following acute unilateral common carotid artery ligation in gerbils. J Neurol Sci 25:75–92
73. Schonfeld AR, Glick SD (1979) Cerebrovascular abnormalities associated with seizure susceptibility in the Mongolian gerbil. Brain Res 173:147–151
74. Mayevsky A, Bar-Sagie D, Levy N, Amosi M (1978) The effects of ischemia on the metabolic response of the cerebral cortex of the gerbil. (Abstract). Fed Proc 37:97
75. Friedli CM, Sclarsky DS, Mayevsky A (1982) Multiprobe monitoring of ionic, metabolic, and electrical activities in the awake brain. Am J Phys 243:R462–R469
76. Mayevsky A, Sclarsky DS (1983) Correlation of brain NADH redox state, $K^+$, $PO_2$ and electrical activity during hypoxia, ischemia and spreading depression. Adv Exp Med Biol 159:129–141
77. Chance B, Mayevsky A, Goodwin C, Mela L (1974) Factors in oxygen delivery to tissue. Microvasc Res 8:276–282
78. Mayevsky A, Lebourdais S, Chance B (1980) The interrelation between brain $PO_2$ and NADH oxidation- reduction state in the gerbil. J Neurosci Res 5:173–182
79. Mayevsky A, Friedli CM, Reivich M (1985) Metabolic, ionic and electrical responses of the gerbil brain to ischemia. Am J Phys 248:R99–R107
80. Mayevsky A, Zarchin N (1987) Metabolic ionic and electrical activities during and after incomplete or complete cerebral ischemia in the Mongolian gerbil. Adv Exp Med Biol 215:265–273
81. Frank KH, Schabert A, Friedl A, Brunner M, Hoper J, Kerl G, Kessler M (1984) Correlation between tissue $PO_2$ and intracapillary Hb spectra. Adv Exp Med Biol 169:811–819
82. Frank KH, Rettig V, Friedl A, Brunner M, Ellermann R, Kerl G, Hoper J (1984) Measurements of intracapillary haemoglobia spectra in the beating heart, skeletal muscle and the liver, using the Erlangen micro-lightguide spectrophotometer. Pfluger Archiv 400(Suppl):219
83. Frank KH, Kessler M, Appelbaum K, Dummler W (1989) The Erlangen micro-lightguide spectrophotometer EMPHO I. Phys Med Biol 34:1883–1900
84. Levy DE, Brierley JB (1974) Communications between vertebro-basilar and carotid arterial circulations in the gerbil. Exp Neurol 45:503–508
85. Mayevsky A, Breuer Z (1990) In: Schurr A, Rigor BM (eds) The Mongolian gerbil as a model for cerebral ischemia. CRC Press, pp 27–46
86. Leão AAP (1944) Spreading depression of activity in cerebral cortex. J Neurophysiol 7:359–390
87. Leão AAP (1944) Pial circulation and spreading depression of activity in the cerebral cortex. J Neurophysiol 7:391–396

88. Leão AAP (1947) Further observations on the spreading depression of activity in the cerebral cortex. J Neurophysiol 10:409–414
89. Mayevsky A, Flamm ES, Pennie W, Chance B (1991) A fiber optic based multiprobes system for intraoperative monitoring of brain functions. SPIE 1431:303–313
90. Mayevsky A, Chance B (1982) Intracellular oxidation-reduction state measured *in situ* by a multicannel fiber-optic surface fluorometer. Science 217:537–540
91. Stern MD, Lappe DL, Bowen PD, Chimosky JE, Holloway GA Jr, Keiser HR, Bowman RL (1977) Continuous measurement of tissue blood flow by laser-Doppler spectroscopy. Am J Phys 232:H441–H448
92. Haberl RL, Heizer ML, Marmarou A, Ellis EF (1989) Laser-Doppler assessment of brain microcirculation: effect of systemic alterations. Am J Physiol Heart Circ Physiol 256:H1247–H1254
93. Dirnagl U, Kaplan B, Jacewicz M, Pulsinelli W (1989) Continuous measurement of cerebral cortical blood flow by laser-Doppler flowmetry in a rat stroke model. J CBF Metab 9:589–596
94. Mayevsky A (1983) Multiparameter monitoring of the awake brain under hyperbaric oxygenation. J Appl Physiol 54:740–748
95. Lanter F, Steiner RA, Ammann D, Simon W (1982) Critical evaluations of the applicability of neutral carrier-based calcium selective microelectrodes. Analitica Chemica Acta:13651–13659
96. Hansen AJ (1985) Effect of anoxia on ion distribution in the brain. Physiol Rev 65:101–148
97. Van Harreveld A, Ochs S (1957) Electrical and vascular concomitants of spreading depression. Am J Phys 189:159–166
98. Mayevsky A, Yoles E, Zarchin N, Kaushansky D (1990) Brain vascular ionic and metabolic responses to ischemia in the Mongolian gerbil. J Basic Clin Physiol Pharmacol 1:207–220
99. Mayevsky A, Weiss HR (1991) Cerebral blood flow and oxygen consumption in cortical spreading depression. J CBF Metab 11:829–836
100. Mayevsky A (1992) Cerebral blood flow and brain mitochondrial redox state responses to various perturbations in gerbils. Adv Exp Med Biol 317:707–716
101. Mayevsky A (1990) Level of ischemia and brain functions in the Mongolian gerbil *in vivo*. Brain Res 524:1–9
102. Wadhwani KC, Rapoport SI, Shepherd AP, Oberg PA (1990) Blood flow in the central and peripheral nervous systems. In: Shephrd AP, Oberg PA (eds) Laser Doppler blood flowmetry. Kluwer Academic Pub, Boston, pp 265–304
103. Breuer Z, Mayevsky A (1992) Brain vasculature and mitochondrial responses to ischemia in Gerbils: II. Strain differences and statistical evaluation. Brain Res 598:251–256
104. Younkin DP, Delivoria-Papadopoulos M, Leonard JC, Subramanian VH, Eleff S, Leigh JS Jr, Chance B (1984) Unique aspects of human newborn cerebral metabolism evaluated with phosphorus nuclear magnetic resonance spectroscopy. Ann Neurol 16:581–586
105. Hilberman M, Subramanian VH, Haselgrove J, Cone JB, Egan JW, Gyulai L, Chance B (1984) *In vivo* time-resolved brain phosphorus nuclear magnetic resonance. J Cereb Blood Flow Metab 4:334–342
106. Mayevsky A, Subramanian VH, Chance B (1988) Brain oxidative metabolism of the newborn dog: correlation between $^{31}P$ NMR spectroscopy and pyridine nucleotides redox state. J CBF Metab 8:201–207
107. Yoles E, Zarchin N, Mayevsky A (1991) Effects of age on brain metabolic ionic and electrical responses to anoxia in the newborn dog *in vivo*. J Basic Clin Physiol Pharmacol 2:297–313
108. Mayevsky A (1993) Biochemical and physiological activities of the brain as *in vivo* markers of brain pathology. In: Bernstein EF, Callow AD, Nicolaides AN, Shifrin EG (eds) Cerebral revascularization. Med-Orion Pub, pp 51–69
109. Cerdan S, Subramanian VH, Hilberman M, Cone J, Egan J, Chance B, Williamson JR (1986) $^{31}P$ NMR detection of mobile dog brain phospholipids. Magn Reson Med 3:432–439

110. Nioka S, Chance B, Smith DS, Mayevsky A, Reilly MP, Alter C, Asakura T (1990) Cerebral energy metabolism and oxygen state during hypoxia in neonate and adult dogs. Pediatr Res 28:54–62
111. Nioka S, Smith DS, Mayevsky A, Dobson GP, Veech RL, Subramanian H, Chance B (1991) Age dependence of steady state mitochondrial oxidative metabolism in the *in vivo* hypoxic dog brain. Neurol Res 13:25–32
112. Zeuthen T, Dora E, Silver IA, Chance B, Kovach AGB (1979) Mechanism of the cerebrocortical vasodilatation during anoxia. Acta Phys Acad Sci Hung 54:305–318
113. Novack RL, LaManna JC, Rosenthal M (1982) Ethanol and acetaldehyde alter brain mitochondrial redox responses to direct cortical stimulation *in vivo*. Neuropharmacology 21:1051–1058
114. Mayevsky A, Zarchin N, Friedli CM (1982) Factors affecting the oxygen balance in the awake cerebral cortex exposed to spreading depression. Brain Res 236:93–105
115. Mayevsky A, Zarchin N, Kaplan H, Haveri J, Haselgrove J, Chance B (1983) Brain metabolic responses to ischemia in the Mongolian gerbil: *in vivo* and freeze trapped redox state scanning. Brain Res 276:95–107
116. Mayevsky A, Zarchin N, Tannenbaum B (1984) Brain responses to experimental oxygen deficiency in the Mongolian gerbil. Adv Exp Med Biol 180:191–202
117. Mayevsky A, Frank K, Muck M, Nioka S, Kessler M, Chance B (1992) Multiparametric evaluation of brain functions in the Mongolian gerbil *in vivo*. J Basic Clin Physiol Pharmacol 3:323–342
118. Mayevsky A, Zarchin N, Sonn J (1993) Brain redox state and $O_2$ balance in experimental spreading depression and ischemia. In: Lehmenkuhler A, Grotemeyer K-H, Tegtmeier F (eds) Migraine – basic mechanisms and treatment. Urban & Schwarzenberg, Munchen-Wier/Baltimore, pp 379–393
119. Mayevsky A, Nioka S, Wang DJ, Chance B (1997) The functioning gerbil brain *in vivo*. Correlation between 31P NMR spectroscopy and the multiparametric monitoring approach. Adv Exp Med Biol 411:41–53
120. Mayevsky A, Chance B (1973) Monitoring of brain NADH redox state using a fiber optic based fluorometer/reflectometer. In: The annual meeting of the international society of oxygen transport to tissue. ISOTT, South Carolina, USA
121. Mayevsky A, Zarchin N (1981) The effects of unilateral carotid occlusion on the responses to decapitation in the gerbil brain. Brain Res 206:155–160
122. Mayevsky A, Frank KH, Nioka S, Kessler M, Chance B (1990) Oxygen supply and brain function *in vivo*: a multiparametric monitoring approach in the Mongolian gerbil. In: Piper J, Goldstick TK, Meyer M (eds) Oxygen transport to tissue XII. Plenum Pub, pp 303–313
123. Mayevsky A, Meilin S, Rogatsky GG, Zarchin N, Sonn J (1995) Multiparametric monitoring of the awake brain exposed to carbon monoxide. J Appl Physiol 78:1188–1196
124. Mayevsky A, Doron A, Manor T, Meilin S, Zarchin N, Ouaknine GE (1996) Cortical spreading depression recorded from the human brain using a multiparametric monitoring system. Brain Res 740:268–274
125. Cummins JT, Elliott HW (1976) Effect of morphine on the potassium-induced change in the level of reduced pyridine nucleotides and cytochromes in brain slices. Biochem Pharmacol 25:893–896
126. Mayevsky A (1992) Interrelation between intracellular redox state and ion homeostasis in the brain *in vivo*. In: Frank K, Kessler M (eds) Quantitative spectroscopy in tissues. Verlasgruppe GmbH, pp 155–168
127. Mayevsky A, Cohen S (1990) Brain ischemic depolarization and vasospasm in the Mongolian gerbil: the dependence on energy depletion levels. Adv Exp Med Biol 277:315–322
128. Cohen S, Mayevsky A (1989) Effects of nimodipine on the responses to cerebral ischemia in the Mongolian gerbil. Adv Exp Med Biol 248:429–438

129. Mayevsky A, Zarchin N (1988) Microcirculatory and ionic events during and after incomplete or complete cerebral ischemia in the Mongolian gerbil. In: Tomita M, Sawada T, Naritomi H, Heiss WD (eds) Cerebral hyperemia and ischemia. Excerpta Medica, pp 157–169
130. Mayevsky A (1988) In: Manabe H, Zweifach BW, Massmer K (eds) Microcirculatory and ionic responses to ischemia in the Mongolian gerbil. Springer, pp 273–276
131. Mayevsky A, Meilin S, Manor T, Ornstein E, Zarchin N, Sonn J (1998) Multiparametric monitoring of brain oxygen balance under experimental and clinical conditions. Neurol Res 20:S76–S80
132. Meilin S, Zarchin N, Mayevsky A (1999) Inter-relation between hemodynamic, metabolic, ionic and electrical activities during ischemia and reperfusion in the gerbil brain. Neurol Res 21:699–704
133. Vatov L, Kizner Z, Ruppin E, Meilin S, Manor T, Mayevsky A (2006) Modeling brain energy metabolism and function: a multiparametric monitoring approach. Bull Math Biol 68:275–291
134. Mayevsky A, Kutai-Asis H, Tolmasov M (2020) Mitochondrial function and brain Metabolic Score (BMS) in ischemic Stroke: evaluation of "neuroprotectants" safety and efficacy. Mitochondrion 50:170–194

# Chapter 4
# Multiparametric Responses of the Gerbil Brain to Oxygen Supply

**Abstract** The Mongolian gerbil (*Meriones unguiculatus*) serves as a useful animal model for studying the effects of ischemia, hypoxia, and anoxia on brain functions. In this model, it is possible to induce two levels of ischemia in the same animal. Thus, monitoring the brain in vivo in real time will provide meaningful information regarding the development of ischemic injury as well as the follow-up during the recirculation period. The aims of the study were as follows: (1) to elucidate the mechanism behind the development of ischemic depolarization under unilateral and bilateral carotid artery occlusion; (2) to exclude the possibility that the removal of the dura mater will affect the results; and (3) to correlate the kinetics of the recovery processes with the level of ischemia. We tested the correlation between energy depletion levels (evaluated by intramitochondrial NADH redox state) and the development of ischemic depolarization (ID) and vasospasm (evaluated by extracellular $K^+$, DC potential, and 366 nm reflectance changes, respectively) under both partial and complete ischemia (induced by unilateral or bilateral carotid artery occlusion) using the multiparametric monitoring system (MPA). In 12 out of 32 gerbils monitored by the MPA, the dura mater remained intact, while in the other 20, it was removed very gently before connecting the MPA to the brain. Two types of responses to unilateral carotid artery occlusion were recorded, and the gerbils were divided into groups according to the development of the ID. In a third group of 5 gerbils, we tested the effect of 1–5 min of bilateral occlusion on the various monitored parameters. The results could be summarized as follows: (1) Under bilateral occlusion, in all gerbils, the ID was recorded within 1–2 min. (2) Under unilateral occlusion, the level of ischemia obtained was significantly lower, resulting in ID in about 60% of the gerbils. (3) The rate of $K^+$ leakage in phase I (before the ID appearance) was significantly lower under unilateral occlusion. In group B where ID did not develop under unilateral occlusion, the level of ischemia was 50% of its maximum or less. (4) The rate of $K^+$ leakage during phase I was energy dependent across all subgroups of gerbils. (5) The $K^+$ leakage during phase II (ID) had an "all or none" nature in terms of maximal $K^+$ levels and time to reach it, and no significant variation was recorded between various groups. (6) The main effect of various lengths of bilateral occlusion was on the recovery time of extracellular $K^+$ level, namely longer occlusion led to a slower recovery.



A. Mayevsky, *The Mongolian Gerbil Brain*,
https://doi.org/10.1007/978-3-031-69549-0_4

**Keywords** Brain hypoxia · Extracellular $K^+$ · Nicotinamide-adenine dinucleotide redox state · Recirculation · Ischemic depolarization

## 4.1 Introduction

This chapter discusses the three neuropathological events, as shown in Fig. 4.1. These three factors (A, B, and C), namely, systemic hypoxia and anoxia, cerebral hyperoxia, as well as cerebral ischemia, affect the availability of oxygen so that energy metabolism is inhibited (blue arrows) or stimulated (purple arrows). All three perturbations affect the same type of processes in the brain, such as CBF, mitochondrial NADH, electrical activities, and ionic homeostasis. Under hypoxia and ischemia, ATP production by the mitochondria decreases and becomes a limiting factor in the system. Due to the inhibition of the ionic pumps, the ionic homeostasis is disrupted and hypoxic or ischemic cortical depolarization is developed. The endpoints of this process are the changes in the electrical activities, recorded as a depression of the ECoG and a negative shift in the DC steady potential. In addition, due to the depolarization event, the extracellular potassium is elevated. If this situation persists for more than a few minutes, irreversible damage to the cortex may develop.

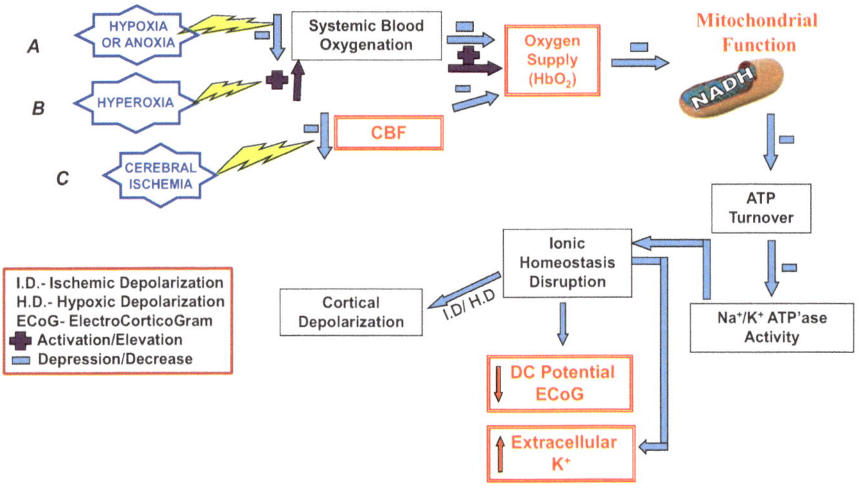


**Fig. 4.1** Effects of various pathological conditions on the energetic mitochondrial function, ionic and hemodynamic state of the cerebral cortex (details are in the text) [1]

## 4.2 Anoxia

In this book, the exposure of a rat or gerbil to anoxia refers to providing the animal with 100% nitrogen. In the late 1950s and early 1960s, Chance and his collaborators reported on the responses of the rat brain and other organs to anoxia [2–5]. Figure 4.2 shows the results [3] of a rat brain and kidney exposed to anoxia in Part A and another rat exposed to anoxia while correlating the EEG activity to the brain NADH in Part B.

The first report describing the effects of anoxia on the mitochondrial NADH redox state in the Mongolian gerbil was published by my laboratory in 1978, as demonstrated in Fig. 3.2 (Chap. 3). In the next phase, I moved from the single or dual parameter monitoring to the multiparametric monitoring approach. The first example is presented in Table 3.1, Protocol 3.3 [6]. The effect of anoxia using this Protocol is shown in Fig. 4.3.

Figure 4.3 shows the response to 1-min anoxia induced by nitrogen respiration. Within 20 s, the extracellular potassium ( $K_e^+$ ) concentration started to change as the log of $K^+$ activity gradually increased. $K_e^+$ returned to the baseline level very shortly after the room air was reintroduced. At the same time, NADH levels (third trace from top) responded as previously described and reached an elevated level under anoxic conditions. The typical decrease of the reflected light is recorded in the upper trace. The ECoG activity became severely depressed during complete anoxia. Very shortly after the full recovery of $K_e^+$ and NADH to their initial levels, a steep increase of $K_e^+$, which peaked at 91 meq/liter, was measured, while the surface potential showed a wave whose negative shift reaching 8.5 mV. The late response of $K_e^+$ is identical to that recorded during spreading depression (SD) [6]. Electrical activity and metabolic response also resembled those found in SD.

Two nitrogen cycles measured in another gerbil are shown in Fig. 4.4. In this animal, $K_e^+$ was measured bilaterally, while NADH level and ECoG were monitored only from the right hemisphere. Of the eight traces shown in Fig. 4.4, only the

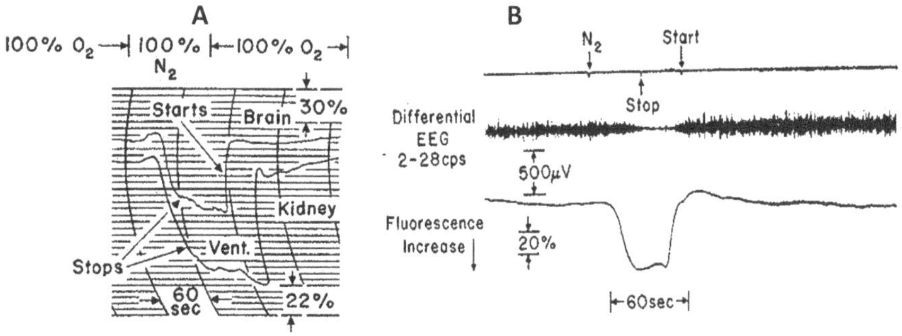


**Fig. 4.2** (**a**) Simultaneous recordings of fluorescence changes in rat brain and rat kidney in a cycle of anoxia. Fluorescence increases are indicated in a downward direction [3]. (**b**) Examples of correlation between the EEG and the fluorescence changes in the rat brain cortex in the presence of energy reserves [3]

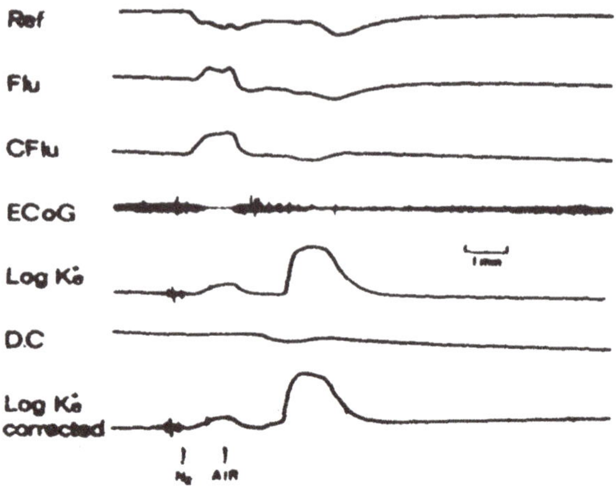


**Fig. 4.3** Effects of anoxia on the metabolic, electrical, and extracellular $K_e^+$ activity in the gerbil brain [6]

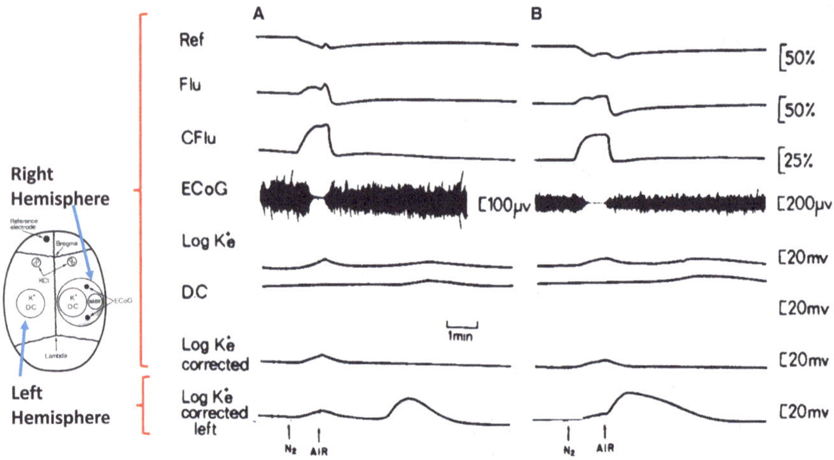


**Fig. 4.4** (**A**) Effects of anoxia on the levels of NADH and $K_e^+$ in the gerbil brain. In **A** and **B**, the interval between the two $N_2$ cycles was 5 min. NADH was monitored only in the right hemisphere, while $K_e^+$ was measured in both the right hemisphere (traces 5–7) and the left hemisphere (trace 8) [6]

one on the bottom was measured from the left hemisphere. The interval between the two nitrogen cycles was 5 min. As can be seen, the responses of the two hemispheres to $N_2$ anoxia are markedly different. As a response to anoxia, the NADH level increased (CFlu) about 30% as compared to the normoxic level on the right side. ECoG became isoelectric when the NADH level reached 80–90% of its maximum reduction level. The response of $K_e^+$ was very small when compared with that shown in Fig. 4.3. Extracellular potassium recovery after the reintroduced air revealed some differences between the two hemispheres of the same animal. In the right hemisphere, NADH was oxidized to its baseline levels rapidly. ECoG and $K_e^+$

returned to their baseline levels in less than 1 min. The response of the right hemisphere was similar in the two $N_2$ cycles. The responses of the left hemisphere measured during the same $N_2$ cycles varied from those of the right side; responses of the left hemisphere were also not identical during two subsequent cycles. The main difference can be seen during the recovery from anoxia, that is, after restoration of energy to the system. In the first $N_2$ cycle, a SD-like response was recorded about 2 min after air respiration, while in the second $N_2$ cycle, the large increase in $K_e^+$ appeared immediately following the anoxia-induced initial increase in $K_e^+$. The reason for the observed differences between cycles and between hemispheres will be discussed later.

In the next study [7], the partial pressure of oxygen was measured together with the mitochondrial NADH. Figure 4.5 shows the effects of anoxia on the slightly awake and anesthetized brain. Parts A and B show the responses to two nitrogen cycles. The reflectance and fluorescence responses are typical, and NADH became fully reduced within less than 1 min. The decrease in $pO_2$ had faster kinetics and stayed at the zero level until rebreathing started. The main difference between the two cycles can be seen in the recovery phase from anoxia. In the first cycle (A), an overshoot of $pO_2$ was higher than the normoxic level and showed a lower level followed by a small overshoot. At the same time, NADH showed an oxidation cycle typical of the SD response. The ECoG trace showed a typical anoxic depression in Parts A and B and a secondary depression in Part B concomitant with the metabolic and hemodynamic responses to the SD wave in B.

The response to anoxia in the anesthetized state is shown in Part C and is followed by a spontaneous SD cycle, as shown in the ECoG trace. The $pO_2$ trace shows the typical decrease cycle during the SD, concomitant with a reduction in NADH. The recovery phase from anoxia, as described in Figs. 4.5a, b, and c, had a special pattern depending on the state of the brain. In Part B, the $pO_2$ trace showed a sharp dip after the fast recovery to the normoxic level. At the same time, the NADH was also above the normoxic level, indicating that it is not an artifact of the

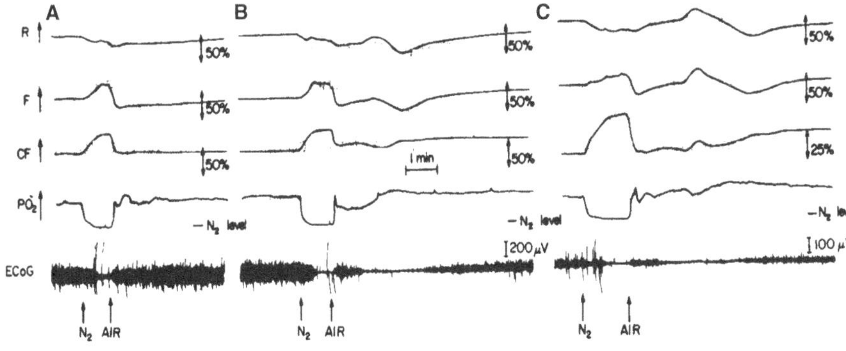


**Fig. 4.5** Effects of anoxia (**a**, **b**, **c**) on NADH, $pO_2$ in the awake (**a**, **b**) and deeply anesthetized (E) gerbil brain [7]

$pO_2$ electrode. This dip in the $pO_2$ is not clear in Fig. 4.5a and c, as the NADH trace also recovers to the normoxic level.

The next protocol (Table 3.1) used in the gerbil brain exposed to anoxia was 3.6 [8, 9]. Figure 4.6 shows the effects of a brief period of anoxia (0.8 min) on the various parameters measured from the gerbil brain. Due to the brief duration of the insult, the main responses are seen in the parameters representing energy metabolism. The NADH level increased by 25% compared with the normoxic level (defined as 100% signal) and returned to normoxic values after rebreathing air or during recirculation, respectively. The increased level of NADH recording during anoxia was maximal, i.e., that all NAD pool was reduced to NADH and those levels were unchanged after death. Due to lack of available energy during insult, extracellular $K^+$ increased (third trace from the top) and then decreased as energy was available during the recovery phase. The pH of the brain during the short ischemia interval became more acidic (signal moving upward). Depression of the ECoG occurred during the anoxic insults. The DC steady potential was not significantly affected in any of the three electrodes used: $DC_{H+}$, $DC_{K+}$, and $DC_F$.

The results of using Protocol 3.9 in Table 3.1 are presented in Fig. 4.7 [11].

In order to illustrate the usage of the multiparametric monitoring system, Fig. 4.7 shows the effects of anoxia. Under complete deprivation of $O_2$, as can be seen also in the $pO_2$ trace, the intramitochondrial NADH (CF) became fully reduced, namely, all NAD accumulates as NADH. During the anoxic episode, a vasodilation occurred due to the autoregulation mechanism, and it is recorded as a decrease in the reflectance trace (R). Due to the short duration of the anoxic cycle, only a small increase in the extracellular $K^+$ level was recorded ( $K_e^+$ ). The effect of the anoxic cycle on

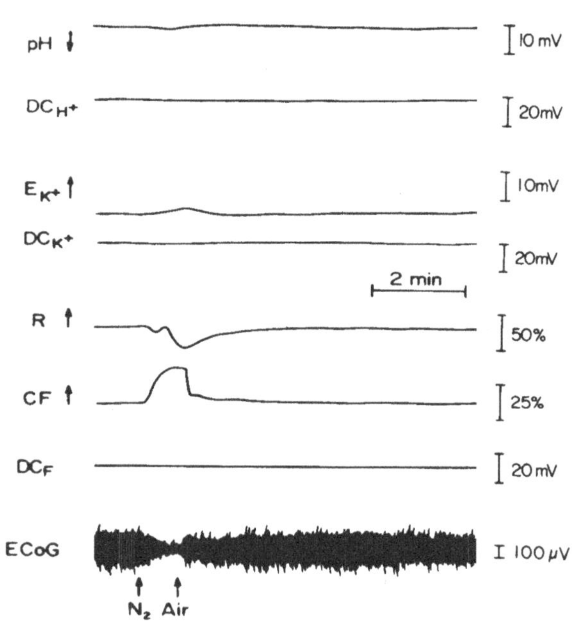


**Fig. 4.6** Effects of anoxia on metabolic, ionic, and electrical activities in right hemisphere of a gerbil brain. ECoG, electrocorticogram; CF, corrected fluorescence; R, reflectance; $EK^+$, extracellular $K^+$ level; $DC_{H+}$—$DC_{K+}$ and $DC_F$, DC steady potential measured near pH electrode, $K^+$ electrode, and light guide, respectively. Arrows show an increase in parameters measured [9]

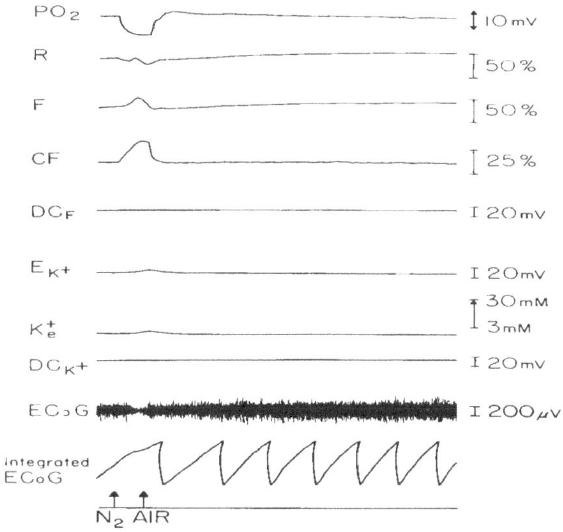


**Fig. 4.7** Effects of anoxia (100% $N_2$) on the metabolic ionic and electrical activities of the gerbil brain. $pO_2$, tissue surface $O_2$ levels; R, 366 nm reflectance; F, 450 nm fluorescence; CF, NADH-corrected fluorescence; $DC_F$ and $DC_K$, DC steady potential around the optical fibers and the $K^+$ electrode, respectively; $EK^+$ and $K_e^+$, uncorrected extracellular $K^+$ electrode potential and corrected (for DC changes) $K^+$ levels, respectively; ECoG, electrocorticogram [10]

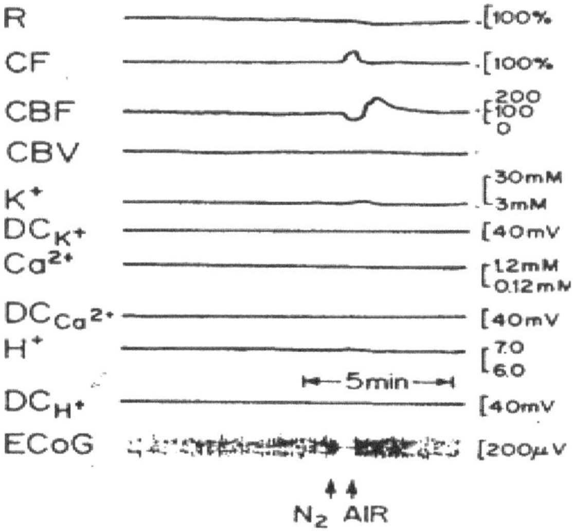


**Fig. 4.8** Metabolic ionic and electrical responses to anoxia in the Mongolian gerbil. CF, corrected NADH fluorescence; ECoG, electrocorticogram; R, reflectance; CBF, CBV, cerebral blood flow and volume; $K^+$, $Ca^{2+}$, $H^+$, extracellular levels of ions; DC, steady potential around the three ion electrodes [12]

the DC steady potential was undetectable, but a small depression of the ECoG was noted.

During the recovery period, the intramitochondrial NADH recovered to the pre-anoxic level without any hyperoxidation as expected from the overshoot in the $pO_2$ trace due to the high blood flow level created during anoxia.

The last example of the response of the gerbil brain to anoxia is presented in Fig. 4.8 [12].

Figure 4.8 shows responses to anoxia recorded from a gerbil brain. All signals recorded were calibrated in relative (NADH and CBF) or absolute (all ions and electrical signals) units. During a short anoxic episode, NADH became more reduced indicating a lack of $O_2$ while CBF showed a marked decrease. The ECoG became isoelectric while only extracellular $K^+$ showed a very small increase. During the recovery period, the CBF showed a clear overshoot, but the NADH did recover to the pre-anoxic level without an undershoot. This indicates that the auto regulatory response to the lack of $O_2$ was clear at the vascular level, but at the mitochondrial level, this effect was not recorded.

## 4.3 Hypoxia

In my studies, I exposed the gerbils to two types of hypoxic protocols. In the first protocol presented in Figs. 4.9, 4.10, and 4.12, the gerbils were exposed to a gradual decrease of $FiO_2$ down to 100% nitrogen. In the second protocol, the gerbils were exposed to the same low level of oxygen for at least 5 min.

The effects of gradual hypoxia are shown in Fig. 4.9 [15]. Due to the decrease in oxygen supply, which was detected by the increase in the intramitochondrial NADH levels, an increase in CBF was recorded (LDF). At the same time, the reflectance (R) trace showed a decrease direction and is correlated to the increase in blood volume. The ionic responses to the decrease in energy availability were similar to those recorded in another gerbil (Fig. 4.10) [13].

The ECoG was affected mainly when the hypoxic level was more severe. During the recovery period, a typical response to SD was also recorded.

Figure 4.10 [13] shows a typical response to gradual hypoxia ranging from 100% to 0% oxygen in the breathing mixture and then a recovery to 100% $O_2$ for another 5 min. As seen, a gradual decrease in hemoglobin saturation ($HbO_2$) was recorded simultaneously with the increase in the intramitochondrial redox state (CF). In parallel, changes were recorded in the reflected light intensity (R) and the Hb concentration, representing the autoregulation response due to hypoxia. The first change in the ECoG was noted at $FiO_2$ level below 9%, was severely decreased to 4.8%, and became isoelectric under $N_2$ breathing.

The resting level of the extracellular $K^+$ level was 2.5 mM, which was increased gradually to a maximum of 3.9 mM during the short complete anoxic period. The resting level of $Ca^{2+}$ was quite low as compared to other brains monitored (due to a film of blood below the electrode) and was decreased from 0.6 mM (normoxia) to 0.5 mM during anoxia. The DC steady potential around the $K^+$ and $Ca^{2+}$ electrodes showed a positive change ranging from 1 to 2 mV during hypoxia but we suspect it is because of a small instability of the reference electrode since it did not recover during the normoxic recovery period.

The recovery period presents a short period of reoxygenation followed by a wave of spreading depression (SD) developed probably by the increased $K^+$ level during the anoxic period. The brain became hyperoxygenated due to the vasodilated state (at the end of anoxia) and the breathing of 100% oxygen. In order to describe the

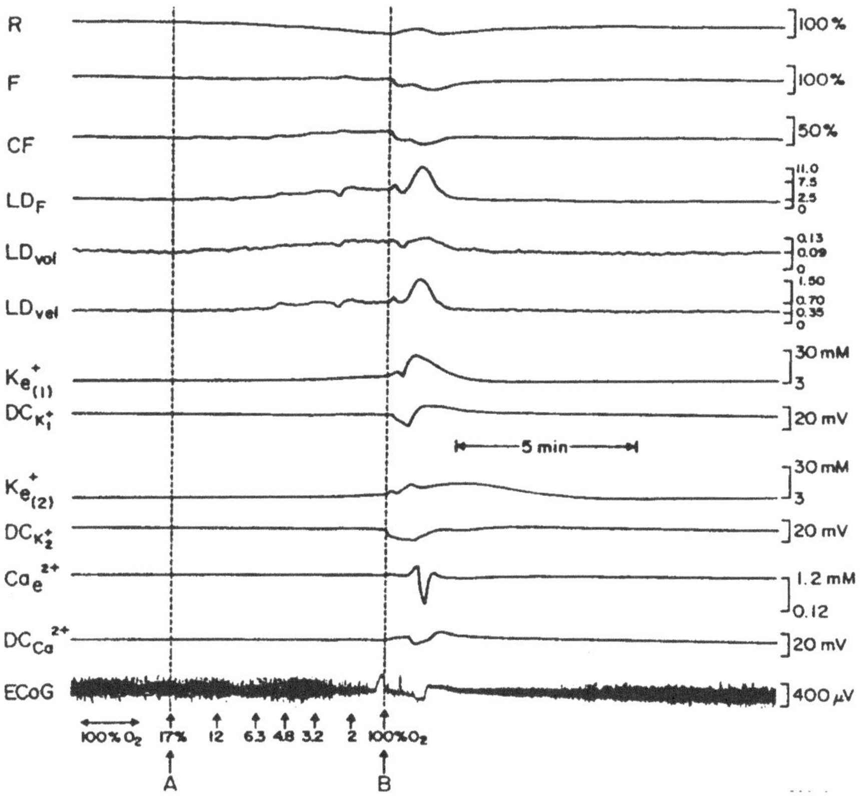


**Fig. 4.9** Hemodynamic metabolic, ionic, and electrical responses to gradual hypoxia in the Mongolian gerbil. At Point **A,** the gerbil was exposed to 17% $O_2$ and recovery started at Point **B** (100% $O_2$). R, 366 nm reflectance; F, 450 nm fluorescence; CF, NADH-corrected fluorescence; $LD_F$ $LD_{vol}$ $LD_{vel}$, Laser Doppler flow, volume, and velocity; $K_e^+$ (1)' $K_e^+$ (2)'; $Ca_e^{2+}$, extracellular potassium (two electrodes) and calcium electrodes; $DCK^+$ 1' DCK 2' $DCCa^{2+}$, DC steady potential measured concentric to the three electrodes; ECoG, electrocorticogram [15]

exact relationship between various parameters, we analyzed the analog signals. These normalized results are presented in Fig. 4.11. The initial value (for each parameter) under 100% $O_2$ was taken as 100% and all the rest of the values were related to those of control. During the first 25 s of recovery, oxygen availability and intramitochondrial redox state were completely recovered. The time for complete recovery of the vascular changes was longer, especially for the recovery of Hb concentration signal.

The SD episode occurred immediately after the reorganization of the brain showing the typical ionic homeostasis disturbances, i.e., an increase in extracellular $K^+$ (26.3 mM) and a decrease in $Ca^{2+}$ (0.22 mM). During the recovery from the SD wave, the $K^+$ showed an undershoot due to the activated $Na^+$-$K^+$-ATPase or the higher permeability of $K^+$ through the plasma membrane. The NADH became more

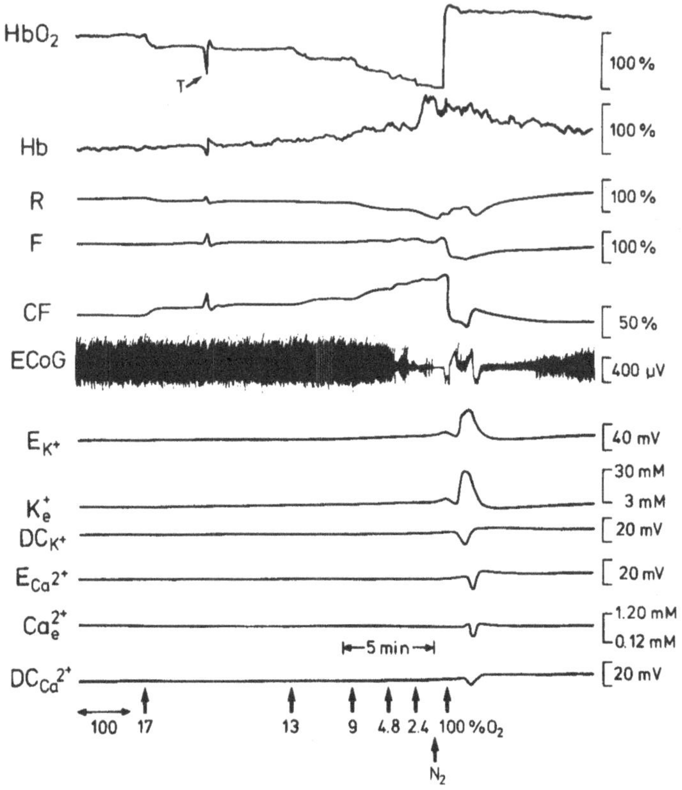


**Fig. 4.10** Metabolic, ionic, and electrical responses to gradual hypoxia induced in the gerbil brain. R, F, and CF, reflectance, fluorescence, and corrected fluorescence; $ECa^{2+}$, $EK^{+}$, uncorrected calcium and potassium potentials; $Ca_e^{2+}$, $K_e^{+}$, corrected ion activities; $DCCa^{2+}$, $DCK^{+}$, DC steady potential measured around the calcium and potassium electrodes; ECoG, electrocorticogram; $HbO_2$, hemoglobin oxygenation; HB, hemoglobin volume or content [13]

**Fig. 4.11** Effect of gradual hypoxia on various parameters monitored from the gerbil brain as shown in Fig. 4.10. Each data point represents the value calculated at the end of the period of lower $O_2$ breathing. $R_1$ and $R_2$ values represent 25 s and 5 min recovery at 100% $O_2$ [13]

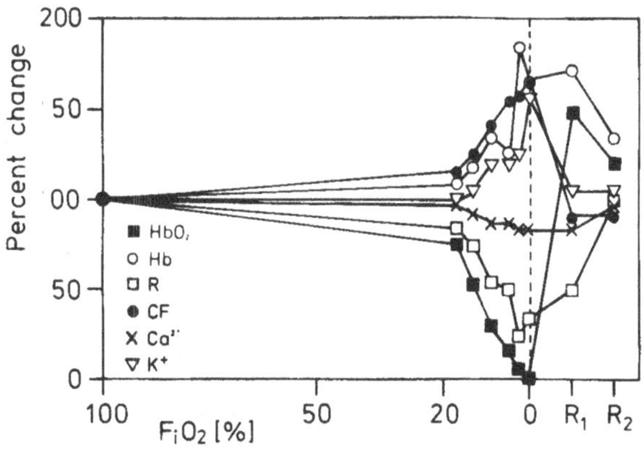

oxidized during the SD cycle, while the blood $HbO_2$ showed a very small, if at all, transient deoxygenation.

The transient change in blood oxygenation and NADH redox state observed during the first step of hypoxia is due to a short bilateral carotid artery occlusion. This procedure enabled us to indirectly determine the $O_2$ uptake rate and will be discussed separately.

Figure 4.12 [14] presents the effects of gradual hypoxia on various monitored parameters. A similar type of response to $O_2$ deprivation was recorded when a gerbil was exposed to systemic hypoxia, as shown in Fig. 4.12 [14]. In order to verify the correlation between $HbO_2$ and NADH redox state, we added an oxygen electrode to the monitoring system (in few experiments) to measure the average oxygen level in tissues, and a typical response to hypoxia, using this system, is shown. The level of $O_2$ was decreased (in steps) in the breathing mixture from 100% to 0% $O_2$ (100%

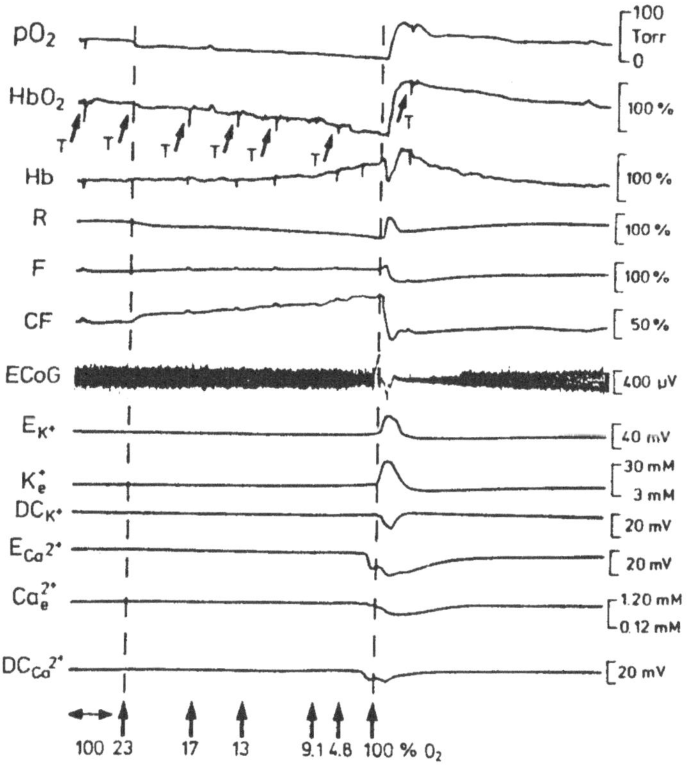


**Fig. 4.12** Responses of the gerbil brain to gradual hypoxia induced by lowering the $FiO_2$. $pO_2$, tissue oxygen levels; $HbO_2$, hemoglobin oxygenation (the pre-ischemic value was taken as 100%); Hb, hemoglobin volume or content; R, F, CF, reflectance, fluorescence, and corrected fluorescence of NADH; ECoG, electrocorticogram; $Ek^+$, $ECa^{2+}$, uncorrected potassium and calcium potential; $K_e^+$, $Ca_e^{2+}$, corrected extracellular potassium and calcium levels; $DC_{k+}$, $DC_{Ca2+}$, DC steady potential measured around the potassium and calcium electrodes; T oxygen consumption test [14]

$N_2$) and then returned to 100% $O_2$. The three indicators of oxygen supply levels changed during hypoxia and correlated to each other. The gradual decrease in $O_2$ delivery, as seen in $HbO_2$ and the CF traces, had a very small effect on ionic homeostasis. Only a small increase in extracellular $K^+$ was noted during most of the hypoxic episodes. The ECoG trace showed changes only when $FiO_2$ was below 5%.

Toward the end of hypoxia, a wave of depolarization developed next to the calcium electrode and was propagated to the rest of the hemisphere during the reoxygenation period of the brain. It can be considered a wave of spreading depression (SD) developed by an intrinsic factor which led to an increase in extracellular potassium and a decrease in extracellular $Ca^{2+}$, as recorded simultaneously with a negative shift in the DC steady potential. The depolarization that occurred during hypoxia led to a large decrease in blood volume, as seen by a transient decrease in the Hb trace and a large increase in the R trace reported by us previously [16]. The SD wave slowed down the ECoG recovery, leading to an oxidation wave of the NADH redox state due to ion movement.

Figures 4.13, 4.14, and 4.15 illustrate the effects of gradual hypoxia on the metabolic, ionic, and electrical activities in the same gerbil. The effects of 5-min exposure to 6% $O_2$ are shown in Fig. 4.13 [17]. Here again, due to a decrease in ATP level (an increase in NADH and a decrease in $pO_2$), $K^+$ increased and a general depolarization occurred (1 in Fig. 4.13). The main differences between the responses to ischemia and hypoxia were that the R trace showed a decrease during hypoxia (increased blood volume due to autoregulation mechanism) and the recovery change of the R was monotonic rather than a two-step decrease. Here, the hypoxic depolarization (HD) led to an increase in R (1 in Fig. 4.13) as found in ischemia. The ECoG trace showed a small decrease during the initial stage of hypoxic episode (as also found in the $K^+$ trace) and a severe depression was recorded during the HD period. In the recovery period, $K^+$ showed a 1-min delay before decreased levels were recorded. A similar kinetic of the $pO_2$ and the R traces was very clear after the hypoxic episode, as found after ischemia. The recovery of the ECoG trace was quite fast and very similar to the partial ischemic episode. The results shown in Figs. 4.14 and 4.15 represent responses to mild (10% $O_2$) and severe (2% $O_2$) hypoxia as compared to 6% as shown in Fig. 4.13. Basically, the qualitative responses to various levels of $O_2$ were the same and the difference was noticed in the intensity of the response; the lower the $pO_2$ given to the animal, the severances of the responses as seen in the time to get the HD or the depression level of the ECoG. The recovery process was also correlated to the level of hypoxia induced.

Figure 4.16 [6] shows the effect of hypoxia. In the first stage (data not shown in the figure), a control response to SD was measured in the subject. $K_e^+$ showed a typical response, while NADH showed an oxidation cycle. A small increase in NADH was recorded in the initial phase of this cycle. When the gerbil was made hypoxic by administration of 5% $O_2$ in 95% $N_2$, an increased NADH level of about 10% was recorded. At the same time, $K_e^+$ began to increase gradually above its baseline level, finally reaching a point at which a fast, large increase was measured. During this phase, NADH level remained high while the gerbil breathed from the 5% $O_2$ mixture. When the gerbil was switched back to a room air breathing mixture,

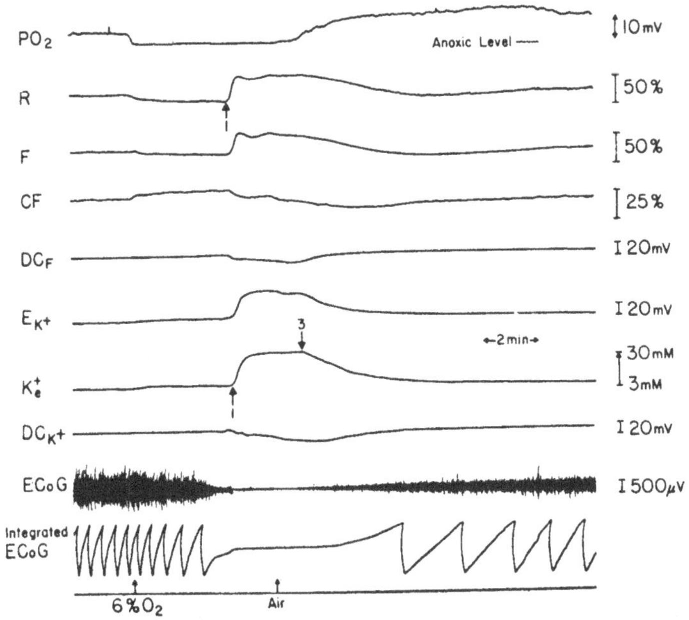


**Fig. 4.13** Cerebral responses to 6% $O_2$ in $N_2$ apply to the gerbil via a nasal mask. Arrow 1 indicating the appearance of the hypoxic depolarization (HD) and arrow 3 shows the beginning of $K_e^+$ recovery [17]. $pO_2$, tissue surface $O_2$ levels; R, 366 nm reflectance; F, 450 nm fluorescence; CF, NADH-corrected fluorescence; $DC_F$ and $DC_K$, DC steady potential around the optical fibers and the $K^+$ electrode, respectively; $EK^+$ and $K_e^+$, uncorrected extracellular $K^+$ electrode potential and corrected (for DC changes) $K^+$ levels, respectively; ECoG, electrocorticogram

a small transient oxidation of NADH occurred, but in this animal only 100% $O_2$ supplied enough oxygen to oxidize NADH to its original level. $K_e^+$ also started to decrease only after 100% $O_2$ was introduced.

In another gerbil, the effect of 6% of oxygen was tested as shown in Fig. 4.17. In this gerbil, the correlation between the decrease in brain oxygen level and mitochondrial NADH was very clear. In this animal, the hypoxic depolarization was not developed.

## 4.4 Partial and Severe (Complete) Ischemia

### 4.4.1 Introduction

The effects of ischemia on the gerbil model were demonstrated in our laboratory in 1978 [19]. In this study, it was clear that the gerbil model is different from the rat model. As shown in Fig. 3.4, in the current book, the occlusion of the two carotid arteries led to a large but not a maximal increase in NADH redox state. This result

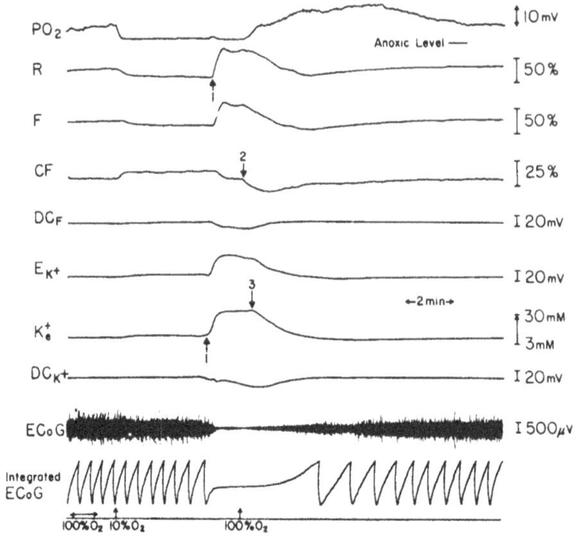


**Fig. 4.14** Metabolic, ionic, and electrical responses to 10% $O_2$ in $N_2$. The animal was breathing 100% $O_2$ before the hypoxia. Arrow 1 indicating the hypoxic depolarization—HD, arrow 2—the oxidation of NADH after rebreathing 100% oxygen and arrow 3—the phase of $K^+$ recovery to the pre-hypoxic level [17]. $pO_2$, tissue surface $O_2$ levels; R, 366 nm reflectance; F, 450 nm fluorescence; CF, NADH-corrected fluorescence; $DC_F$ and $DC_K$, DC steady potential around the optical fibers and the $K^+$ electrode, respectively; $EK^+$ and $K_e^+$, uncorrected extracellular $K^+$ electrode potential and corrected (for DC changes) $K^+$ levels, respectively; ECoG, electrocorticogram

was different from the response measured in the rat under similar conditions [20, 21].

The effects of ischemia on the gerbil brain were studied, which has an incomplete circle of Willis. The light guides were implanted in four locations as shown in Fig. 4.18. We tested the variability between anterior and posterior areas in the same hemisphere by comparing the two hemispheres. In the gerbil, one can obtain a unilateral incomplete or complete ischemia by occluding the ipsilateral common carotid artery. By means of ligation of the two carotids, a complete bilateral cerebral ischemia is achieved. Figure 4.18 shows the corrected fluorescence response to anoxia (Fig. 4.18a) and to unilateral or bilateral occlusions performed in various gerbils (Fig. 4.18b–e). The upper two traces were recorded from the right hemisphere (RA + RP) and the other traces from the left one (LA + LP). In each hemisphere, we compared the responses recorded in the anterior part to that of the posterior area. In Fig. 4.18a, the effects of anoxia are presented. Although each probe was located in different fields of the cortex, the responses are quite identical. The small variability between the four responses is probably due to the differences in the tissue properties as well as geometry of the light guide implantation. The response of the brain to anoxia could serve as a reference point to calculate other responses and then they were normalized for better comparison. The responses to unilateral and bilateral occlusion enable us to determine the type of communication

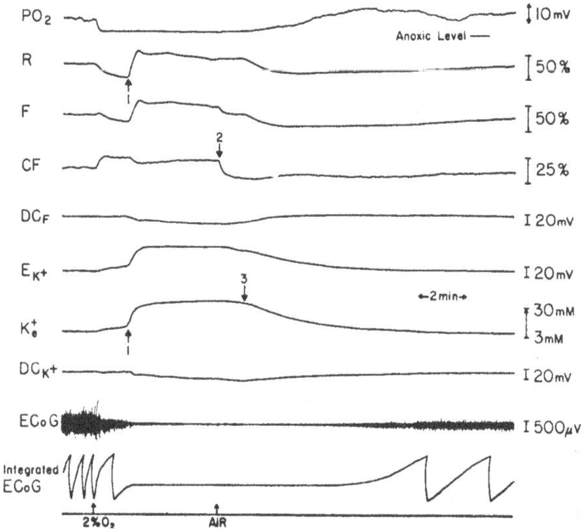


**Fig. 4.15** Effects of severe hypoxia $-2\%$ $O_2$ in $N_2$ on various parameters measured in the gerbil brain. Arrow 1 indicating the hypoxic depolarization—HD; arrow 2, the oxidation of NADH after rebreathing 100% oxygen and arrow 3—the phase of $K^+$. $pO_2$, tissue surface $O_2$ levels; R, 366 nm reflectance; F, 450 nm fluorescence; CF, NADH-corrected fluorescence; $DC_F$ and $DC_K$, DC steady potential around the optical fibers and the $K^+$ electrode, respectively; $EK^+$ and $K_e^+$, uncorrected extracellular $K^+$ electrode potential and corrected (for DC changes) $K^+$ levels, respectively; ECoG, electrocorticogram recovery to the pre-hypoxic level [17]

**Fig. 4.16** Effects of hypoxia (5% $O_2$) on $K_e^+$ and metabolic activity in the gerbil [6]

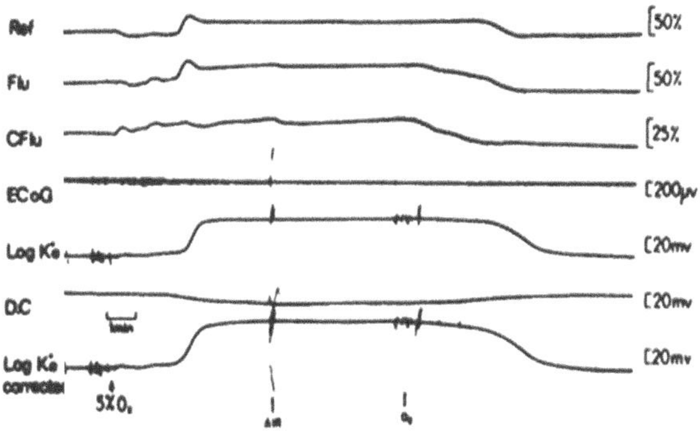


between the two hemispheres through the anterior cerebral arteries. In Fig. 4.18e, one can see that unilateral occlusion (left or right) did not affect the NADH redox state in the occluded as well as in the opened hemisphere. In this gerbil, the contralateral carotid artery did supply the oxygen needed for the two hemispheres. Only bilateral occlusion induced a change and complete ischemia was achieved in the two hemispheres. After reopening, the NADH returned to its normoxic level in all the four sites of measurement. In Fig. 4.18d, a complete separation between the two hemispheres was observed. After occlusion of the right artery, the right hemisphere

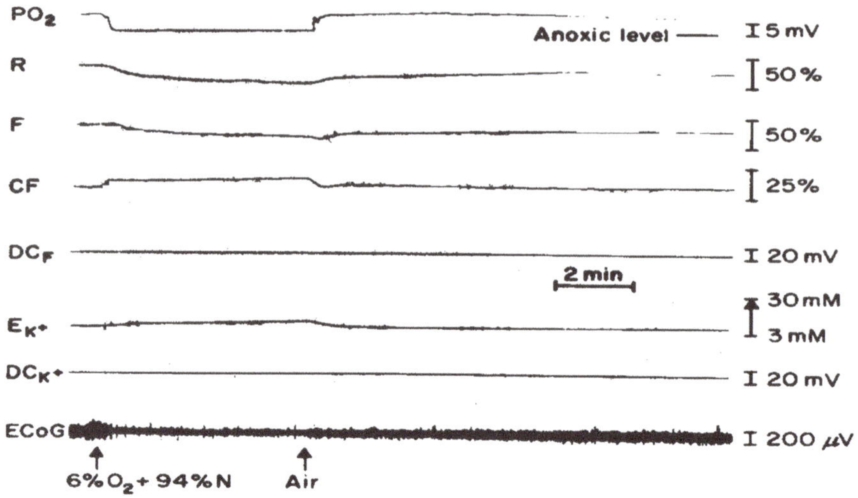


**Fig. 4.17** Responses of the gerbil brain to hypoxia on various parameters measured from the gerbil brain. R, F, CF, reflectance, fluorescence, and corrected fluorescence, respectively; $DC_F$, $DC_K+$, DC steady potential measured around the light guide and the potassium electrode, respectively; ECoG, electrocorticogram [18]

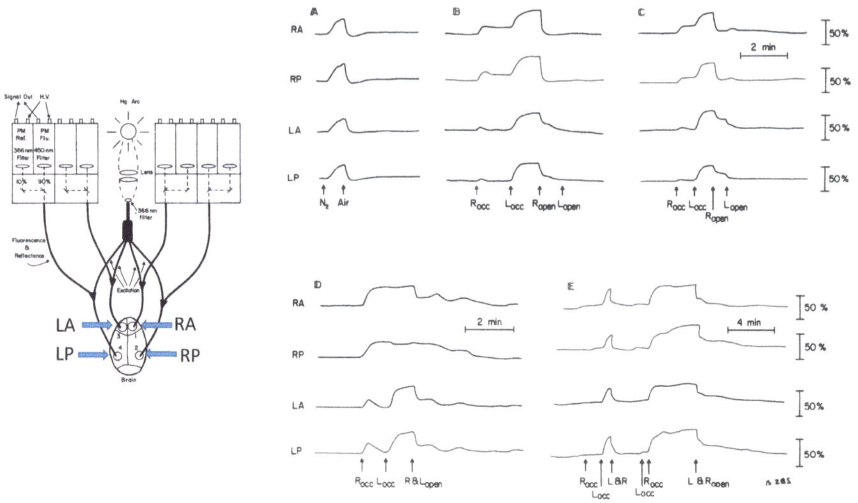


**Fig. 4.18** Effects of anoxia (**a**), partial and complete ischemia (**b–e**) on the corrected fluorescence (right side) measured from four different sites in the gerbil brain (as shown in the left side of the figure [22])

became fully ischemic (maximum NADH level), while the left one showed a transient response which returned to the baseline level in a very short time. The left hemisphere also became ischemic after occlusion of the left carotid artery. The two responses to ischemia presented in Fig. 4.18b and c were observed in gerbils that had a partial connection between the two hemispheres via the anterior cerebral arteries. In Fig. 4.18b, the occlusion of the right artery induces a change of about 25% of the maximum response obtained after bilateral occlusion. In Fig. 4.18c, the response to unilateral occlusion was about one-third of its maximum. In Fig. 4.18b and c, the reopening of the carotid arteries was carried out in two steps and the NADH recovery was performed accordingly.

### 4.4.2 Basic Responses to Ischemia in Gerbils

To test the effects of ischemia on the slightly anesthetized gerbil brain, two types of experiments were performed. The first experiment included unilateral and bilateral carotid artery ligations (Fig. 4.19). Figure 4.19 shows a typical response to unilateral carotid artery occlusion. The occlusion of the left carotid artery ($L_{occl}$) did not affect the level of NADH in the contralateral hemisphere; the NADH trace that was obtained from the contralateral hemisphere during occlusion showed only a very small change. When the ipsilateral carotid artery was occluded ($R_{occl}$), a 25% increase in NADH (CFlu) was recorded within 10 s of the occlusion. At the same time, $K_e^+$ showed a gradual increase, which continued until the reopening of the artery ($R_{open}$). After the reopening of the right carotid, NADH returned to its baseline level within a few seconds. Electrical activity (ECoG) responded only to ipsilateral occlusion with a decrease in amplitude that required about 5 min of reperfusion for complete recovery. The DC potential level did not respond to either ipsilateral or

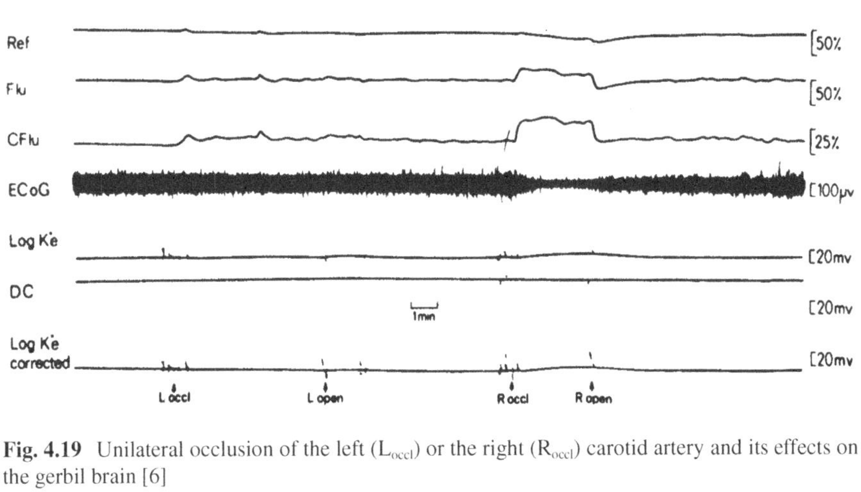


Fig. 4.19 Unilateral occlusion of the left ($L_{occl}$) or the right ($R_{occl}$) carotid artery and its effects on the gerbil brain [6]

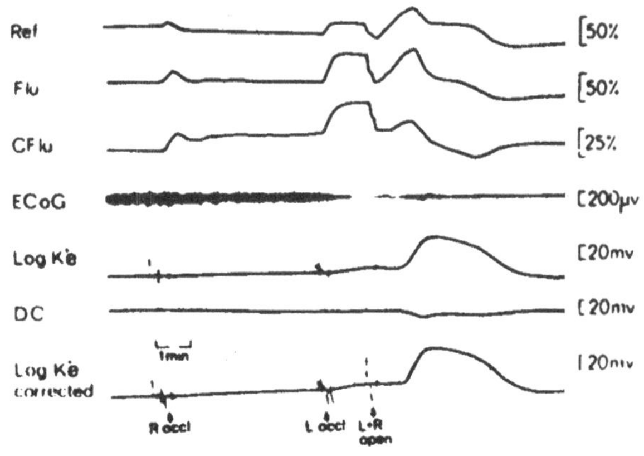


**Fig. 4.20** Effect of partial and complete ischemia on the metabolic and $K_e^+$ responses of the brain [6]

contralateral occlusion. A small decrease in response in the reflected light channel was observed only during the ipsilateral (right) carotid occlusion.

Using another gerbil, the effects of a short unilateral and bilateral occlusion on the response of various parameters were tested. When the right carotid was occluded, the NADH level increased to 10–15%, and $K_e^+$ showed a very slow gradual increase (Fig. 4.20). When the left carotid and right carotid arteries were occluded, a large increase in NADH, which reached 40% above the control level, was observed. At the same time, $K_e^+$ continued to increase until the reopening of the two carotids. After recirculation, a spreading depression cycle developed spontaneously, and typical $K_e^+$ and metabolic responses are shown in Fig. 4.20.

### 4.4.3 *Monitoring of Extracellular $K^+$ in Two Hemispheres*

The results shown in Figs. 4.21 and 4.22 were taken from an experiment in which $K_e^+$ was monitored bilaterally, while NADH and ECoG were measured only from the right hemisphere [6]. When the right carotid was occluded, an increase of 10% in NADH level was followed by a gradual small response of $K_e^+$ in the right hemisphere, but without any change in $K_e^+$ of the left hemisphere (Fig. 4.21).

When the left carotid was occluded ($L_{occl}$), the right hemisphere did not show any metabolic or ionic changes (Fig. 4.21). In the ipsilateral hemisphere, two phases in the $K_e^+$ changes were measured: the first one shows a gradual slow increase for about 2 min followed by a rapid large increase remaining at a plateau until the reopening of the artery.

The response to a bilateral carotid occlusion is shown in Fig. 4.22 [6]. Immediately after the occlusion ($L + R_{occl}$), $K^+$ started to increase gradually, with levels in both hemispheres following similar response curves. When the carotid arteries were reopened, a large increase in $K_e^+$ was measured in each hemisphere. The duration of

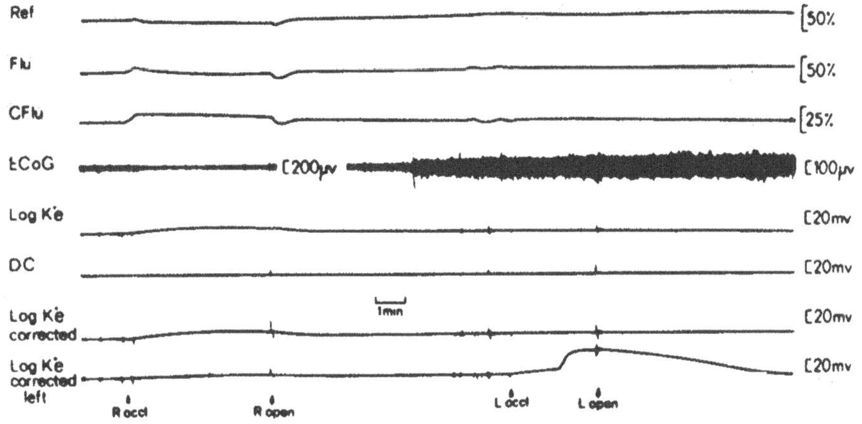


**Fig. 4.21** Unilateral carotid artery ligation in the gerbil. $R_{occl}$, right carotid ligation; $L_{occl}$, left carotid ligation [6]

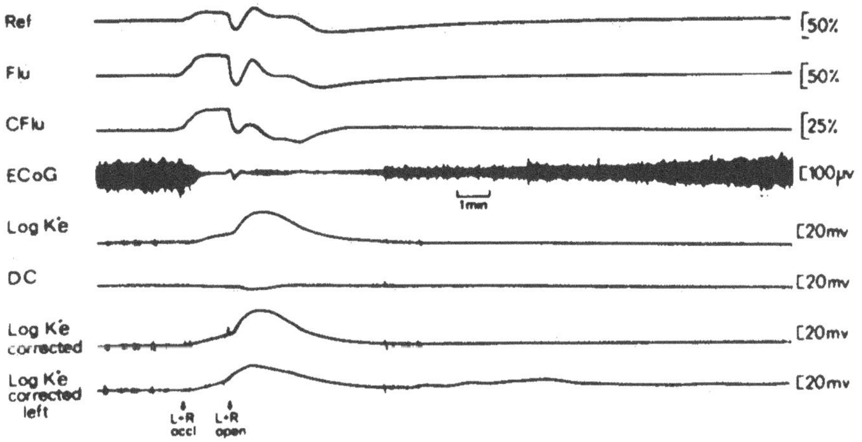


**Fig. 4.22** Bilateral carotid artery ligation in the gerbil. L + $R_{occl}$, left and right carotid ligation; L + $R_{open}$, left and right carotid open [6]

ischemia was 90 s. The metabolic response to complete ischemia is shown in the upper three traces. During the period of occlusion, NADH was reduced by 20% as compared to the normoxic level. Following restoration of flow in the carotid arteries, a large oxidation cycle of NADH was recorded. This oxidation cycle appeared to be a spreading depression cycle as indicated by the large transient increase in $K_e^+$ and the depression of the electrical activity (ECoG).

### 4.4.4 Discussion

The data presented in Figs. 4.18, 4.19, 4.20, 4.21, and 4.22 demonstrate various typical responses in the brain to partial or complete cerebral ischemia. The advantage of the gerbil brain model is that in the same animal model, the effects of complete cortical ischemia (bilateral occlusion) and those of various levels of partial ischemia (unilateral occlusion) can be studied. As we described previously, the relationship between basilar and carotid arteries is probably absent in the gerbil [23–25], while the relationship between the two anterior cerebral arteries differs among individual gerbils [19, 25, 26]. These findings have been reconfirmed in the present study. The responses of the brain to partial or complete ischemia depend very much upon the animal and the metabolic state of the tissue at the site of measurement. If, for one of a number of reasons, a local area of brain has suffered trauma, then its response to ischemia may differ from that of a normal site, as demonstrated in Fig. 4.21 in which the left side is shown to be more sensitive to ipsilateral carotid artery occlusion. However, bilateral occlusion induced similar changes in the two hemispheres (Fig. 4.22). In few animals, unilateral occlusion had a very small effect on $K_e^+$ [Figs. 4.19 and 4.21 (the right hemisphere)], while in other cases [Fig. 4.19 (left hemisphere)], $K_e^+$ accumulated after unilateral occlusion and led to spreading depression. The same sensitivity was found in the same gerbil under anoxic condition. The variability among gerbils with respect to the relationship between the two anterior cerebral arteries causes the varied responses of $K_e^+$ to ischemia.

### 4.4.5 Monitoring of NADH Together with $pO_2$, ECoG, Extracellular $K^+$, and DC

#### 4.4.5.1 Introduction

In the present study [7], we correlated for the first time the changes in NADH fluorescence, changes in the reflected light, and the changes in $pO_2$ measured with a surface electrode. The interrelation between the three parameters measured was tested under various conditions, such as anoxia and ischemia.

Figure 4.23a presents the results obtained from a different gerbil. The response to anoxic episode is basically similar to the one shown in other gerbils, but at the recovery phase the $pO_2$ response to SD has a different nature. One minute after breathing was started, a biphasic response was recorded, which was inversely related to the reflected light changes, i.e., when the reflectance showed an increase, the $pO_2$ trace showed a parallel decrease, and when the reflectance showed a large decrease (increased blood volume), the $pO_2$ showed three- to fourfold increase and then gradually recovered to the normoxic level in parallel to the gradual increase in the reflected light.

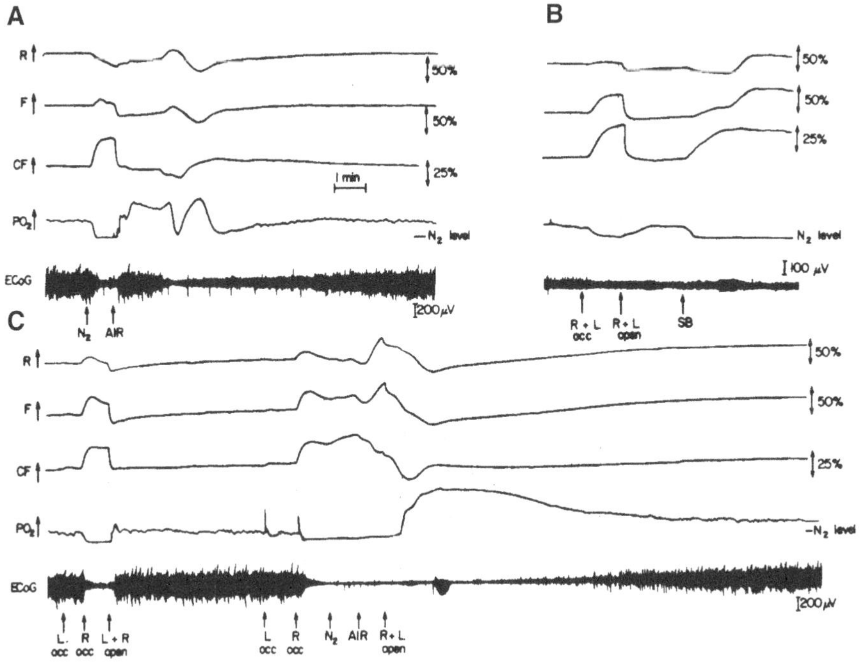


**Fig. 4.23** Effects of anoxia and carotid arteries occlusion on the metabolic and electrical activity of the awake (**a**, **b**) and anesthetized (**c**) gerbil brain [7]

The biphasic response in the $pO_2$ trace is in correlation to the response of the NADH fluorescence trace and we assume that it is the typical response to SD, as described previously [27], after anoxic cycle. Figure 4.23b presents the effects of unilateral or bilateral carotid artery occlusion. In this animal, the occlusion of the contralateral carotid ($L_{OCC}$) had a very small effect on NADH or $pO_2$. Occlusion of the right carotid (while the left one was closed) induced a large fall in $pO_2$ and an increase in NADH with a similar kinetics. The recovery after bilateral recirculation was fast, without any significant overshoot in $pO_2$. In the second ischemic episode, when $N_2$ was applied to the ischemic brain, a large overshoot in $pO_2$ was recorded after recirculation. Again, the large increase in $pO_2$ was parallel to the large decrease in reflectance with the same kinetics toward the normoxic level. NADH shows an oxidation cycle after recirculation due to SD, as can be seen in the ECoG trace. Figure 4.23c shows the response of the anesthetized brain to bilateral carotid artery occlusion. During the recovery period, $pO_2$ returns to its normoxic level and then decreases after nitrogen application together with the large increase in NADH.

#### 4.4.5.2 The First Study

The typical response to partial ischemia is shown in Fig. 4.24 [17]. After occlusion of the right carotid artery, $pO_2$ dropped simultaneously with the increased level of intramitochondrial NADH. The occlusion time was 5 min and the responses were observed in two steps during this period. The occlusion led to partial ischemia due to the compensation of blood flow from the left carotid artery through the anterior part in the Circle of Willis. A small increase in $pO_2$ and oxidation of NADH were observed 30 s after the occlusion. During this stage, ECoG was somewhat depressed as seen in the integrated ECoG. About 2 min after the occlusion, a second phase, "ischemic depolarization," started, with the property to start in the most sensitive area of the cortex and spread through the entire hemisphere. This event is labeled as 1 in Fig. 4.24 and it started near the light guide and propagated to the $K^+$ electrode area. During the ischemic depolarization (ID), few events occurred. We believe that the increased level of $K^+$ due to the decreased energy availability is the main factor leading to the development of the ID. As a result, $K^+$ accumulated and reached a

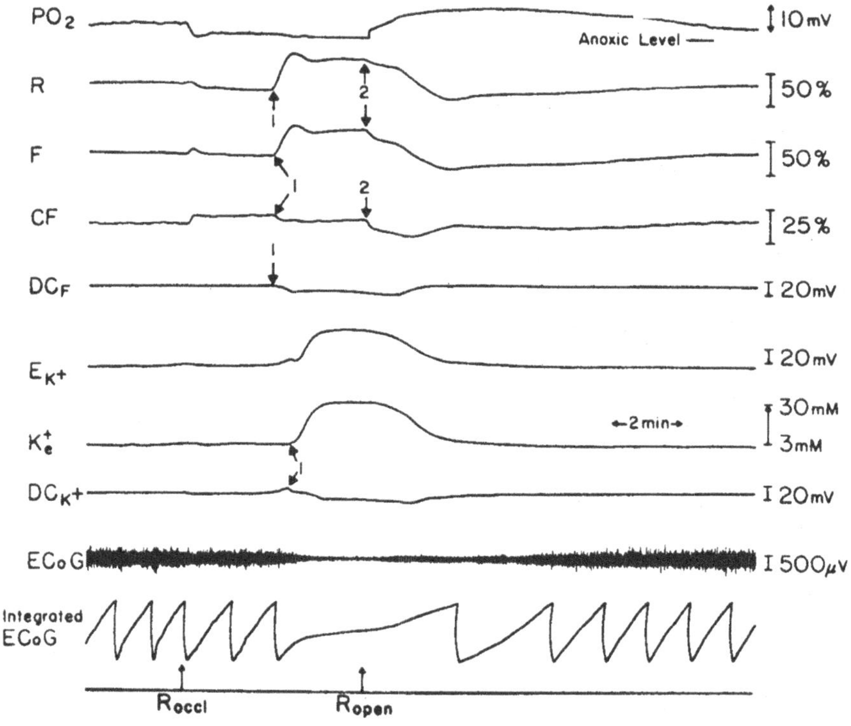


**Fig. 4.24** Metabolic ionic and electrical responses to partial ischemia induced by right carotid artery occlusion in the Mongolian gerbil. Arrow 1 represents the beginning of the ischemic depolarization (ID) and arrow 2 represents the recovery of blood flow after reopening of the carotid artery. $pO_2$ values are not an absolute one and the anoxic level obtained under $N_2$ breathing is indicated [17]

maximum level, which was more than ten-fold in this gerbil as compared to the baseline level. Due to the ID phenomenon which was also recorded as a negative shift in the DC traces and ECoG depression ($DC_F$ and $DC_K^+$), a large increase in the reflection signal was recorded leading to an apparent oxidation of NADH, but we showed that this is an artifact of the correction factor used for the F signal [28]. In many gerbils, we observed a drop in $pO_2$ during the ID period as shown in Fig. 4.24, which suggested the occurrence of constriction of blood vessels.

During the recovery period (after the reopening of the artery), a strong correlation was noted between the intensity of the ischemic insult and reestablishment of normal cortical activities. Simultaneously, with the opening of the vessels, a decrease in the R signal was recorded together with an increase in $pO_2$ (2 in Fig. 4.24). A decrease in the R signal was again recorded after 1 min. These two changes of the R signal are probably due to increased blood volume in the tissue and absorption of more light by the hemoglobin. The recovery of R signal to the pre-occulation level was very slow and parallel to the decreases of the overshoot $pO_2$ levels. The recovery of the $K^+$ and DC traces occurred as soon as energy was available and it took about 6–7 min in this occlusion (partial ischemia only). The ECoG also recovered very fast after this ischemic episode.

When complete ischemia was induced by bilateral carotid artery occlusion (Fig. 4.25), the kinetics of the changes was faster and the severity of the insult was evaluated by the duration of the recovery period. The increases in NADH as well as the decrease in $pO_2$ reached maximal and animal levels, respectively. The ID (1 in Fig. 4.25) appeared much faster than the partial ischemic episode (Fig. 4.24). Also, the ECoG disappeared very fast even before the ID occurred. The R change led to an artifact in the CF trace, and the CF reached on the high level later in this episode as recorded before the appearance of the ID (CF at 1 = CF at 2). The recovery period after the complete ischemia was very long as compared to the previous one (Fig. 4.24). Here, one can see very clearly the two stages in the R changes, i.e., (1) a fast initial decrease (2) followed by a secondary long-lasting slower decrease (3) and then a recovery to the pre-ischemic level. The $pO_2$ trace recovery also showed a two-step increase in parallel to the decrease in the R trace. A delay of more than 2 min was recorded in the decrease of $K^+$ and the large decrease started soon after the large decrease in the R trace (3 in Fig. 4.25). The recovery of the ECoG was very slow after this insult, and even after 1 h the ECoG was depressed as compared to the pre-ischemic level.

#### 4.4.5.3 The Second Study

##### 4.4.5.3.1 Introduction

This study was performed according to the protocol [17] described in Chap. 3. It appears in the study by Mayevsky and Zarchin [29].

The purpose of the present study was (1) to correlate the degree of ischemia developed under unilateral and bilateral carotid artery occlusions with the

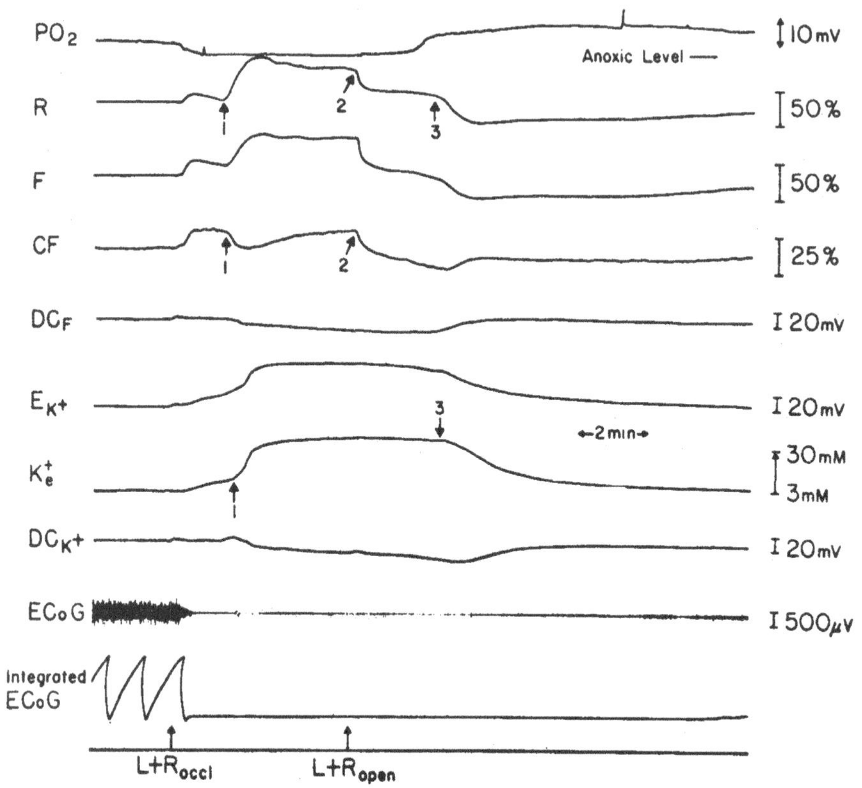


**Fig. 4.25** Effects of complete cerebral ischemia (left and right carotid artery occlusions -L + $R_{occl}$) on the functional state of the brain and the recovery after reopening of the two blood vessels (L + $R_{open}$). Arrow 1 indicates the beginning of the ischemic depolarization period (ID) and arrows 2 and 3 show the two steps of blood flow recovery after the ischemic insult removal [17]

impairment of ionic homeostasis developed under these conditions and (2) to test the recovery rate of the extracellular $K^+$ level ( $K_e^+$ ) under the same conditions.

In order to evaluate the functional state of the brain, a multiprobe assembly was used [8, 29]. Energy metabolism was evaluated by monitoring the intramitochondrial NADH redox state simultaneously with tissue surface $pO_2$. The reflected light (366 nm) measured simultaneously with the fluorescence signal (450 nm) served as an indirect measure of blood volume changes in the observed area. The ionic activity was monitored by measuring the extracellular $K^+$ levels using the valinomycin type of electrode. Electrical activities were evaluated by bipolar electrocorticography (ECoG) and DC steady potential.

#### 4.4.5.3.2 Methods

The multiprobe assembly (MPA) used in the present study was described in detail in our previous publication [8]; hence, only a few major points are described here. The MPA has the following main features: (1) all components have the same type of noninvasive surface contact with the cortex; (2) it can be removed very easily from the skull at the end of the experiment without any damage, so that repetitive experiments can be performed in a very low sensitivity to movements of the animal, even while exposed to hyperbaric pressure.

Figure 3.11 shows schematically the relationship between the MPA and the cerebral cortex of the gerbil brain. The probes are mounted in a lucite cannula (c) which can be cemented directly to the skull. The holes accommodating the specific sensing electrodes ($K^+$, $pO_2$) converge at the implantation site to occupy less space in the brain and diverge at the top of the cannula to leave maximum room for mounting and replacing the electrodes. Each sensor chamber is pierced approximately half way down by a hole drilled obliquely from the top which houses an Ag/AgCl wire to record DC potential around the sensing electrode. One more hole drilled horizontally into the DC wire chamber allows for filling with saline and for cleaning ECF and blood around the electrodes at the end of an experiment. A lucite sleeve (s) placed over the entire assembly is held in place with a set of screws. This sleeve holds the assembly together as 1 unit, protects the electrodes and wiring, and also provides shielding for the high impedance $K^+$ electrode. The sensing electrodes, Ag/AgCl wires for DC measurement, stainless steel wires for ECoG measurement, and a polyethylene-sleeved temperature probe are held in and sealed to the cannula with 5-min epoxy or by melting wax around them. Another block of lucite attached to the outer shield holds the reference assembly and the tubes (f) for filling the DC chambers.

**Surface Fluorometry**

The intramitochondrial NADH redox state was monitored by a light-guide DC fluorometer/reflectometer [22]. The source for the 366 nm light was split in a 90:10 ratio for the measurement of the NADH fluorescence and reflectance. The common part of the light guide was cemented inside the multiprobe assembly in the same way as the electrodes. A branch of the excitation and emission fiber bundle was connected to the common part.

**Animal preparation**

The gerbil was anesthetized by i.p. injection of Equithesin (0.3 ml/100 g body weight) (each ml contains pentobarbital, 9.72 mg; chloral hydrate, 42.51 mg; magnesium sulfate, 21.25 mg; propylene glycol, 44.34% w/v; alcohol, 11.5%; and water). The skull was exposed through a midline incision and a hole of 6.0 mm diameter was drilled in the right parietal bone. Four holes (with a diameter of 1.5 mm) were drilled in various points of the skull and stainless steel screws were used to hold the cement to the skull. The dura mater was removed carefully, only from the area where the $K^+$ electrode was placed. The MPA was held in a micromanipulator and located over the cortex without creating any extra pressure on the brain. The ground and reference electrodes were placed below the skin and were cemented to the skull, together with the MPA and the screws, by dental acrylic cement.

#### 4.4.5.3.3 Results

Figure 4.26 shows the effects of complete ischemia induced by bilateral carotid artery occlusion ($L + R_{occl}$) on various parameters monitored from the gerbil brain. Upon occlusion, tissue oxygen level decreased to the anoxic level and stayed low until reperfusion started ($L + R_{open}$). As oxygen became limited, the NADH level (CF) increased and reached its maximum soon after the occlusion. During the occlusion, a large increase in the reflectance signal (R) was recorded, which led to an almost similar change in the uncorrected fluorescence (F). The NADH level recovered to its pre-ischemic value within 2 min after the reperfusion started. Due to the

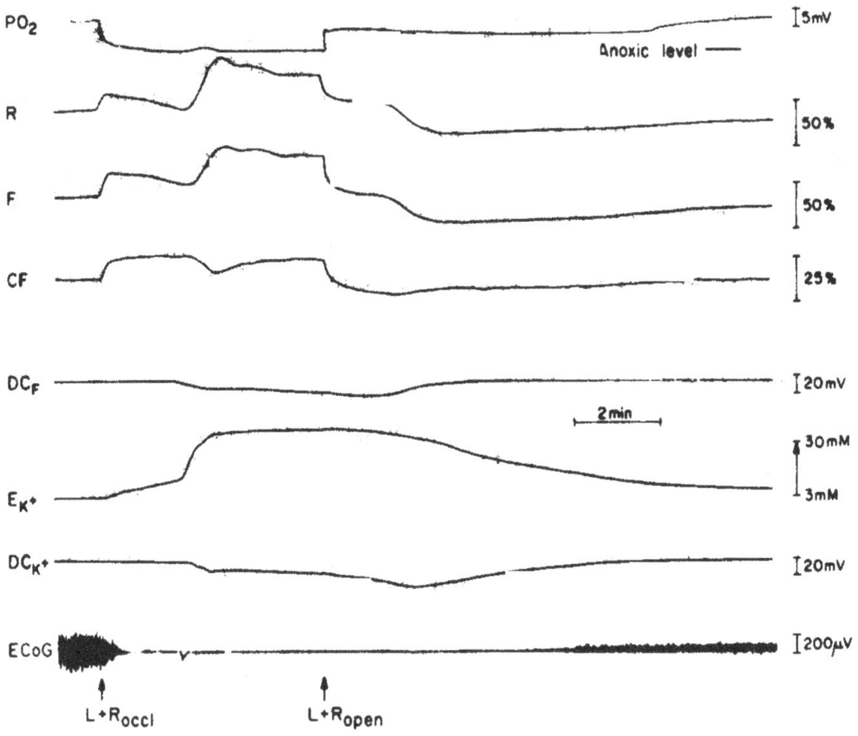


**Fig. 4.26** Effects of bilateral carotid artery occlusion ($L + R_{occl}$) on metabolic, ionic, and electrical activities measured from the right hemisphere of the gerbil brain. $pO_2$, tissue oxygen partial pressure; R, reflectance; F, fluorescence; CF, corrected fluorescence; $E_K$+, extracellular $K^+$ level; $DC_K$+ and $DC_F$, DC steady potential measured around the $K^+$ electrode and the light guide respectively; ECoG, electrocorticogram. The $pO_2$ anoxic level was determined by exposing the gerbil to 100% $N_2$ for 1 min [29]

secondary reflectance increase (SRI), the NADH trace showed no apparent oxidation although the $pO_2$ was at the anoxic level at this time. The recovery of the reflectance signal exhibits two-step kinetics: a fast decrease was recorded immediately after reopening the blood vessels, followed by a further slow decrease which recovered afterward to the pre-ischemic level. The effect of the occlusion on the extracellular $K^+$ level ($E_K$+) shows clear two-step kinetics. Soon after the occlusion, a slow gradual increase in $E_K$+ was recorded followed by a rapid large increase. During the first phase, $E_{K+}$ changed from approximately 3 to 6 mM, whereas during the second phase it reached the level of 40–50 mM.

The rapid increase in $E_{K+}$ was concomitant with the negative shift in the DC steady potential measured around the potassium electrode ($DC_{K+}$) as well as around the light guide ($DC_F$). Both the $E_K$+ and the DC steady potential recovered to pre-ischemic levels within 10 min after reopening the two carotid arteries. The ECoG trace shows that the EEG disappeared within 30 s after bilateral carotid occlusion. The recovery of the EEG (data not shown in the figure) was much slower than any other parameter measured and took about 30 min in this particular ischemic episode.

Figure 4.27 shows the response of a gerbil brain to unilateral carotid artery ligation for 3 min. The main difference between the responses in this ischemic episode

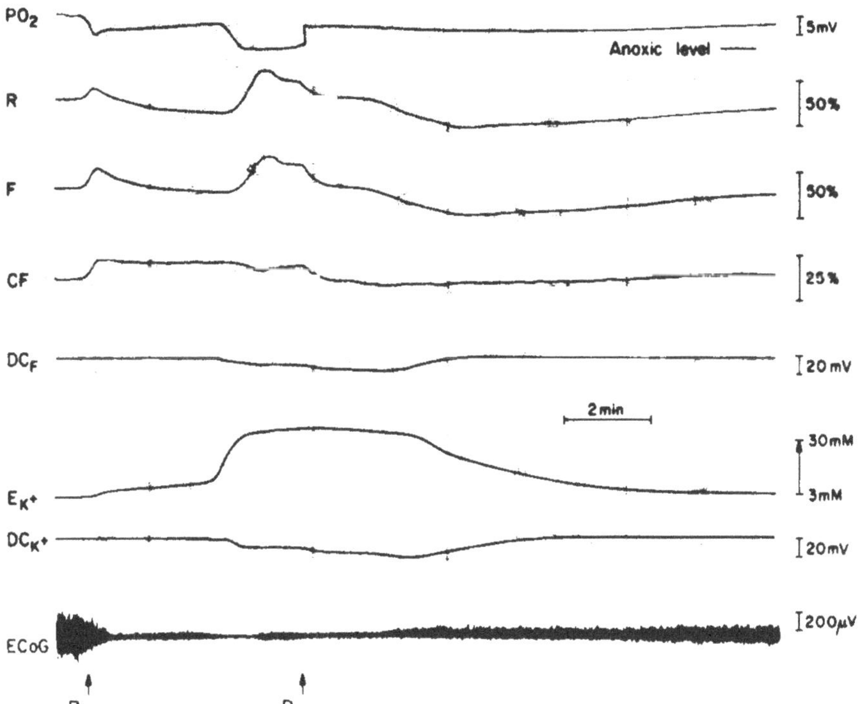


**Fig. 4.27** Responses of various parameters, measured in the same gerbil as in Fig. 4.26, to unilateral carotid artery occlusion ($R_{occl}$) [29]

and those shown in Fig. 4.26 is due to the lower level of ischemic insult. The $pO_2$ trace shows an initial decrease followed by a slow recovery due to compensatory flow through the contralateral anterior cerebral artery. The same pattern of events is seen in the NADH redox state. As soon as the large and rapid increase in $E_K+$ occurred, the SRI was recorded in the reflectance trace simultaneously with a sharp decrease in $pO_2$.

Leakage of $K^+$ from the cells in phase I of the insult was slower and smaller compared with that produced by bilateral occlusion. The effect of partial ischemia on the ECoG activity was also smaller and recovery was faster. Figure 4.28 shows a different type of response to unilateral occlusion found in 40–50% of the gerbils used. The level of ischemia achieved was smaller and, as a result, probably the changes in $E_K+$ were more gradual and did not reach the second phase of large $K^+$ leakage. The ECoG trace shows only a very small change, if any. The analog signals recorded from 11 gerbils included in this study were analyzed for quantitative changes in NADH redox state and $E_K+$ and are presented in Tables 4.1 and 4.2. As

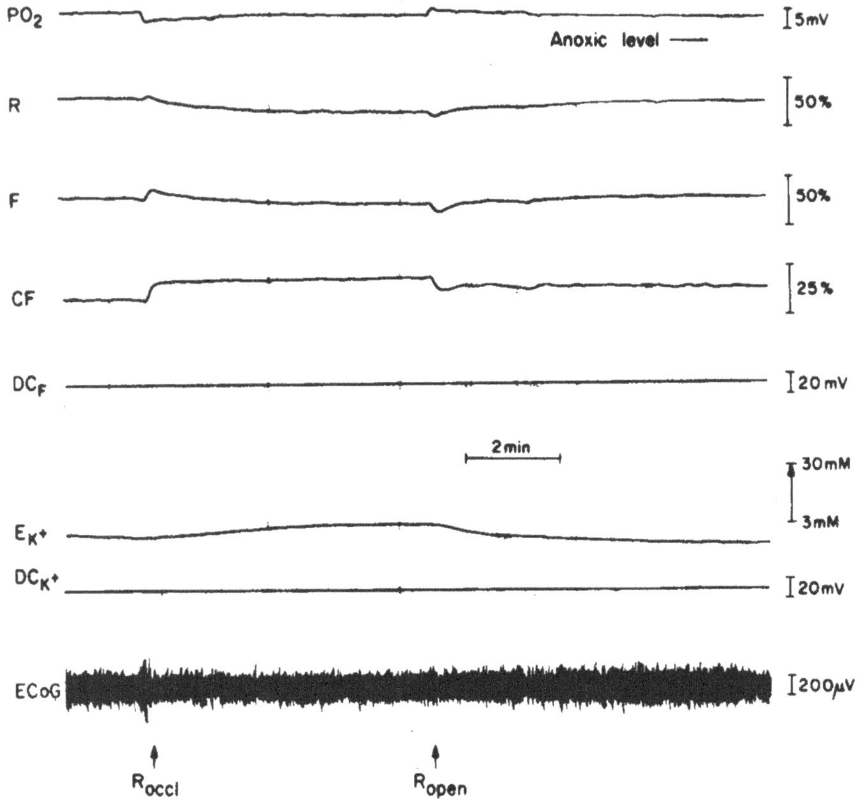


**Fig. 4.28** Effects of unilateral carotid artery occlusion ($R_{occl}$) on the metabolic, ionic, and electrical activities measured in a different gerbil brain from that in Figs. 4.26 and 4.27 [29]

**Table 4.1** Effects of partial or complete ischemia on NADH redox state and extracellular $K^+$ levels during occlusion [29]

| Treatment | Duration | NADH | | $K_e^+$ | During Phase I | | $K_e^+$ | During Phase II | Time to |
|---|---|---|---|---|---|---|---|---|---|
| ($n = 11$) | of Ischemia (min) | % increase | % ischemia | Baseline (mH) | mM increase | Duration (min) | Max increase (mM) | Time to max min | Max/2 (min)% ischemia |
| R Occlusion | 5.1 ±0.04 | 8 ±0.7 | 66 ±6 | 3.1 ±0.3 | 4.5 ±0.4 | 2.1 ±0.3 | 41 ±2.6 | 2.8 ±0.3 | 0.62 ±0.05 |
| L + R Occlusion | 5.1 ±0.06 | 13*** ±0.5 | 100 | 3.3 ±0.4 | 6.1** ±0.7 | 1.4* ±0.1 | 49* ±3.8 | 3.6 ±0.2 | 0.61 ±0.05 |

Data represent means ± S.E.M. *$p < 0.05$; **$p < 0.01$; ***$p < 0.001$

**Table 4.2** Recovery of extracellular $K^+$ levels after 5 min partial or complete ischemia [29]

| Treatment | Reflectance | $K_e^+$ Time (min) to: | | | $K_e^+$ New baseline |
|---|---|---|---|---|---|
| ($n = 11$) | Time to second decrease (min) | Initial decrease | Minimum | Minimum/2 | level (mM) |
| R Reopening | 2.8 ±0.7 | 1.58 ±0.44 | 13.2 ±1.7 | 3.9 ±0.6 | 2.95 ±0.34 |
| L + R Reopening | 3.8 ±0.8 | 2.8 ±0.9 | 21.6** ±2.8 | 6.5 ±1.3 | 3.06 ±0.38 |

Data represent means ± S.E.M. $^{**}p < 0.01$

presented in Table 4.1, the level of ischemia produced by unilateral artery occlusion was significantly lower than that produced by bilateral occlusion. During phase I of potassium leakage, two significant differences were found between partial and complete ischemia as well as differences in the maximum level of $E_K$+.

Table 4.2 presents the recovery of the reflectance trace and the $E_K$+ after ischemic episodes. The time to reach the second reflectance decrease was longer but was not significant. All parameters represented the slower recovery of $E_K$+ in the completely ischemic episodes but only the time to reach the $E_K$+ minimum level was significantly longer. The new $E_K$+ level after recovery was not different from the pre-ischemic values with either insult.

#### 4.4.5.3.4 Discussion

The Mongolian gerbil provides a useful model for studying brain responses to partial or complete ischemia (refer to "Introduction"). In our previous study [9], we compared the effects of 5- and 30-min complete cerebral ischemia induced by bilateral carotid artery occlusion. In the present experiments, we compared the events that occurred in the brain under partial or complete cerebral ischemia in the same animal. The effects of bilateral occlusion were as expected from previous work, but with unilateral carotid artery occlusion, two very different responses were recorded. The reason for the variation between animals is not only that the level of ischemia induced differs in different animals, but it also involves other factors such as membrane permeability to $K^+$. The type of response is very much dependent on the level of ischemia induced. Due to the ischemic depolarization that developed in about 50% of the gerbils exposed to unilateral carotid artery occlusion, the functional state of the brain will be completely different as shown in Figs. 4.27 and 4.28. The results can be summarized as follows: (1) bilateral occlusion led to the development of complete ischemia, while after unilateral carotid occlusion the level of ischemia (as evaluated by the NADH redox state) was only 50–60%; (2) extracellular $K^+$ level under bilateral occlusion showed a two-step increase reaching its maximal level (40–50 mM) within 3.5 min; (3) under partial ischemia, two types of responses in $K_e^+$ were noted: the first was similar to that recorded under complete ischemia, and in the second, only a small monotonic increase in $K_e^+$ was seen; (4) during the recovery phase, the kinetics of $K_e^+$ toward the base line level were significantly faster after partial ischemia as compared with complete ischemia. We concluded that after incomplete ischemia, the recovery processes were faster due to the recovery of the ECoG traces.

#### 4.4.5.4 The Third Study

##### 4.4.5.4.1 Introduction

In order to evaluate the functional state of the brain, both during and after the ischemic insult, we used a multiparametric monitoring system [8, 30] that provides information about the brain metabolic, ionic, and electrical activities. Metabolic activity was assessed by monitoring the tissue surface $O_2$ tension with surface electrodes. The intramitochondrial NADH redox state was evaluated by the surface fluorometry/reflectometry technique described in detail previously [30]. The reflected light at the excitation wavelength (366 nm) measured simultaneously with the fluorescence light (450 nm) was used to correct the fluorescence signal for hemodynamic artifacts. Extracellular $K^+$ activity was measured with surface valinomycin electrodes located in the multiprobe assembly.

##### 4.4.5.4.2 Methods

All probes used were located in a multiprobe assembly (MPA) constructed according to the principles described previously [8, 31, 32]. The probes were mounted in a lucite holder (cannula) cemented directly to the skull. The holes accommodating the specific sensing electrodes converge at the implantation site to occupy less space on the brain and diverge at the top of the cannula to leave maximum room for mounting and replacing the electrodes. Another hole was drilled obliquely from the top to intersect the first chamber approximately half way down. This hole houses an Ag/AgCl wire to record the DC potential concentrically to the sensing electrodes. One more hole was drilled horizontally into the chamber of the DC wire for filling with saline and for cleaning extracellular fluid and blood around the electrodes at the end of an experiment. A lucite sleeve was placed over the entire assembly and was held in place with a set of screws. This sleeve holds the assembly together as one unit, protects the electrodes and wiring, and also provides shielding for the $K^+$ electrode's high impedance. The sensing electrodes, Ag/AgCl wires for DC measurement, stainless steel wires for electrocorticogram (ECoG) measurement, were held in and sealed to the cannula with 5-min epoxy, or by melting wax around them. Another block of lucite was attached to the outer shield, which holds the reference assembly and the tubes for filling the DC chambers.

**Surface Fluorometry**

The intramitochondrial NADH redox state was monitored with a light-guide DC fluorometer/reflectometer [30]. The source for the 366 nm light was a 100-W mercury arc cooled by air. The emitted light was split in a 90:10 ratio for measurement of the NADH fluorescence and reflectance, respectively. The common part of the Y-shaped light guide was cemented inside the multiprobe assembly in the same way as the other electrodes.

**Animal Preparation**

The gerbil was anesthetized by intraperitoneal injection of Equithesin (0.3 ml/100 g body weight) (each ml contains pentobarbital, 9.72 mg; chloral hydrate, 42.51 mg; magnesium sulfate, 21.24 mg; propylene glycol, 44.34% w/v; alcohol, 11.5%; and water). After making a midline incision of the skin, the skull was exposed and a hole of 5.0 mm in diameter was drilled in the right parietal bone. Four holes (1.5 mm in diameter) were drilled at various points on the skull and stainless steel screws were used to hold the cement to the skull. The dura mater was removed very gently only from the area where the $K^+$ electrode was placed. The MPA was held by a micromanipulator and located above the cortex without creating any extra pressure on the brain. The ground and reference electrodes were placed below the skin and were cemented to the skull together with the MPA and the screws using dental acrylic cement.

#### 4.4.5.4.3 Results

In order to illustrate the usage of the multiparametric monitoring system, Fig. 4.7 shows the effects of anoxia (left traces) and unilateral carotid artery ligation (right traces). Under complete deprivation of $O_2$, as can be seen also in the $pO_2$ trace, the intramitochondrial NADH (CF) became fully reduced, i.e., all NAD accumulates as NADH. During the anoxic episode a vasodilation occurred due to the autoregulation mechanism, and it is recorded as a decrease in the reflectance trace (R). Due to the short duration of the anoxic cycle, only a small increase in the extracellular $K^+$ level was recorded ( $K_e^+$ ). The effect of the anoxic cycle on the DC steady potential was undetectable but a small depression of the ECoG was noted. During the recovery period, the intramitochondrial NADH recovered to the pre-anoxic level without any hyperoxidation as expected from the overshoot in the $pO_2$ trace due to the high blood flow level created during the anoxia.

During the occlusion of the right carotid artery (Fig. 4.29), only partial ischemia was induced as can be seen from the $pO_2$ trace and the NADH recording. Very small changes were found in the extracellular $K^+$ level, and the DC steady potential or ECoG was recorded during 5 min of partial ischemia.

In the same gerbil, occlusion of the two carotid arteries (Fig. 4.30) led to complete ischemia, as can be seen from the $pO_2$ and the CF traces. During the initial phase of the ischemia, $K_e^+$ increased gradually as the energy was depleted due to the decrease in oxygen supply.

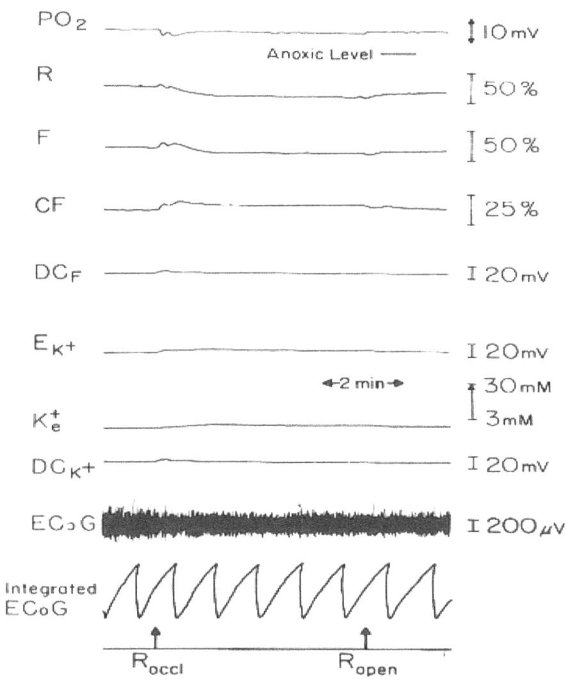


**Fig. 4.29** Effects of partial ischemia (right carotid artery occlusion) on the metabolic, ionic, and electrical activities of the gerbil brain. $pO_2$, tissue surface $O_2$ levels; R, 366 nm reflectance; F, 450 nm fluorescence; CF, NADH-corrected fluorescence; DCF and DCK, DC steady potential around the optical fibers and the $K^+$ electrode, respectively; $EK^+$ and $K_e^+$, uncorrected extracellular $K^+$ electrode potential and corrected (for DC changes) $K^+$ levels, respectively; ECoG, electrocorticogram [11]

At a certain point, a massive depolarization occurred and the $K_e^+$ levels exceeded the 30 mM range (ten-fold more than the resting pre-ischemic level). During depolarization, the optical properties of the tissue were changed and a large increase in the reflectance trace (R) was recorded. This change in the R signal was secondary to the initial response to the occlusion of the arteries itself and we defined it as the secondary reflectance increase (SRI) [33]. The change in the R signal led to an apparent decrease in the NADH-corrected fluorescence trace due to the inability of the correction factor to handle such a large change that also includes optical changes in the tissue besides the hemodynamic ones. The DC steady potential showed a negative shift of about 20 mV during the massive depolarization event. The ECoG trace was depressed as soon as energy was depleted and it became isoelectric 20 s after the occlusion.

Following recirculation, the tissue $pO_2$ as well as the intramitochondrial NADH redox state recovered to the pre-ischemic level within less than 1 min. The extracellular $K^+$ recovery showed a two-stage response, i.e., immediately after recirculation a slow decrease was recorded followed by a fast recovery to the baseline levels. During the recovery of the $K^+$ and DC potential, a secondary large decrease in the reflectance trace was recorded, suggesting a hyperemic event that gradually recovered to the pre-ischemic blood volume levels.

In order to examine the mechanism of $K^+$ disappearance from the extracellular space during the recovery from ischemia, we exposed gerbils to a short moderate

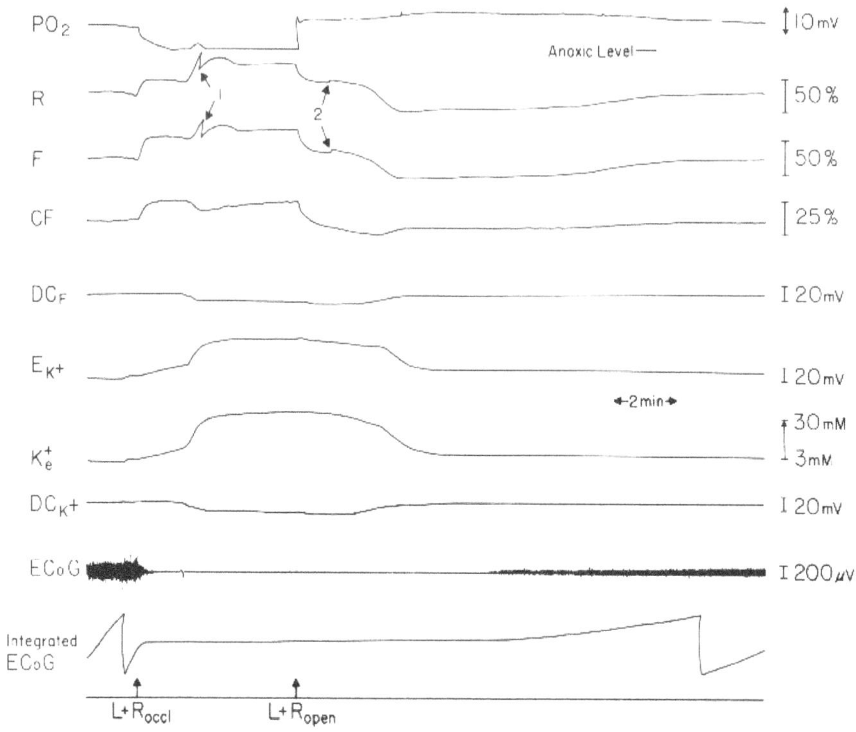


**Fig. 4.30** Responses to bilateral carotid arteries occlusion (L + $R_{occl}$) and post-ischemic recovery after recirculation (L + $R_{open}$) of the gerbil brain. The anoxic level line represents the level of $pO_2$ during complete deprivation of oxygen as shown in Fig. 4.29. Arrow 1 indicates a change in scale sensitivity which was returned to the original one at Arrow 2 (abbreviations as in Fig. 4.29) [11]

hypoxia during the recirculation phase. As shown in Fig. 4.31, a gerbil was exposed to partial ischemia by occlusion of the right carotid artery. Only partial ischemia was induced, as can be seen from the $pO_2$ level after the occlusion.

When massive depolarization occurred, the large increase in the R trace (SRI) was concomitant with the decrease in $pO_2$ to reach the minimum anoxic levels. Also, the CF level increased after a short artifact transient decrease during the depolarization phase. These results suggested that the large increase in the R trace was associated with a vasoconstriction of blood vessels leading also to a decrease in $pO_2$ to the cells under those conditions. During this recovery from the ischemia, when the energy level was recovered ($pO_2$ and CF traces), exposing the gerbil to hypoxia (6% $O_2$ in N2) led to a decrease in $pO_2$ and an increase in NADH (CF), but at the same time the R trace showed a decrease suggesting a further vasodilation although the R trace was already very low. This increase in blood flow due to hypoxia had no beneficial effect on the rate of $K^+$ disappearance which was even inhibited. Only when 100% $O_2$ was given to the animal did $K^+$ recover to the pre-ischemic levels.

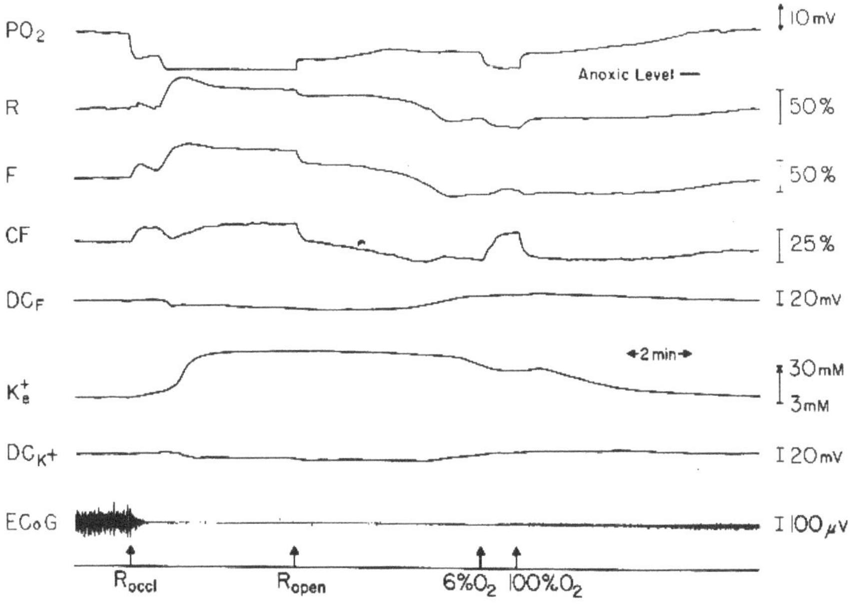


**Fig. 4.31** Effects of hypoxia induced during the post-ischemic recovery phase on the pumping of $K^+$ into the cellular compartment. Partial ischemia was induced by the occlusion of the right common carotid artery ($R_{occl}$). Hypoxia was induced by exposing the head of the gerbil to a 6% $O_2$ and 94% $N_2$ mixture (abbreviations as in Fig. 4.29) [11]

In order to study the kinetic effects of ischemia on the brain, we exposed a group of gerbils to 1–5 min of bilateral carotid artery occlusion. After connecting the animal to the measuring system, a short $N_2$ cycle was applied and ischemia was induced in the same animal for different durations as shown in the next series of figures.

Figure 4.32 illustrates the effects of 1-min (left) and 2-min (right) bilateral occlusion on the metabolic, ionic, and electrical activities of the same gerbil. In this gerbil, we found that occlusion of the left artery had no effect on the measured hemisphere, i.e., the right one. Therefore, the left common carotid artery was occluded permanently and only the right one was reopened during the recirculation episode after the ischemia. As shown, the 1-min ischemia led to a small increase in extracellular $K^+$ which recovered to the baseline level shortly after the recirculation. The DC steady potential was not affected, while the depression of the ECoG began to recover very soon after the recirculation. During the 2-min occlusion, the same responses obtained during the 1-min occlusion were recorded, but during the recovery period a spontaneous wave of spreading depression (SD) was observed. A negative shift in the DC potential as well as a transient depression of the ECoG was recorded. This wave of SD was induced by internal elevation of the extracellular $K^+$ levels during the ischemia, but was pronounced during the recirculation phase due to the short period of ischemia. During this SD event, a secondary increase in $K^+$

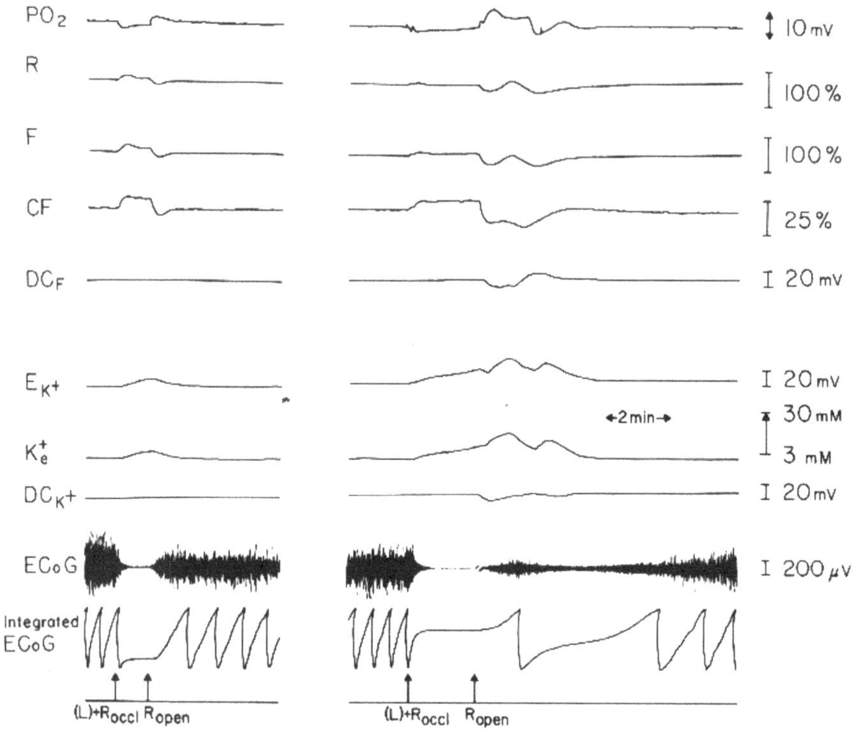


**Fig. 4.32** Metabolic, ionic, and electrical responses to 1 min (left) or 2 min (right) bilateral carotid artery occlusion in the gerbil. The left common carotid artery was permanently occluded and therefore was not opened during the recirculation phase (abbreviations as in Fig. 4.29) [11]

level was recorded together with a negative shift in the DC steady potential and the depression wave of the ECoG.

The NADH redox state reveals an oxidation wave during the SD event simultaneously with the increase in the $pO_2$ levels. Figure 4.33 illustrates the effects of 3-min bilateral occlusion on various parameters measured. About 2 min after the occlusion, ischemic depolarization (ID) occurred and a large efflux of $K^+$ was recorded simultaneously with the large reflectance increase (SRI). Also, a further decrease in $pO_2$ was noted suggesting that the level of ischemia was below 100% and the vasoconstriction led to complete ischemia. The recovery of the $K^+$ and DC steady potential were very fast compared to the ECoG which were partially depressed for a longer period of time. The $pO_2$ trace exhibited an overshoot during the recovery period correlating to the large blood volume in the tissue as seen by the large decrease in the R trace. The effects of 4-min occlusion (data not shown) were basically the same as those in the case of 3 min.

The effects of 5-min bilateral carotid occlusion are illustrated in Fig. 4.34. SRI was recorded after 2 min and the corrected fluorescence showed a transient apparent oxidation which returned to the maximal reduced level after 3 min. ECoG was

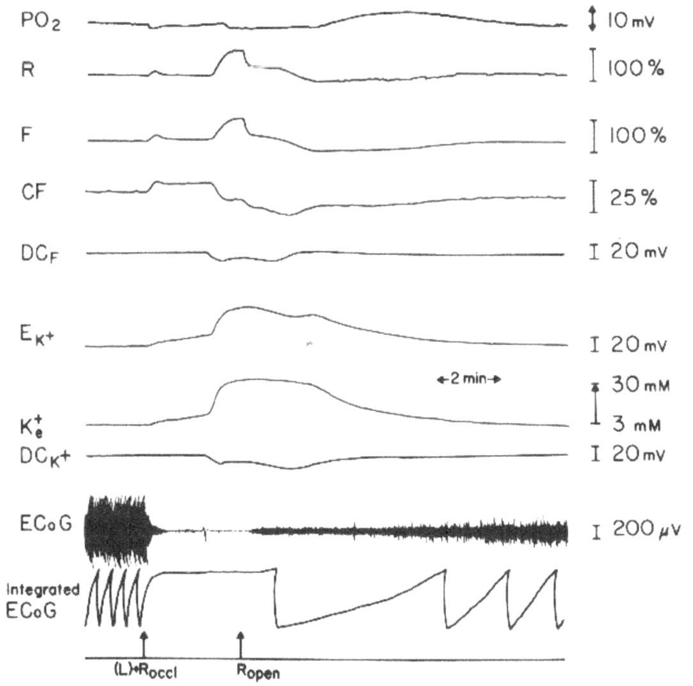


**Fig. 4.33** Effects of 3-min bilateral carotid artery occlusion on various parameters measured in the gerbil brain (abbreviations as in Fig. 4.29) [11]

depressed for a longer period of time correlating with the slower recovery of other parameters. The NADH redox state revealed a large undershoot during the recovery period due to the extra pumping activity done by the $Na^+$-$K^+$-ATPase to shift $K^+$ into the cellular compartment. Figure 4.35 shows the responses of $K_e^+$ and ECoG to 5-min (Fig. 4.34a) and 1- to 5-min (Fig. 4.34b–f) bilateral carotid artery occlusion in the same gerbil as shown in Figs. 4.32, 4.33, and 4.34. A very small difference was found between the first (A) and second (F) 5-min occlusion in the same animal suggesting a very small deterioration of the brain during the experiment. The effects of the duration of the insult on the recovery of $K^+$ as well as ECoG are illustrated clearly in Fig. 4.35. Only Fig. 4.35b shows that under 1-min ischemia, ECoG *recovered* to the pre-ischemic levels very shortly after the recirculation.

#### 4.4.5.4.4 Discussion

As demonstrated by various investigators, the Mongolian gerbil is a good model for studying the effects of partial and complete ischemia on the brain. One of the problems, using any model of ischemia, is to evaluate the level of the ischemic insult developed. In each gerbil, we can induce complete ischemia or anoxia obtained by the 100% level of insult. The level of insult developed under unilateral carotid artery occlusion is calculated in relation to the maximal level of insult determined.

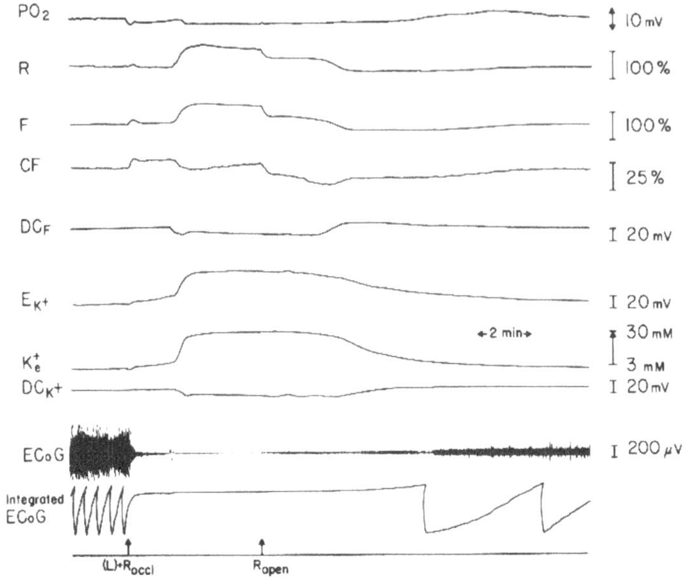


**Fig. 4.34** Brain responses to 5-min bilateral carotid artery occlusion in the gerbil (abbreviations as in Fig. 4.29) [11]

Employing this approach, one can correlate in a better way the various events that occur during the ischemic episode. In the present study, we found that during unilateral carotid occlusion, the primary insult was proportional to the degree of communication between the two hemispheres' circulation via the anterior communicating artery. In about 50% of the gerbils, a secondary response was developed due to the massive depolarization which occurred under the partial ischemic conditions. This secondary response led to a further decrease in oxygen supply as demonstrated in Figs. 4.31, 4.33, 4.34, and 4.35. It is not yet clear whether the secondary response is due to the level of ischemia developed during the unilateral occlusion or dependent on other factors such as removal of the dura mater or the operation itself.

As shown in Figs. 4.29 and 4.31, under unilateral occlusion (right artery), the difference in the responses of the two gerbils led to variations in the extracellular $K^+$ levels during the ischemia, and thus the activation of the $Na^+$-$K^+$-ATPase was different using various amounts of energy for the restoration of ionic homeostasis.

Under bilateral carotid artery occlusion, all gerbils developed complete or close to complete cerebral ischemia in which NADH reached its maximal level and was accompanied by ischemic depolarization (ID) in all animals. A high level of ischemia under bilateral occlusion has been described by all investigators using this model [19, 22, 25, 34–38]. Under unilateral ischemia, the only parameter employed to evaluate the degree of ischemia was the behavioral modification which is not a quantitative one. Use of the NADH increase response due to ischemia is most adequate for this purpose and indeed provides valuable information.

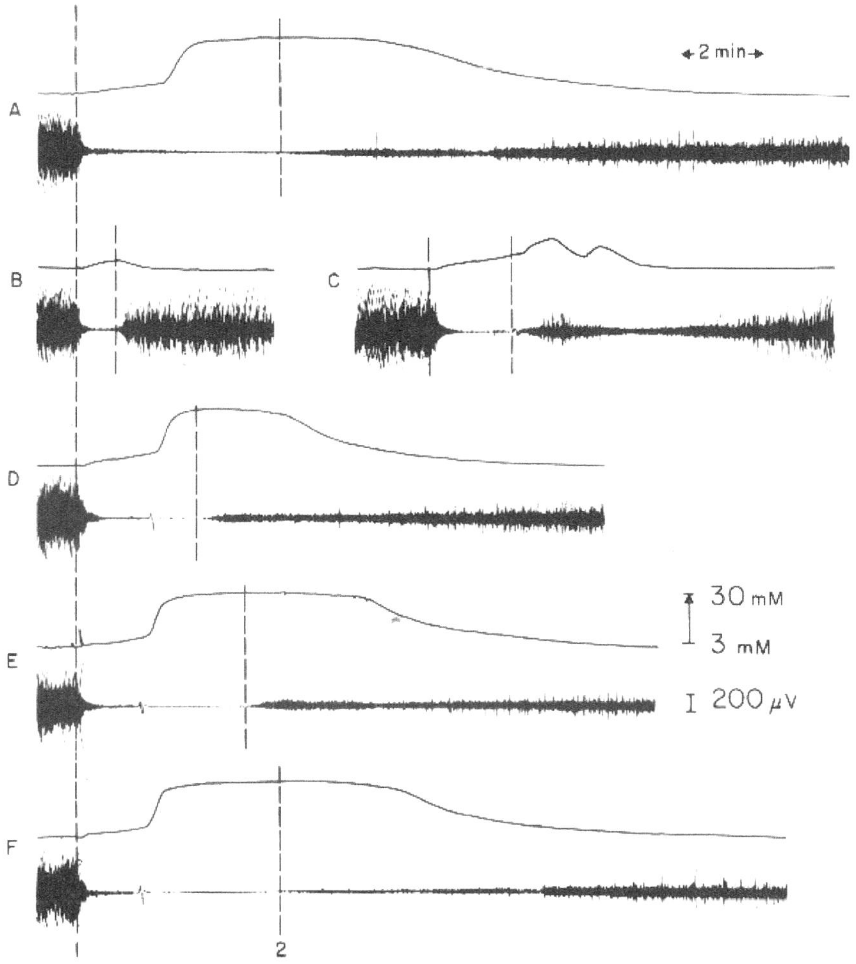


**Fig. 4.35** Extracellular $K^+$ level and ECoG responses to various durations of bilateral carotid artery occlusion in the gerbil brain. (**a**) shows the first 5-min occlusion as compared to the second 5 min (**f**). (**b–e**) show the responses of 1–4 min occlusion. The first dashed vertical line (1) represents the occlusion point, while the second one (2) represents the reopening of the right carotid artery [11]

The main contribution of the present study is its demonstration of the timing factor in ischemia studies. As shown in Figs. 4.32, 4.33, 4.34, and 4.35, ID appeared at a certain point (about 2-min after occlusion) and led subsequently to its secondary effects. Therefore, the recovery processes will be correlated to the appearance of the ID. If the ID occurs during the transition to the recirculation phase, a similar response to spreading depression will take place as shown in Fig. 4.32.

The present results can be summarized as follows:

1. Bilateral occlusion led to the development of complete ischemia, while after unilateral carotid occlusion the level of ischemia (as evaluated from the NADH redox state) was only 50–60%.
2. The extracellular $K^+$ level ( $K_e^+$ ) under bilateral occlusion revealed a two-step increase, reaching its maximal level (40–50 mM) before 3.5 min.
3. Under various durations of complete ischemia, two types of responses in $K^+$ were recorded: one was similar to that recorded under 5-min complete ischemia, while in the second group only a small monotonic increase in $K^+$ was noted.
4. During the recovery phase, the kinetics of $K^+$ toward the baseline levels were significantly faster after short ischemia as compared to the longer one. We conclude that after short ischemia, the recovery processes were regained faster as also evaluated from the recovery of the ECoG traces.
5. During the initial phase of the ischemic insult, massive vasoconstriction occurred and, as recorded by the reflectance trace, it was an indication of further damage to the microcirculation.

#### 4.4.5.5 The Fourth Study

##### 4.4.5.5.1 Introduction

The development of ischemic depolarization is a well-known and documented phenomenon [17, 39, 40], but the mechanism behind it is not yet clear (for review, see [41]). Also, the reason for the change in $K^+$ leakage during ischemia is not yet clear. One possibility is that the energy storage depletion reached a critical low level or the level of $K^+$ in the extracellular space reached a certain point. This hypothesis has not yet been tested in the gerbil model. Loss of calcium ion homeostasis was also suggested as a mechanism behind cellular damage (for review, see [42]). In order to shed some light on the question of ID development mechanism, we have designed this study. To evaluate the functional state of the brain during ischemia as well as after reperfusion, we used our multiparametric monitoring assembly (MPA) developed several years ago [6, 8, 31, 43, 44]. The level of ischemia was determined by monitoring the intramitochondrial redox state using surface fluorometry/reflectometry [30, 45]. Tissue $pO_2$ was used only as a parameter of qualitative type for comparison with the mitochondrial redox state. The extracellular $K^+$ level was determined by surface mini-electrode surrounded by a DC steady potential electrode. Electrocortical surface spontaneous activity (ECoG) was measured by bipolar electrodes. The specific questions we aimed to answer by this study [18] are as follows:

1. Is the energy depletion level during the ischemic insult a critical factor in the development of ID (complete loss of ion homeostasis)?
2. What is the effect of $Na^+$-$K^+$-ATPase inhibition level during early ischemia (rate of $K^+$ leakage) on the development of ID?
3. Is the rate of ion homeostasis recovery dependent on the level as well as the duration of the ischemic insult?

4. What is the contribution of removal of the dura mater to the development of the ID event?

In order to obtain answers to these questions, we performed the following experimental protocols:

1. The effects of two types of ischemic insult, i.e., two levels of ischemia (unilateral or bilateral carotid occlusion), were compared to a constant ischemia level imposed for different durations (bilateral occlusion for 1–5 min).
2. We evaluated the differences between gerbil brains that developed ischemic depolarization under 5 min of unilateral occlusion and those that did not.
3. A comparison of the recovery processes from various types of ischemia induced was done by calculating the rate of K/recovery to the intracellular compartment.
4. We determine the effects of dura mater removal on the responses to ischemia using unilateral or bilateral carotid artery occlusion for 5 min. Preliminary results of this study have been published earlier [11, 29].

#### 4.4.5.5.2 Methods

The MPA (multiprobe assembly) developed by our group [8] was changed and modified [31] to be used in the present study as a standard monitoring device. In the MPA device, the probes do not penetrate the tissue as is done using microelectrodes. The main reason for this is that we wanted to compare the $K^+$ levels to the NADH redox state using a surface light guide touching the surface of the brain rather than penetrating the tissue. According to Crowe et al. [46] surface mini potassium electrodes gave similar results of $K^+$ changes during spreading depression as recorded by the usual microelectrodes. Therefore, all our probes had a direct contact with the surface of the brain [11, 29]. The mitochondrial redox state was measured by surface fluorometry/reflectometry [30, 45]. A 100 W Hg arc was used as a light source which passed through a 366-nm filter toward one branch of a Y-shaped light guide. The emitted light returned from the brain via the other branch and was split in a 90:10 ratio to provide NADH levels and 366 nm reflectance traces. The common part of the light guide was located inside the MPA next to the $K^+$ and $pO_2$ electrodes.

**Animal preparation**

The gerbils (50–60 g) obtained from Tumblebrook Farms (MA) were anesthetized by i.p. injection of Equithesin (0.3 ml/100 g body weight). Each ml contained pentobarbital 9.72 mg; chloral hydrate, 42.51 mg; magnesium sulfate, 21.25 mg; propylene glycol, 44.34% w/v; alcohol, 11.5%; and water. The skull was exposed through a midline incision and a hole of 6.0 mm diameter was drilled in the right parietal bone. Three holes (1.5 mm diameter) were drilled in various points of the skull and stainless-steel screws were used to hold the cement to the skull. The dura mater was removed carefully, only from the area where the $K^+$ electrode was placed. The MPA was held in a micromanipulator and located over the cortex without creating any extra pressure on the brain. The reference electrode was placed below the skin and was cemented to the skull, together with the MPA and the screws, by dental acrylic cement. The cementation procedure protects the brain from drying since the interface between the MPA and the *brain* was completely sealed. Figure 3.17 shows the longitudinal section of the MPA connected to the brain used in the present study.

**Experimental Protocols**
Each gerbil was exposed to two sessions of 5 min of unilateral and bilateral carotid artery occlusions with a 30-min interval. In 50% of the animals, we started with unilateral and continued with the bilateral occlusion, while in the other 50% the order was reversed. By using this approach, we minimized the effects of the early episode on the results of the later one in the same experiment. Only animals that showed a complete recovery of various parameters to the pre-ischemic levels were exposed to the next episode. The results presented were calculated from the animals which were exposed to the two types of ischemia and therefore the variations were smaller as compared to one occlusion per animal approach used by other investigators.

#### 4.4.5.5.3 Results

Before presenting various quantitative results obtained in this study, an introductory qualitative part will show various responses recorded under unilateral or bilateral carotid artery occlusion. Under complete or near-complete ischemia (bilateral carotid artery occlusion), a uniform type of response was recorded in all gerbils with or without dura mater, as shown in Fig. 4.36b. The initial response was a decrease in energy supply as can be seen by the decrease in $pO_2$ and an increase in the NADH redox state (CF trace). In this gerbil, the $pO_2$ was decreased to almost the anoxic level. It seems that a small posterior communication artery to the basilar artery supplied a small amount of $O_2$ to the monitored site even under bilateral occlusion. At a certain point (II), ischemic depolarization occurred and led to a change in the slope of the $K^+$ accumulation and a negative shift in the DC steady potential. The DC negativity was recorded also near the light guide ($DC_F$) indicating that the ischemic depolarization is a general event in all cortical areas. As detected by the light guide, a large increase in light reflectance was recorded and affected by the NADH signal, but as shown by the $pO_2$ levels the apparent oxidation of NADH seen in the CF signal is only a transient artifact. When partial ischemia was induced, two different types of responses were recorded as shown in Figs. 4.36a and 4.37 when only the right carotid artery was occluded. The responses shown in Fig. 4.36a are very similar qualitatively to those recorded under bilateral occlusion (Fig. 4.36b) but with different kinetics (slower) as shown by the appearance of line II. The other type of response to unilateral occlusion shown in Fig. 4.37a has less dramatic effects on the $K^+$ leakage pattern. After an initial increase, the $K^+$ reaches a plateau level until the reperfusion time and the ID response were not recorded in this type of

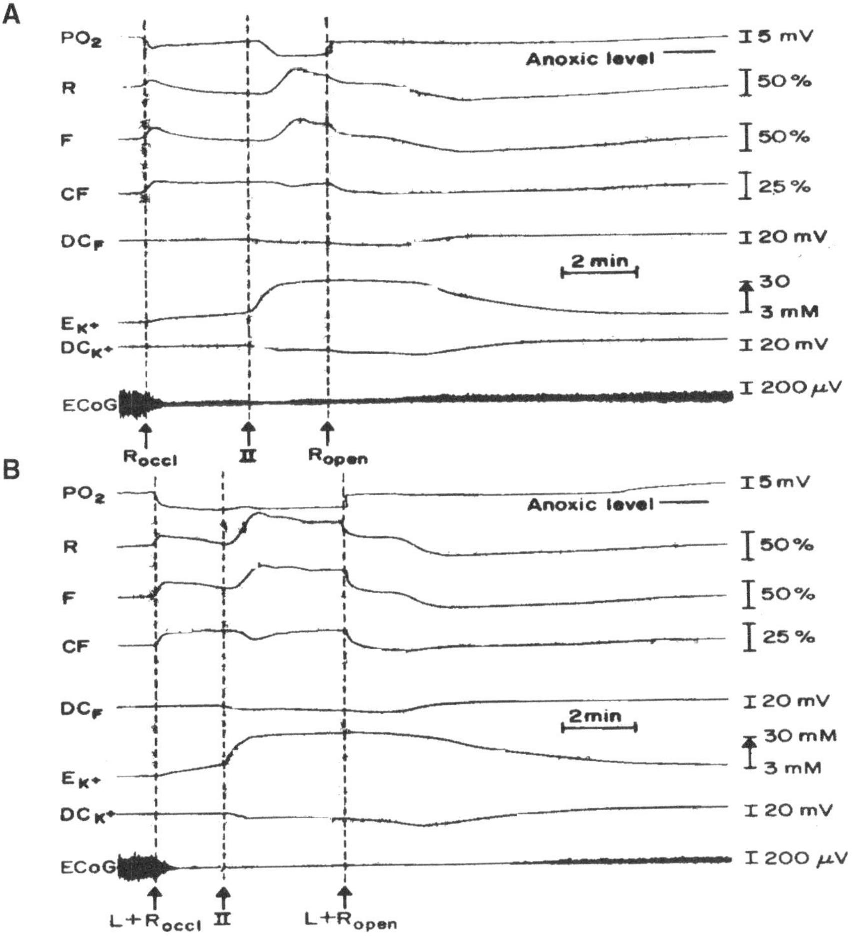


**Fig. 4.36** Effects of unilateral (**a**) and bilateral (**b**) carotid artery occlusion on the metabolic, ionic, and electrical activities in the gerbil brain. Line II represents the appearance of the ischemic depolarization event. $pO_2$, partial pressure of oxygen; R, F, CF, reflectance, fluorescence, and corrected fluorescence, respectively; $DC_F$, $DC_K+$, DC, steady potential measured around the light guide and the potassium electrode, respectively; ECoG, electrocorticogram [18]

gerbil. The same type of response was recorded when energy deficiency was induced by lowering the $FiO_2$ to 6% (Fig. 4.17). The level of ischemia created in each occlusion was calculated by the ratio between the percent change in the NADH redox state achieved under unilateral occlusion and the maximal level measured under bilateral occlusion in the same animal. In the first set of experiments, a total of 32 gerbils were used. In 12 animals the dura mater remained intact, while in the remaining 20 measurements were taken after gentle removal of the dura mater. In each group, we separated the gerbils into two subgroups according to the responses to unilateral occlusion. Less than 50% of the gerbils with or without the dura mater responded, as shown in Fig. 4.37, i.e., ischemic depolarization was not recorded under unilateral occlusion. Each extracellular $K^+$ response to occlusion was analyzed for changes in $K^+$ levels as well as various time events according to the scheme shown in Fig. 4.38. The changes in the shape of the $K_e^+$ curves are very similar when MV or mM levels were used. In some animals at T3-time of the end of ischemia, $K_e^+$ did not reach its maximal level ($K^+_{max}$). In each animal, we compared various parameters analyzed under unilateral occlusion to those under bilateral occlusion, and the results are presented in Table 4.3 (dura mater remaining intact) and Table 4.4 (after removing dura mater), respectively. Table 4.3 presents results collected from 12 gerbils with intact dura mater exposed to 5 min of right carotid and bilateral carotid occlusions. In 5 gerbils, (group B) unilateral occlusion did not induce the typical ischemic depolarization and the second phase of $K^+$ leakage was not recorded. The main difference between the two groups (A and B) is the level of ischemia (77% vs. 45%) which was found to be statistically significant ($P < 0.05$). As a result of the lower level of ischemia and probably other factors, the rate of $K^+$ leakage was much slower (0.06 vs. 0.6 mM/min). Table 4.3 demonstrates significant differences between unilateral and bilateral carotid occlusions regarding the level of ischemia and the slope of $K_e^+$ leakage. Table 4.4 shows that the removal of the dura mater did not affect the baseline levels of $K_e^+$ as well as the difference in the level of ischemia between unilateral and bilateral occlusions. The slope of $K^+$ under right occlusion was five times smaller in the group, which did not develop ischemic depolarization (0.16 vs. 0.83 mM/min). The level of ischemia under right occlusion was smaller but not significant in the group where ischemic depolarization was not recorded (52.9% vs. 65.5%). The baseline $K^+$ levels before ischemia and after recovery from partial or complete ischemia were not different between all groups. It is very interesting that $T_2$ that represents the slope of $K^+$ leakage during the ischemic

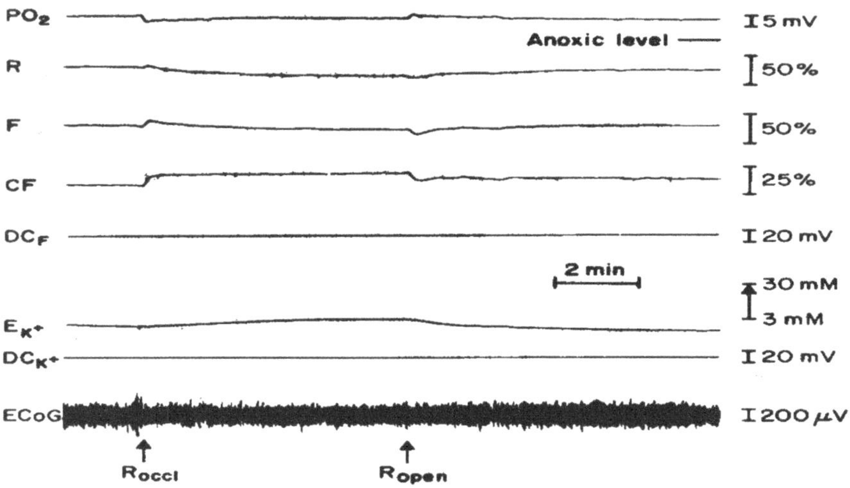


**Fig. 4.37** Responses of the gerbil brain to unilateral carotid occlusion on various parameters measured from the gerbil brain. All abbreviations are as in Fig. 4.36 [18]

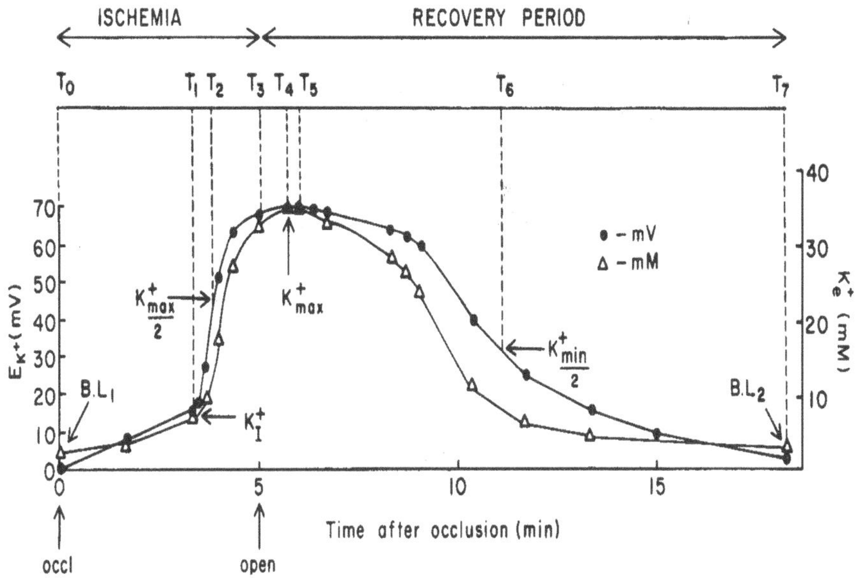


**Fig. 4.38** Time course of the changes in extracellular $K^+$ levels under ischemia and various parameters used for the analyses. The definition of various parameters is as follows: $BL_1$, $BL_2$, baseline $K^+$ levels before and after the ischemic episode; $K_I^+$– level of $K^+$ at the end of phase I; $K^+_{max}$ maximum level of $K^+$ during the ischemic episode; $K^+(max/2)$, $K^+(min/2)$, 50% of the $K^+$ level during phase II of ischemia and recovery period; $T_0$, $T_3$, time of occlusion and reopening of the carotid artery (arteries); $T_1$, end of phase I in $K^+$ leakage; $T_2$, $T_6$, time to reach 50% of $K^+$ level during phase II and recovery, respectively; $T_4$, time to reach maximum level of $K^+$ during the episode; $T_5$, time to initial decrease of $K^+$ after reperfusion started; $T_7$, time to achieve a new steady baseline level of $K^+$ [18]

depolarization phase was the same in all four groups of gerbils. Also the rate of $K^+$ recovery (time to minimum/2) was the same, suggesting that when energy was restored, the initial rate of $K^+$ recovery in the fast kinetic phase was the same although the $T_7$ (time to restore a new baseline $K^+$ level) was significantly different. Figure 4.39 shows the difference in responses of three parameters as calculated in the four groups of gerbils exposed to bilateral carotid occlusion. The time to reach maximum level of $K^+$ during phase II (T) was similar in all groups. Also, the duration of phase I (D) was very similar in all groups although in the DR, A group it was shorter. The main difference was found in the $K^+$ leakage slope (S). In the dura mater intact group, it was significantly slower due to the barrier produced by the dura mater itself. Student's t-test was used to compare various groups of gerbils and the results are shown in Fig. 4.39.

In the second set of experiments (five gerbils), we induced the same level of ischemia in all gerbils (bilateral occlusion) but the effects of ischemia duration were tested. Figure 4.40 shows the effects of 1–5 min of bilateral occlusion (B–F) on the DC steady potential, extracellular $K^+$ level ( $K_e^+$ ), and electrical activity (ECoG). In Fig. 4.39a, the effect of 5-min unilateral occlusion is shown for comparison

**Table 4.3** Effects of unilateral and bilateral carotid artery occlusion of NADH and extracellular $K^+$ responses in the gerbil brain (dura mater removed) [18]

| | Duration | NADH | | $K_e^+$ during phase I | | | | $K_e^+$ during phase II | | | $K_e^+$ during recovery | | | |
|---|---|---|---|---|---|---|---|---|---|---|---|---|---|---|
| | of ischemia | % Increase | % Ischemia | Baseline (BL) | Increase (mM) | Duration (T1) | Slope (mM, min) | Max incr. (mM) | Time to | | Time to | | | New $K^+$ level (BL2) |
| | | | | | | | | | Max incr ($T_1$) | Max ($T_2$) | Init. Decr. (T1) | Min (T2) | Min/2 (T3) | |
| Group A | | | | | | | | | | | | | | |
| R | 5.0 ±0.05 | 12.6 ±2.0 | 77.1 ±5.9 | 3.1 ±0.3 | 4.7 ±0.5 | 2.3 ±0.3 | 0.6 ±0.1 | 20.7 ±2.9 | 3.8 ±0.7 | 0.7 ±0.1 | 1.6 ±0.48 | 16.2 ±3.0 | 4.4 ±1.2 | 3.5 ±0.31 |
| L + R (*n* = 11) | 5.21 ±0.11 | 16.0 ±1.9 | 100 – | 3.0 ±0.3 | 4.9 ±0.6 | 1.96 ±0.2 | 0.9 ±0.2 | 22.3 ±4.0 | 3.5 ±0.3 | 1.01 ±0.13 | 2.87 ±0.53 | 20.0 ±2.1 | 5.5 ±1.3 | 3.31 ±0.35 |
| Level of significance *p*< | n.s | 0.01 | – | n.s | n.s | n.s | 0.05 | n.s | n.s | 0.05 | n.s. | n.s | n.s. | n.s. |
| Group B | | | | | | | | | | | | | | |
| R | 5.06 ±0.05 | 6.6 ±1.6 | 44.8 ±9.3 | 2.0 ±0.3 | 2.3 ±0.2 | 5.1 ±0.1 | 0.06 ±0.02 | – – | – – | – – | – – | 3.3 ±2.2 | – – | 2.0 ±0.1 |
| L + R (*n* = 9) | 5.44 ±0.19 | 16.6 ±3.9 | 100 – | 2.2 ±0.1 | 3.5 ±0.4 | 1.9 ±0.3 | 0.76 ±0.3 | 25.1 ±6.5 | 3.4 ±0.3 | 0.7 ±0.1 | 1.04 ±0.5 | 13.4 ±4.2 | 3.7 ±0.7 | 2.48 ±0.27 |
| Level of significance *p*< | n.s | 0.05 | – | n.s. | 0.05 | 0.01 | 0.05 | – | – | – | – | 0.05 | – | n.s. |

All time values presented are minutes; *n.s.* not significant

**Table 4.4** Effects of unilateral and bilateral carotid artery occlusion of NADH and extracellular K⁺ responses in the gerbil brain (dura mater removed) [18]

| | Duration | NADH | | $K_e^+$ during phase I | | | | $K_e^+$ during phase II | | | $K_e^+$ during recovery | | | |
|---|---|---|---|---|---|---|---|---|---|---|---|---|---|---|
| | of ischemia | % Increase | % Ischemia | Baseline (BL) | Increase (mM) | Duration (T1) | Slope (mM, min) | Max incr. (mM) | Time to | | Time to | | | New K+ level (BL2) |
| | | | | | | | | | Max incr ($T_1$) | Max ($T_2$) | Init. Decr. (T1) | Min (T2) | Min/2 (T3) | |
| Group A | | | | | | | | | | | | | | |
| R | 5.1<br>0.04 | 8.4<br>±0.7 | 65.5<br>±6.3 | 3.1<br>±0.3 | 4.5<br>±0.4 | 2.1<br>±0.3 | 0.83<br>±0.2 | 41.0<br>±0.6 | 2.8<br>±0.3 | 0.62<br>±0.05 | 1.58<br>±0.44 | 13.2<br>±1.7 | 3.9<br>±0.6 | 2.95<br>±0.34 |
| L + R<br>(*n* = 11) | 5.07<br>±0.06 | 13.0<br>±0.5 | 100<br>– | 3.3<br>±0.4 | 6.1<br>±0.7 | 1.4<br>±0.1 | 2.1<br>±0.3 | 48.9<br>±3.8 | 3.6<br>±0.2 | 0.61<br>±0.05 | 2.8<br>±0.86 | 21.6<br>±2.8 | 6.5<br>±1.3 | 3.06<br>±0.38 |
| Level of significance $p<$ | n.s | 0.001 | – | 0.5 | 0.01 | 0.05 | 0.001 | 0.05 | 0.05 | n.s. | n.s. | 0.01 | n.s. | n.s. |
| Group B | | | | | | | | | | | | | | |
| R | 5.2<br>±0.1 | 6.7<br>±0.9 | 52.9<br>±8.7 | 2.7<br>±0.3 | 3.6<br>±0.3 | 5.2<br>±0.1 | 0.16<br>±0.03 | –<br>– | –<br>– | –<br>– | –<br>– | 3.2<br>±1.8 | –<br>– | 2.5<br>±0.3 |
| L + R<br>(*n* = 9) | 5.06<br>±0.05 | 13.90<br>±1.6 | 100<br>– | 2.8<br>±0.3 | 5.4<br>±0.6 | 1.8<br>±0.1 | 1.5<br>±0.3 | 45.3<br>4.6 | 3.4<br>±0.1 | 0.8<br>±0.1 | 2.94<br>±0.71 | 28.7<br>±7.8 | 6.2<br>±1.2 | 2.8<br>±0.4 |
| Level of significance $p<$ | n.s | 0.01 | – | n.s. | 0.01 | 0.001 | 0.001 | – | – | – | – | 0.01 | – | n.s. |

All time values presented are minutes; *n.s.* not significant

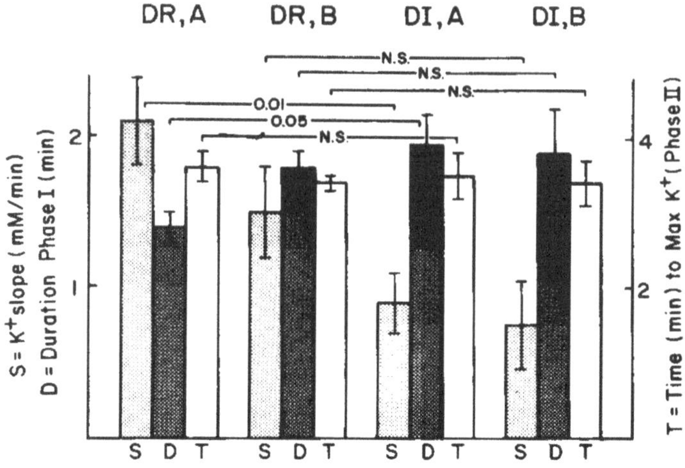


**Fig. 4.39** Comparative changes in three parameters measured from four groups of gerbils exposed to bilateral carotid artery occlusion. DI dura intact, DR dura removed. A and B are groups of gerbils in which the ID was or was not developed under unilateral occlusion, respectively [18]

purposes. The distance between lines 1 and 2 represents the duration of ischemia. The ischemic depolarization appeared clearly in 3, 4, and 5 min, while it started together with the opening of the blood vessels in 2 min. Table 4.5 summarizes the quantitative analysis of the 1- to 5-min ischemic experiments. The baseline level of $K^+_e$ was the same before and after the ischemic episodes suggesting that the $Na^+$-$K^+$-ATPase was able to restore the ionic homeostasis. The duration of the first phase of $K^+$ leakage was shorter only in the 1-min group due to the reopening of the blood vessels before the end of the first phase. The slope of $K^+$ leakage was the same in all groups, ranging from 1.1 to 1.3 mM/min. Figure 4.41 shows the effect of ischemic duration on the maximal $K^+$ level during phase II and on the time to reach the new baseline level of $K^+$ after reperfusion. After 3 min of ischemia, the level of $K^+$ remained in the same range. The time for the complete recovery was correlated very clearly to the duration of the ischemic insult.

#### 4.4.5.5.4 Discussion

Since the gerbil brain model became so popular in studying cerebral ischemia, we performed the present study in order to answer a few basic questions regarding the on-line monitoring of the metabolic, ionic, and electrical activities of brain during and after ischemia. One of the key events occurring under brain ischemia is the development of the complete depolarization associated with the massive ion shifts [41] and with vasospasm which may induce secondary damage under incomplete ischemia. In the present study, we aimed to understand the factors that may lead to the development of the ischemic depolarization (ID) in the gerbil model. Our

**Table 4.5** Effects of 1–5 min of bilateral carotid occlusion on various parameters measured from the gerbil brain (five animals) [18]

| Duration of ischemia | NADH % | $K_e^+$ during phase I | | | | $K_e^+$ during phase II | | | $K_e^+$ during recovery | | | |
|---|---|---|---|---|---|---|---|---|---|---|---|---|
| | | Baseline ($BL_1$) | Increase (mM) | Duration ($T_1$) | Slope (mM/min) | Max. incr. (mM) | Time to | | Time to | | | New $K^+$ level ($BL_2$) |
| | | | | | | | Max. ($T_4$) | Max/2 ($T_2$) | Init. Decr. ($T_5$) | Min ($T_7$) | Min/2 ($T_6$) | |
| 1.1 | 8.5<br>±1.0 | 2.3<br>±0.2 | 3.6<br>±0.4 | 1<br>±0 | 1.3<br>±0.2 | –<br>– | –<br>– | –<br>– | –<br>– | 1.2<br>±0.1 | –<br>– | 2.2<br>±0.2 |
| 2 | 9.8<br>±1.1 | 2.3<br>±0.1 | 4.3<br>±0.4 | 1.8<br>±0.1 | 1.1<br>±0.1 | 22.6<br>±5.3 | 1.1<br>±0.1 | 0.6<br>±0.0 | 1<br>±0.3 | 3.6<br>±0.3 | 1.6<br>±0.2 | 2.4<br>±0.2 |
| 3.1 | 11.4<br>±1.5 | 3.2<br>±0.6 | 5.3<br>±0.6 | 1.7<br>±0.2 | 1.3<br>±0.1 | 35.9<br>±2.9 | 1.4<br>±0.1 | 0.6<br>±0.1 | 2<br>±0.9 | 9.6<br>±1.5 | 2<br>±0.9 | 3.4<br>±1.1 |
| 4 | 12.0<br>±1.5 | 3.2<br>±0.2 | 5.6<br>±0.6 | 2.0<br>±0.3 | 1.3<br>±0.3 | 35.8<br>±6.6 | 2.2<br>±0.3 | 0.6<br>±0.1 | 1.4<br>±0.3 | 12.5<br>±1.8 | 4.5<br>±1.1 | 3.1<br>±0.7 |
| 5 | 13.8<br>1.7 | 2.5<br>±0.3 | 4.5<br>±0.5 | 2.0<br>±0.3 | 1.2<br>±0.5 | 32.2<br>±5.9 | 3.1<br>±0.2 | 0.8<br>±0.1 | 2.2<br>±0.5 | 17.7<br>±1.6 | 4.9<br>±0.8 | 2.2<br>±0.3 |

All time values presented are minutes

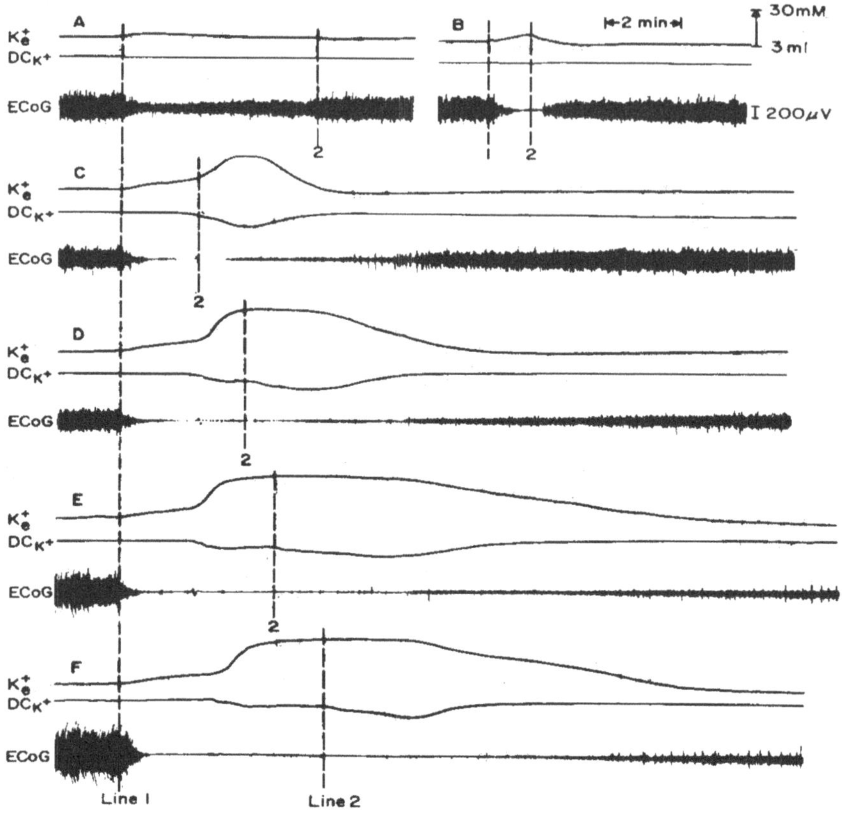


**Fig. 4.40** Effects of 1- to 5-min bilateral carotid occlusion (**b-f**) and unilateral 5-min occlusion (**a**) on the extracellular K⁺, DC potential, and ECoG. Time of occlusion presented as line 1 and reopening done on line 2 [18]

**Fig. 4.41** Ischemic duration effects on the accumulated $K_e^+$ and the time for recovery of K⁺ to the new baseline level. The 3-min point was significantly higher than the 2- and 1-min groups [18]

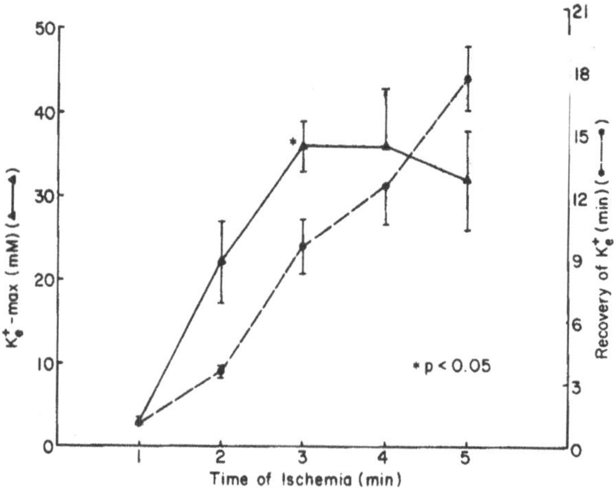

multiparametric monitoring approach is a unique one and enabled us to correlate the ionic changes during the ischemic episode to the level of ischemia created by the unilateral or bilateral carotid artery occlusion in the same brain. The range of extracellular $K^+$ levels in all groups of gerbils tested were comparable to those published by other researchers for the rat brain [47, 48] or the cat brain [49], i.e., around the 3 mM level. Only one group of gerbils showed a lower baseline $K_e^+$ level (Table 4.3, group B). This group also did not develop the ischemic depolarization but the linkage between these phenomena needs further investigation. The main problem in conducting an on-line in vivo monitoring of ion homeostasis experiments is the need for dura mater opening or removal of $K^+$ measurements. It seems that the removal of the dura mater itself (if done in a very delicate way) did not damage the brain thanks to the fact that the location of the MPA on the brain and the cementation to the skull closed the brain and protected it from swelling. In recent studies, we were able to keep this kind of preparation vital for a period of 8–10 h post-operation maintaining normal extracellular $K^+$ and $Ca^{2+}$ levels as well as the normoxic intramitochondrial redox state and electrical activities (Mayevsky, unpublished results). As shown in Tables 4.3 and 4.4, the slope of $K^+$ leakage during phase I was higher when the dura mater was removed but not always statistically significant. The insignificant difference in the duration of phase I suggests that the dura mater served as a barrier for the $K^+$ ions from reaching the electrode. The same pattern was found when the maximal level of $K^+$ during phase II was compared but the time to reach those different values was the same in all bilaterally occluded groups. As concluded recently by Erecinska and Silver [47], "under normal conditions, the maintenance of ionic gradients requires 50–60% of the total $O_2$ consumed of which the major fraction is used by the $Na^+$-$K^+$-ATPase." As indicated in the present study, once the blood supply to the brain was diminished, an immediate accumulation of extracellular K+ was recorded. The main question was "Is this $K^+$ leakage coupled to the level of energy depletion?" To answer this very basic question under in vivo conditions, we compared the two ischemic levels created in the gerbil under unilateral or bilateral carotid artery occlusion. Since our measurements were done in the same animal preparation, we eliminated the effects of the variations between individual gerbils as well as the operation itself. Indeed, the rate of $K^+$ leakage (slope) was significantly different between the R and R + L occlusions in all groups of gerbils (Tables 4.3 and 4.4). As presented in Tables 4.3 and 4.4, the difference in slopes between the two levels of ischemia in the A groups was much smaller than in the B groups. There are at least two causes for this difference, i.e., the level of ischemia created in the two groups under unilateral occlusion and variation in the sensitivity to oxygen deficiency due to other factors. Indeed, the depletion of $O_2$ in the B groups (Tables 4.3 and 4.4) was smaller under R occlusion, but we believe that both factors mentioned play a role in determining the rate of $K^+$ leakage. One of the consequences of the different slopes is the development of the ischemic depolarization (ID) in various groups. The only groups in which ID was not developed under unilateral occlusion were the two B groups in which the slope was 0.06 or 0.16 mM/min. The practical meaning of it is that those brains in which the development of ID will be avoided (e.g., by drugs) under clinical situations, the beneficiary effect will be evidently significant [50]. As we indicated in Fig. 4.36a, under partial ischemia the ID developed led to a large increase in the reflected light trace (SRI). Using a photography technique (Shaked and Mayevsky, unpublished data) or the laser Doppler flowmetry approach (Mayevsky, unpublished data), we found that during the SRI event, a large vasoconstriction was noted. This vasoconstriction led to a secondary damage to the brain by decreasing $O_2$ availabilities as well as a massive depolarization which by itself has a negative effect on the brain. In our recent work [51], we showed that the Ca antagonist, Nimodipine, delayed the appearance of the ID in gerbils under ischemic conditions. Thus, the understanding of the mechanism behind the development of ID has a significant meaning to its prevention by drugs. This concept was also discussed by other groups very recently [52, 53]. Once the ID started to develop, the effects of the level of ischemia on the development of phase II was minimal. It seems that it is an "all or none" phenomenon. The same type of phenomenon was found when the effect of ischemia duration was studied. Table 4.5 shows that $T_2$ was the same in all groups, i.e., once the ID developed, the rate of leakage of $K^+$ was the same. Regarding the recovery processes, it seems that the level of $K^+$ during ischemia is not the only factor to determine the time to reach a new $K^+$ level after the recovery. As the levels in Fig. 4.41 and Table 4.5 show, the time to achieve a complete recovery of $K^+$ ($T_7$) was correlated to the duration of the ischemia. As shown in Tables 4.3 and 4.4, the time to complete recovery of $K_e^+$ was shorter under unilateral occlusion, suggesting that various levels of ischemia have the same effect as compared to different durations of the same level of ischemia. In the present study, we used repeated ischemic episodes in the same animal and we assumed a minimal residual effect of the previous episodes. Kato et al. [54] showed that repeated ischemia episodes led to morphological damage in the selective vulnerable regions. In our study, we tested only the cortical region and only gerbils that showed a complete recovery of all parameters were exposed to another episode. Also, we changed the sequence of the R and R + L occlusions in various animals so that any residual effect was canceled out by the statistical procedure. The level of ischemia was calculated by the changes in the intramitochondrial NADH redox state. This is the best approach that exists for the real-time monitoring and, in a separate study, we found a very good correlation between the NADH redox state responses and the blood flow monitored by a laser Doppler flowmeter (Mayevsky, unpublished results). Also, the cerebrovascular anatomy was tested in the gerbils and was found to be significantly correlated to the NADH redox state changes under unilateral and bilateral carotid artery occlusions [55]. It seems that the main effects of having the dura mater intact during the measurements are on the parameters that represent the rate of diffusion of $K^+$ to the electrode area. No damage to the brain was created by removing the dura mater as indicated by the same baseline $K^+$ levels in all groups of animals. Therefore, we concluded that removal of the dura mater will provide a stable and reliable model which is also comparable to other studies when microelectrodes are in use. In conclusion, it seems that the gerbil model could provide significant information regarding the development of ischemic damage and recovery. The removal of the dura mater did not change the behavior of the system and provided more accurate measures of $K_e^+$. Since the main event to be further investigated is the mechanism of ID development, we are now using the $Ca^{2+}$ electrode which together with the $K^+$ electrode and redox state monitoring will provide better understanding of brain mechanisms of ischemia as well as recovery. The combination of various probes in the MPA and the monitoring in real time of various parameters open up new possibilities for pharmacological as well as clinical applications.

#### 4.4.5.6 The Fifth Study

This study was performed using Protocol 3.6 described in Chap. 3, Table 3.1 [8].

##### 4.4.5.6.1 Introduction and Methods

Our aim was to provide a new holding system allowing an easy implantation of surface probes in various combinations on the brain of small mammals and to reevaluate the potential of such surface electrode assemblies in tracing physiological and pathological events. In this chapter, we describe the basic assembly adopted and discuss its application to the measurement of the extracellular tissue pH and $K^+$ activities in ischemia, spreading depression, and other parameters, such as reduced pyridine nucleotide fluorescence, DC potential, electrocorticogram (ECoG), and local temperature. Ideally, the multiprobe assembly designed for our project had to be rugged enough to withstand routine use by no specialists and yet to be miniaturized enough to fit the limited surface available for implantation on gerbil skull (about 6–7 mm in diameter on each hemisphere). For ischemic studies, interest was focused in monitoring mean values over large areas, suggesting the use of larger electrodes. In contrast, following propagated events such as spreading depression the tip of multiprobe assembly would have required to be concentrated in a tiny space observing the same phase of the wave unless each sensor could be located with respect to the wave front by an auxiliary signal. The second option was retained by recording the wave of the DC potential concentrically to each sensor. In addition, the complete assembly had to provide adequate protection and shielding of the high-impedance ion-sensitive stable enough reference electrodes ($K^+$ and pH) and a junction for use in un-anaesthetized animals. With $K^+$ surface electrodes, concentric DC potential has been proposed as the best approximation for the DC bipotential component to be subtracted from the sensor signal [46]. In this study, we also attempted to check the accuracy of the method by recording DC potential differentially between a central barrel and the peripheral slit and to reevaluate its usefulness in correcting surface $K^+$ and pH measurements under various conditions.

#### 4.4.5.6.2 Results

The results presented here were selected from a set of 40 gerbils used in our study. We show typical results of the new assembly connected to the gerbil brain exposed to various perturbations. Brain temperature was kept between 35° and 36 °C with an automatic heating system controlled by the temperature probe located in the assembly. Figure 4.42 shows a typical response of the gerbil brain to anoxia (left) and bilateral carotid occlusion (right). The initial response to anoxia is an increase of the NADH level as shown in the corrected fluorescence (CF) trace (a 30% increase compared with the normoxic level). As a result, the extracellular level of $K^+$ increased slightly and returned to the baseline level after restoration of air to the breathing mixture. The pH showed an increased response under anoxia. The DC potential measured in all three electrode sites ($DC_{H+}$, $DC_{k+}$, and $DC_{F}$) did not show any change during the anoxic cycle. ECoG showed a typical depression during the anoxic episode. The effect of ischemia was tested in two steps. After occlusion of the right carotid artery ($R_{occ}$), the NADH increased by 10% compared with the normoxic level and reached only one-third of the maximal NADH level measured during the anoxic cycle or bilateral carotid occlusion. The partial ischemia induced by $R_{occ}$ did not cause any other significant changes during the short period of occlusion. After complete ischemia was induced by occlusion of the left artery ($L_{occ}$ while the right one was occluded), the NADH showed a 30% increase and returned to the baseline level after both left and right arteries were reopened. During the period of

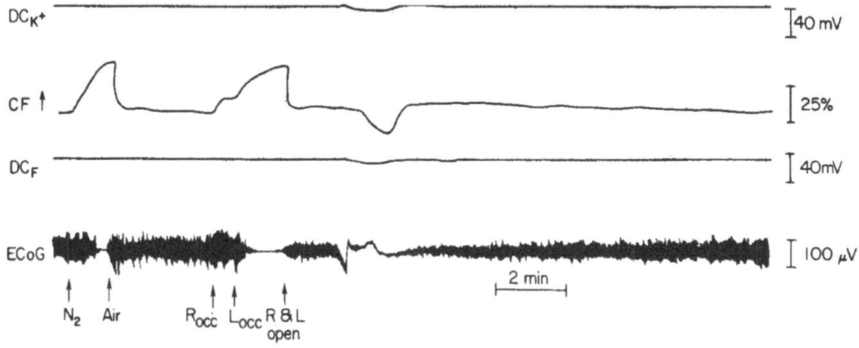


**Fig. 4.42** Effects of anoxia (left) and ischemia (center) on metabolic, ionic, and electrical signals measured from a slightly anesthetized gerbil. Spontaneous spreading depression cycle was developed after ischemic episode. In this experiment, we recorded corrected fluorescence alone (CF) without uncorrected fluorescence and reflectance. For abbreviations, see text [8]

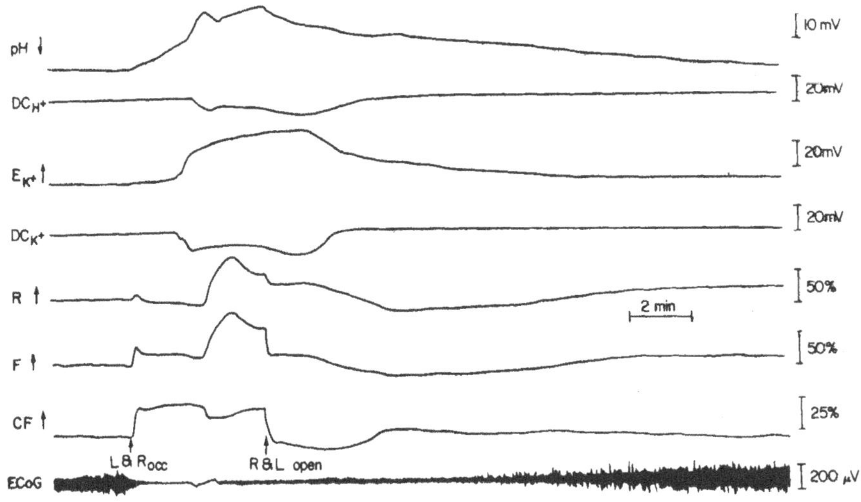


**Fig. 4.43** Responses of gerbil brain to about 4 min of complete ischemia and recovery after reopening of both carotid arteries. Arrows near signal label on left represent an increase in level of parameter. For abbreviations, see text [8]

complete ischemia, extracellular $K^+$ level increased and the pH decreased (more acidic). A spontaneous SD was developed during the complete ischemia, and the wave was propagated to the measuring site after the two carotid arteries were reopened. During the SD event, typical responses were recorded, i.e., $K^+$ was elevated and then pumped back into the cells, and as a result the NADH became more oxidized (oxidation cycle). The pH showed an acidification response to the SD. As expected, the three DC potential signals showed a transient negative shift during SD. We found that the resting level of extracellular $K^+$ was 3 ± 1 mM in all undamaged brains. When the initial levels were higher, the animal was not included. The effect of complete ischemia was also tested in another gerbil in which the carotids were occluded for more than 4 min (Figs. 3.12 and 4.43). As a result, the NADH reached its maximal level (CF) in about 2 min, and then a transient apparent oxidation was recorded on the corrected trace. This "oxidation" may be an artifact due to the changes in light absorption by the tissue and blood present in it and was not corrected by the 1:1 correction factor used. One can see a large increase in R as well as in the uncorrected F during this episode, which is a general depolarization occurring during the complete ischemia. The NADH (CF) showed a fast recovery after the reopening of the carotids. The extracellular $K^+$ showed a two-step response to the complete ischemia. Immediately after the occlusion, a small and slow increase in extracellular $K^+$ was recorded, and after 1.5 min a sharp and large increase occurred simultaneously with the negative shift in the DC potential surrounding the $K^+$ electrode. The pH showed a monotonic acidification response until the reopening, but 1.5–2 min after the occlusion a large DC negative shift was recorded near the pH electrode, as can also be seen in the corrected pH trace. In this specific gerbil,

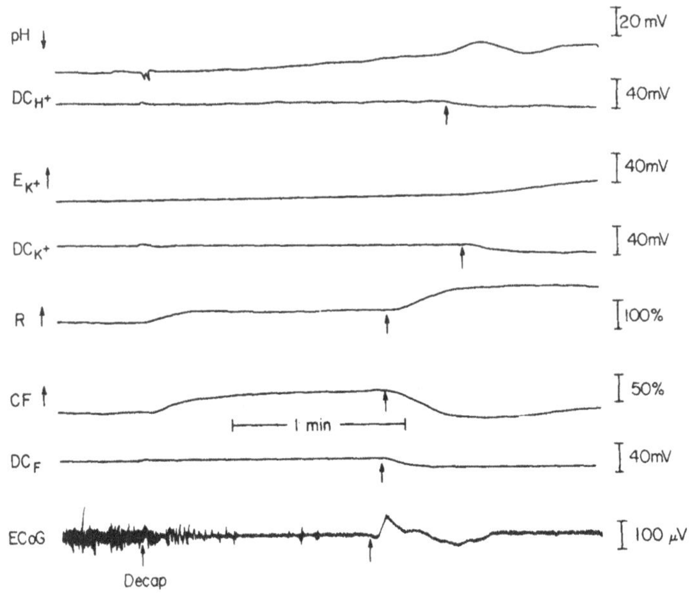


**Fig. 4.44** Effects of complete ischemia—decapitation on various parameters measured from the brain of awake gerbil. Arrows indicate points where complete depolarization started in each area of the brain. For abbreviations, see text [8]

the recovery of the pH started immediately after the reopening of the blood vessels, while the extracellular $K^+$ showed a delay of about 2 min before it started to decrease. The time needed for the complete recovery of the pH and $K^+$ was much longer than the NADH recovery time. The ECoG showed a silent period during the occlusion and gradually recovered after the reperfusion of the brain. The effects of complete irreversible ischemia on the brain were tested in the decapitation model (Fig. 4.44) in which blood supply and oxygen delivery were terminated at once by the decapitation. The initial effect of the decapitation is the increased level of NADH (after 2–3 s of delay) due to the decreased level of oxygen in the tissue. ECoG was the second parameter to respond and it became isoelectric within 30 s. The pH shows a 30 s delay before becoming more acidic, probably due to the accumulation of lactic acid. The extracellular $K^+$ initially increased after about 2 min and then reached its maximal level later on (data not shown in Fig. 4.44). General depolarization of the cerebral cortex started in this gerbil after about 75 s, as indicated by the DC shift measured by the DC electrodes. This depolarization propagated through the entire cortex, as can be seen in the three DC electrodes, i.e., $DC_F$, $DC_{H+}$, and $DC_{K+}$. The R trace showed a large increase during this depolarization event preceded by the DC change near the light guide ($DC_F$). This change in R typical to complete ischemia was described by us previously [20], and we called it "secondary reflectance increase" (SRI). The correction for the DC changes in the pH electrode was not perfect in this specific animal. The pH trace, which is the corrected one (after

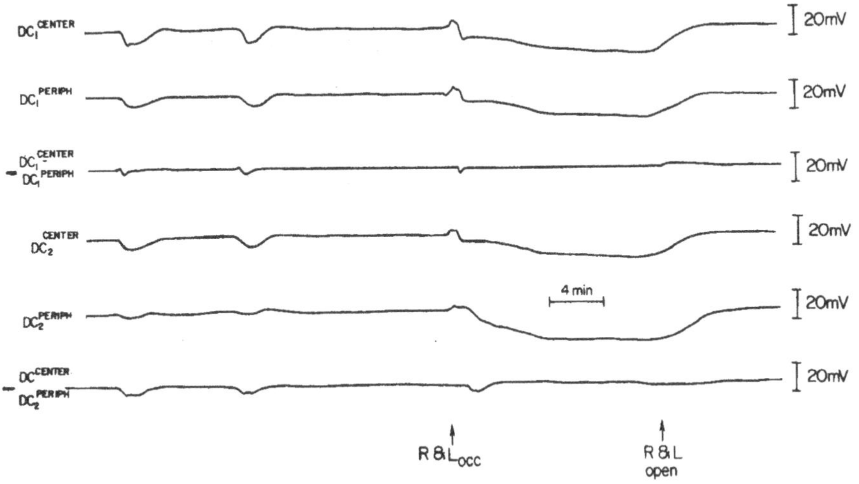


**Fig. 4.45** Comparison between responses of two shapes of DC electrodes to spreading depression and complete ischemia. Two pairs of electrodes were placed in the same hemisphere. Each pair included a center electrode and a peripheral one [8]

subtraction of the DC changes), showed an abrupt increase when the depolarization occurred and it is an artifact of the correction. The geometry of the pH electrode (center circle) is not identical to that of the DC one (peripheral ring), and the cortical field measured by the two electrodes is not the same. To clarify this point, we performed five experiments in which the DC potential was measured both in the center and the periphery (two such electrodes were located in the same brain and the results are shown in Fig. 4.45). The brain was exposed to SD by washing the surface of the brain with 0.5 M KCl (the washing point is not shown in Fig. 4.45), and two SD negative cycles were recorded in all the electrodes. Due to the propagation nature of this phenomenon (SD), the two electrodes are not synchronized in time in relation to the propagation, and as a result the corrected trace (third and sixth traces) also showed a cycle. When complete ischemia was induced ($R + L_{occ}$), the two uncorrected DC traces showed similar kinetics and the corrected trace showed only a transient change and later on remained at its normal level. The artifact in the corrected $DC_1$ was smaller compared with that of corrected $DC_2$.

The last example of the potential use of the multiprobe assembly is shown in Fig. 5.10, in which the repetitive effect of SD was tested. The brain was washed with 0.5 M KCl, which remained on the brain. As a result, six SD cycles were recorded in all the parameters measured. The ECoG was depressed during the entire period of the SD, and toward the end the beginning of epileptic activity was recorded. The three DC channels also showed cycles of negative shift. The main change in the pH was an alkalinization response. The extracellular $K^+$ showed the typical increase–decrease cycle, and as a result the NADH showed the oxidation cycles.

#### 4.4.5.6.3 Discussion

It is well documented and accepted that for understanding of the mechanism of brain pathology development, as many parameters as possible must be measured from the same site in the brain. For this reason, several groups of investigators developed a multiprobe approach for studying and evaluating the functional state of the brain. All the published materials have dealt with anesthetized or paralyzed animals [56–63]. We previously described experimental setups by which metabolic, ionic, and electrical activities were correlated in the awake animals [6]. In the present study, we extended the method so that more parameters were measured simultaneously from the awake animal. The multiprobe assembly can also be very easily modified for use in a hyperbaric chamber in order to study the possible mechanism of brain oxygen toxicity (Mayevsky, unpublished observations). In 1980, we described the correlation between NADH fluorescence extracellular $K^+$ and ECoG in the awake gerbil [6]. We later combined the $pO_2$ electrode with NADH measurements [7]. Although valuable information has been accumulated, to understand the mechanisms of brain pathology in detail, more parameters must be measured at the same time. The multiprobe assembly described in the present study gives information on the metabolic (NADH), hemodynamic (R), ionic ($K^+$ and $H^+$), and electrical activities (ECoG and DC potential). The main problem in developing the multiprobe assembly was the limitation in the size of the brain in rats and gerbils. This was solved by the conical shape of the cannula holding various probes. In our previous cannula (cylindrical), we were able to put together only a $K^+$ electrode, a light guide, and two stainless steel rods for ECoG measurements [6]. A more theoretical problem is the heterogeneity of the tissue analyzed by different probes placed on the cerebral hemisphere. In addition to the intrinsic heterogeneity existing in such measuring systems, the effects of the operation procedure itself is a critical factor. When removing the dura in rats and mainly in gerbils, it is very easy to cause unseen damage to the underlying tissue. For this reason, not all the experiments were successful, i.e., not all the parameters were obtained from each animal. In general, only about 50% of the experiments were perfect and they provided information as expected. To test the functional state of the brain after the operation, we exposed the brain to complete anoxia for a very short time (30–40 s). As shown in Fig. 4.42, the recovery from such an episode is very fast in all parameters measured. The size of the NADH fluorescence increase during the anoxia is a good indicator of the intactness of the mitochondrial activity. The resting levels of the $K^+$ and $H^+$ as well as the amplitude of the ECoG are also significant indicators. Another indicator of the functional state of the brain is its response to activation of the $Na^+$-$K^+$-ATPase system. As can be shown in the right side of Fig. 4.42, when extracellular $K^+$ level was elevated during the ischemic episode, a spontaneous spreading depression developed and the NADH showed an oxidation cycle simultaneously with the decrease in extracellular $K^+$ to its resting level. This type of correlation was shown by us previously [6, 64] and by others [57, 58]. In our recent publication [65], we showed that the type of NADH response (oxidation or reduction) to spreading depression is a good indication of the "oxygen balance" in the tissue at any stage of the experiment. The problem of DC correction in $K^+$ and $H^+$ measurement was discussed by Urbanics et al. [66], and as a result we decided to monitor DC potential near each of the probes that we used. In our case, the correction is not simple due to the large size of the electrodes and the light guide, and indeed we demonstrate the problem in Fig. 4.45. In our setup, it is very difficult to express the results in real pH units, and we prefer to analyze them in relative terms only. On the other hand, we found in the previous study and also in the present study that the correction for DC changes in the $K^+$ electrode is almost perfect. The effect of neuronal activation on the metabolic and ionic responses is a subject discussed by many investigators [57–60, 67]. Most of the results published showed a direct correlation between pumping activity of the ATPase system (due to increased extracellular $K^+$) and the oxidation of NADH. Leniger-Follert et al. [58] found that due to direct electrical stimulation of the cortex, the NADH showed an oxidation while the pH showed a biphasic response, i.e., alkalosis followed by acidotic shift [68]. The effects of ischemia on the multiprobe results are shown in Figs. 4.43 and 4.44. Though the blood vessel occlusion effects were described previously, this is the first time that the effect of decapitation on various parameters has been described. In our previous studies, we showed the effects of decapitation on NADH fluorescence and reflectance as well as ECoG. The response of the reflectance trace was unexplained until now: the origin of SRI was only a speculation until now. As shown in Fig. 4.44, the general depolarization of the cortex occurred at the same time, illustrated by the DC negative shift and later on by the $K_e^+$. We do not know why general depolarization caused a large increase in the reflected light signal. Two mechanisms could be involved, i.e., the change in absorption properties of the tissue due to water and ions movement or the movement of the remaining blood from the vessels due to the depolarization effects on the wall of the vessels. Several reports (for details, see Ref. [63]) support the first possibility mentioned, i.e., during SD and ischemia, a decrease in the extracellular space was found while the cells swelled. The same response of the reflectance was recorded when bilateral carotid ligation was performed (Fig. 4.43). In conclusion, a new multiprobe approach described in the present study opens up new possibilities for studying mechanisms of brain pathology in a very detailed manner. We now use this approach in our laboratory to study hyperbaric oxygen toxicity, long-term ischemia, and recovery.

#### 4.4.5.7 The Sixth Study

This study was performed using Protocol 3.6 described in Chap. 3, Table 3.1 [8, 9].

#### 4.4.5.7.1 Introduction and Methods

To evaluate the functional state of the brain both during and after complete ischemia, it is desirable to measure several parameters from the same site in the brain. We monitored the brain continuously before, during, and after the ischemic insult using a recently developed multiprobe assembly [8].

Three different types of surface probes were used in the present study:

1. Energy metabolism was evaluated by monitoring the intramitochondrial NADH redox state by surface fluorometry. The reflected light measured simultaneously with the fluorescent signal served as an indirect measure of hemodynamic changes.
2. Ionic activities were measured by $K^+$ and pH surface electrodes.
3. Electrical activity was measured by electrocorticogram (ECoG) and the DC steady potential.

In the normoxic brain, the electrical activity of neurons leads to an efflux of $K^+$ and an influx of $Na^+$. The restoration of the ion distribution is accomplished by the adenosine triphosphatase (ATPase) system, which requires ATP. This stimulates the electron transport system of the mitochondria, leading to increased $O_2$ consumption and oxidation of intramitochondrial NADH. During complete ischemia (maximal NADH level), mitochondrial ATP production is inhibited, and $K^+$ accumulates in the extracellular space to its maximal level within a few minutes. The hypothesis tested was that the duration of recovery is correlated to the duration of the ischemic insult.

The aims of the present study were to compare the effects of long-term complete ischemia (20–30 min) with those of a relatively short (4–5 min) ischemic insult during both ischemic and recovery periods. Spreading depression was used as a test stimulus for evaluating the completeness of recovery from ischemia.

Each gerbil was exposed to a short anoxic episode (30–60 s) by flushing 100% N through a face mask. Under these conditions, the gerbil stopped breathing, and the NADH increased to its maximal level. In another study, we found that under the same method of $N_2$ application the maximal increase of NADH was correlated with the decrease in tissue $pO_2$ to its minimal level. The changes in the NADH redox state are determined in terms of percent change of the signal measured in the normoxic brain (100%, full scale). In all gerbils, the percent change in NADH during complete ischemia was similar to that obtained during anoxia (a 25% increase). Gerbils in which changes below 15–18% were observed were excluded from the study due to the high NADH redox state before the experiments. After recovery of the animal from the operation and stabilization of the signals, complete ischemia was induced by occlusion of the two common carotid arteries. A total of 24 gerbils were studied, but only 12 animals provided data on all parameters monitored. We compared the effects of long-term ischemia (20–30 min) with those of short-term ischemia (4–5 min). The interval between the ischemic episodes depended on complete recovery of all measured signals. Only two gerbils (28 and 55) were exposed to short-term ischemia after complete recovery of all parameters from the 30-min ischemia. In a few gerbils, we also tested the response of the cortex to spreading

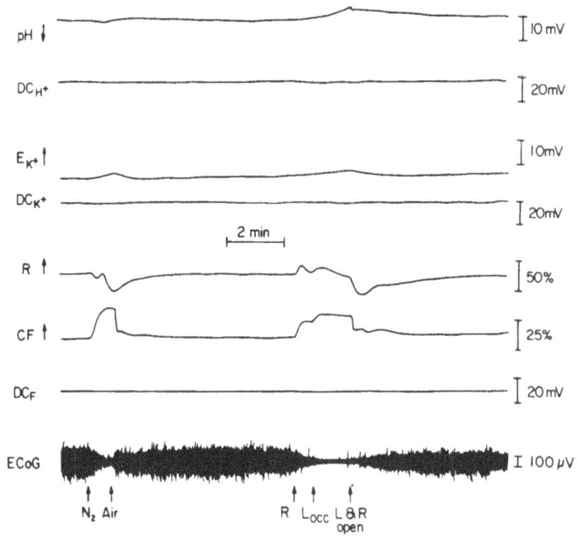


**Fig. 4.46** Effects of anoxia (left side) and short-term unilateral and bilateral carotid artery occlusions on metabolic ionic and electrical activities in right hemisphere of a gerbil brain. $L_{occ}$, left carotid occlusion; L & $R_{open}$, reopening of two carotid arteries; ECoG, electrocorticogram; CF, corrected fluorescence; R, reflectance; $EK^+$, extracellular $K^+$ level; $DC_{H+}$—$DC_{K+}$ and $DC_F$, DC steady potential measured near pH electrode, $K^+$ electrode, and light guide, respectively. Arrows—increase in parameters measured [9]

depression before and after the ischemia to evaluate the completeness of recovery from the ischemic insult. For statistical analysis, we used Student's t-test for comparison between the short- and long-term ischemic groups.

#### 4.4.5.7.2 Results

Figure 4.46 shows the effects of a brief period of anoxia (0.8 min) and unilateral and bilateral carotid artery occlusions on various parameters measured from the gerbil brain. Due to the brief duration of the insult (anoxia or ischemia), the main responses are seen in the parameters representing energy metabolism. The NADH level increased by 25% compared with the normoxic level (defined as 100% signal) and returned to normoxic values after rebreathing air or during recirculation, respectively. The increased level of NADH recording during anoxia or ischemia was maximal, i.e., that all NAD pool was reduced to NADH and those levels were unchanged after death. Due to lack of available energy during insult, the extracellular $K^+$ level increased (third trace from top) and then decreased as energy was available during the recovery phase. The pH of the brain during the short ischemic interval became more acidic (signal moving upward). Depression of the ECoG occurred during both anoxic and ischemic insults. The DC steady potential was not affected significantly

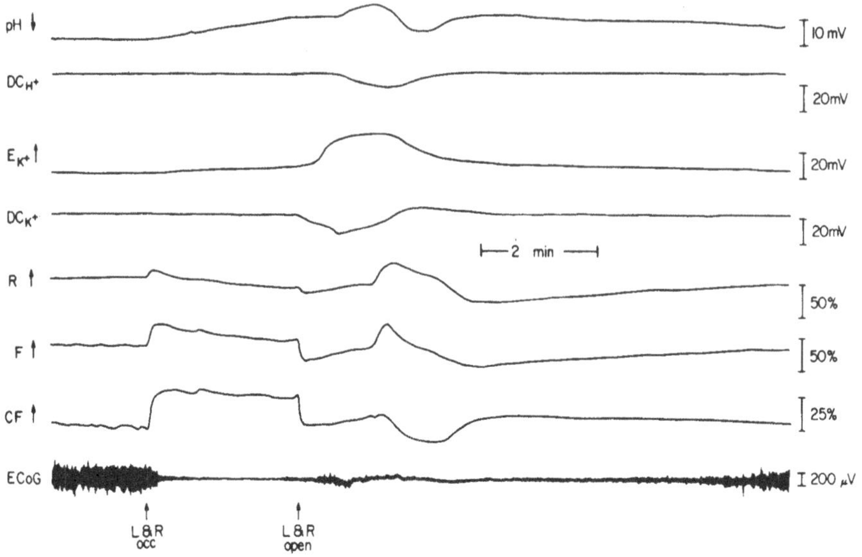


**Fig. 4.47** Appearance of a spreading depression phenomenon after short-term complete ischemia. After reopening of left and right carotids $E_{K+}$ increased, and NADH (CF) showed an oxidation cycle. Uncorrected fluorescence signal (F) was also recorded instead of $DC_F$ recorded in other figures [9]

in any of the three electrodes used [$DC_{H+}$, $DC_{K+}$, and $DC_F$ (fluorescence)]. When the duration of the ischemic insult was longer, a different pattern of responses was recorded. The effects of 2.6-min complete ischemia are shown in Fig. 4.47 and can be compared with the 1.3-min complete ischemia shown in Fig. 4.46. Brain temperature during the ischemic period was 32 °C. During insult, the following changes occurred: NADH was elevated to its maximal level; extracellular $K^+$ was accumulated; pH became acidic, and ECoG became isoelectric. After reopening the carotid arteries, NADH was oxidized immediately. Recovery of the other parameters was carried out as shown in Fig. 4.46. However, during this ischemic insult, the high level of extracellular $K^+$ apparently initiated the depolarization of a large number of cortical cells such that a spreading depression wave (SD) passed through the cortex simultaneously with the opening of the carotid arteries. The wave of SD started near the DC electrode of the $K^+$ channel ($DC_{K+}$) as shown by the negative shift recorded and propagated toward the pH electrode ($DC_{H+}$). The wave of SD was associated with the typical events [8, 22, 26, 46, 63, 65], i.e., an increase in extracellular $K^+$, a biphasic change in pH, and an oxidation wave of NADH. As usual, the ECoG was also depressed during the SD and started to recover 4–5 min later. When the brain is more sensitive to the elevated extracellular $K^+$ level, general depolarization may start during the ischemic insult, and the responses will have a different pattern (Fig. 4.49). In this gerbil, occlusion of the right carotid artery ($R_{occ}$) induced a large increase in NADH, and the later occlusion of the left carotid artery ($L_{occ}$) did not add significantly to the insult. Brain temperature before the occlusion was 36.8 °C, and

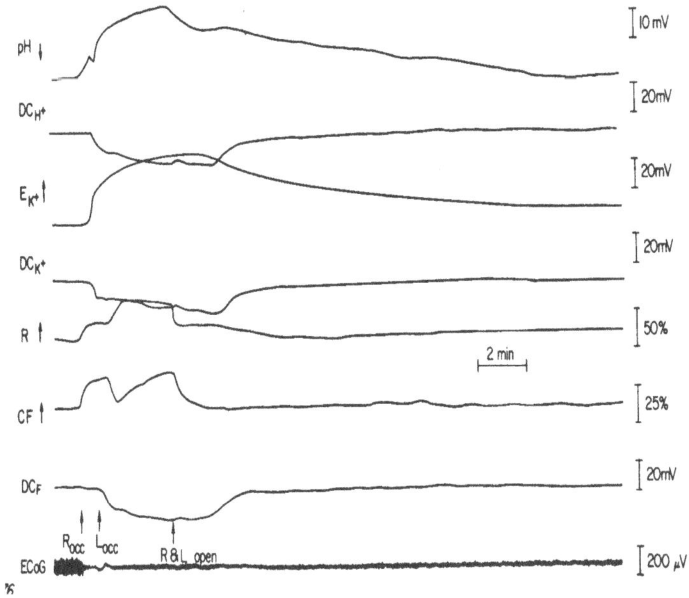


**Fig. 4.48** Metabolic, ionic, and electrical responses to complete cerebral ischemia and recovery kinetics. $R_{occ}$, right carotid occlusion. Other abbreviations as in Fig. 4.47 [9]

this higher temperature probably led to the general fast depolarization starting near the $K^+$ electrode 30 s after occlusion, then passing near the pH electrode, and ending near the light guide ($DC_F$). During ischemia and the following general depolarization, extracellular $K^+$ reached very high levels while the pH became acidic. It can be seen in the reflectance and the NADH-corrected fluorescence traces that general depolarization ($DC_F$) led to an apparent oxidation of NADH in the absence of oxygen. This has been shown to be an artifact of the reflectance changes [8, 20]. The ischemic period lasted 3.8 min, after which the two carotid arteries were opened. Tissue NADH became oxidized as soon as recirculation began, and the NADH reached the pre-ischemic baseline within 1.5 min. The pH started to recover after reopening the two blood vessels. Extracellular $K^+$ level showed a delay of 1.4 min before the beginning of recovery. The record shown in Fig. 4.48 was obtained from gerbil 56, and the detailed analysis of various parameters is presented in Tables 4.6 and 4.7. The effects of long-term ischemia (30 min) are shown in Fig. 4.49. Although the pattern of the responses to a 30-min ischemic insult is very similar to that observed during a short-term ischemic insult (Fig. 4.48), the kinetics of the recovery phase were slower. General depolarization appeared 2 min after the occlusion as can be seen in the three DC traces and the large increase in the reflectance trace. The upward swings seen in the three DC traces during recovery are probably an artifact originating from the reference electrode. The corrected $K^+$ and the pH traces were unaffected by this artifact. All traces recorded from the 12 gerbils were analyzed, and the results are presented in Tables 4.6 and 4.7. Since ischemia induced by

**Table 4.6** Effects of short- and long-term complete ischemia on extracellular $K^+$ activity, pH, and DC steady potential in gerbil brain [9]

| | | Extracellular K+ | | | | | pH | | | DC, mV | |
|---|---|---|---|---|---|---|---|---|---|---|---|
| Gerbil No. | Ischemia | I | | II | | | | | | | |
| | Duration min | mV | Duration min | mV max | Time to mV max min | Time in mV max/2, min | mV max | Time to mV max min | Time in mV max/2, min | $K^+$ | pH |
| 47 | 30 | 5 | 2.2 | 67 | 28 | 3 | 27 | 17.5 | 2.5 | 8 | 6 |
| 26 | 30 | 6 | 0.6 | 65 | 18.8 | 2.1 | 13 | 17.2 | 3.2 | 15 | 13 |
| 28 | 30 | 10 | 2.4 | 54 | 25 | 2.9 | 61 | 19.3 | 4.2 | 32 | 30 |
| 55 | 30 | 5 | 0.7 | 80 | 27 | 1.5 | 100 | 129 | 2.2 | 32 | 30 |
| 45 | 30 | 8 | 3.0 | 55 | 26.5 | 3.6 | 17 | 12 | 2.1 | 21 | 21 |
| 41 | 20 | 2 | 2.5 | 61 | 18.5 | 3.5 | 20 | 13 | 3.0 | 20 | 15 |
| 29 | 21 | 10 | 1.3 | 47* | 11.4 | 1.6 | 27 | 11.2 | 2.1 | 36 | 40 |
| 44 | 20 | 5 | 1.0 | 55 | 15.5 | 1.6 | 11 | 19.5 | 3.5 | 20 | 17 |
| Mean | | 6.4 | 17 | 60.5 | 21.3 | 2.5 | 34.5 | 17.3 | 2.9 | 23 | 21.5 |
| ±SE | | 0.98 | 0.33 | 3.6 | 2.2 | 0.3 | 10.9 | 2.0 | 0.27 | 3.4 | 3.9 |
| 28 | 5 | 6 | 2.2 | 24 | 5 | 2.5 | 35 | 4.8 | 2 | 20 | 16 |
| 30 | 5 | 6 | 2.3 | 48 | 5 | 1.6 | 23 | 5 | 1.6 | 16 | 14 |
| 50 | 4.6 | 9 | 1.6 | 40 | 4.5 | 2.2 | 24 | 4.5 | 2.1 | 15 | 9 |
| 51 | 3.8 | 8 | 1.6 | 41 | 3.8 | 2 | 10 | 3.8 | 2.4 | 15 | 15 |
| 55 | 4.6 | 10 | 0.7 | 51 | 4.6 | 0.7 | 13 | 4.5 | 1.2 | 25 | 19 |
| 56 | 3.8 | 4 | 0.5 | 43 | 3.8 | 1.0 | 23 | 3.8 | 1 | 20 | 22 |
| Mean | | 7.2 | 1.48 | 41 | 4.5 | 1.7 | 21.3 | 4.4 | 1.7 | 18.5 | 15.8 |
| ±SE | | 0.91 | 0.3 | 3.8 | 0.2 | 0.28 | 3.6 | 0.2 | 0.2 | 1.6 | 1.8 |
| Significance | | NS | NS | ‡ | ‡ | † | NS | ‡ | ‡ | NS | NS |

Changes in the level of extracellular $K^+$ were divided into two stages (I and II) according to the slope of the analog signal. In each stage, two types of parameters were calculated, i.e., amplitude of change (in $K^+$ or pH) in mV values and time to reach a certain point in kinetic changes. Parameter time to maximal change in $K^+$ or pH during ischemia (mV max) divided by 2 represents slope of curve. * Initial $K^+$ level higher than normal. † $p < 0.05$; ‡ $p < 0.001$

bilateral carotid occlusion is complete, the NADH response in all gerbils tested reached a maximal level. The ECoG disappeared in all gerbils, and the recovery after short- or long-term ischemia was incomplete at the time when other parameters recovered to their baseline values. There were significant differences between the two groups in the other parameters measured, i.e., extracellular $K^+$, pH, and DC potential. The extracellular $K^+$ and pH data are expressed in terms of millivolts. The sensitivity of the $K^+$ electrode was 50–60 mV for a ten-fold increase in $K^+$ in the calibration solution (3–30 mM). Most brains used in the experiments showed a resting level of 3–5 mM $K^+$. It was difficult to calibrate the pH resting level as discussed previously [8]. The changes that occurred during the ischemic period are presented in Table 4.6, whereas those of the recovery period are presented in Table 4.7. Long-term ischemia was induced in eight gerbils; the occlusion lasted 30 min in five

**Table 4.7** Kinetic changes during the recovery period from complete ischemia in gerbil brain [9]

| | | Recovery Time, min | | | | | | | |
|---|---|---|---|---|---|---|---|---|---|
| Gerbil No. | Ischemia Duration, min | $K^+$ | | | pH | | | DC | |
| | | Initial change | Baseline | 50% | Initial change | Baseline | 50% | $K^+$ | pH |
| 26 | 30 | 2.5 | 120 | 71 | 0.1 | 101 | 70 | 116 | 117 |
| 28 | 30 | 12 | 62 | 27 | 6.5 | 87 | 50 | 69 | 71 |
| 55 | 30 | 1 | 101 | 52 | 0.1 | 97 | 7 | 52 | 52 |
| 45 | 30 | 7.5 | 92 | 35 | 0.5 | 94 | 11 | 40 | 42 |
| 41 | 20 | 1 | 33 | 13 | 0.5 | 29 | 8 | 24 | 22 |
| 29 | 21 | 4.3 | 42 | 10 | 3.9 | 45 | 13 | 54 | 30 |
| Mean<br>±SE | | 4.7<br>1.8 | 75<br>14 | 35<br>10 | 1.9<br>1.1 | 76<br>12 | 27<br>11 | 59<br>13 | 56<br>14 |
| 28 | 5 | 2 | 10.8 | 3.8 | 0.3 | 24 | 2.1 | 10.8 | 26 |
| 30 | 5 | 2.5 | 41.3 | 6.3 | 0.7 | 42.5 | 20 | 9.5 | 27.5 |
| 50 | 4.6 | 1.3 | 13.5 | 3 | 0.1 | 17.3 | 3.1 | 2.8 | 5 |
| 51 | 3.8 | 1.3 | 10.8 | 3 | 0.2 | 3.8 | 1.5 | 5 | 5 |
| 55 | 4.6 | 2.6 | 25.5 | 9.5 | 0.1 | 16 | 4.1 | 18.3 | 18.3 |
| 56 | 3.8 | 1.4 | 26.2 | 5.8 | 0.1 | 21.3 | 5 | 9.3 | 10.1 |
| Mean<br>±SE<br>Significance | | 1.9<br>0.2<br>NS | 21.4<br>4.9<br>* | 5.2<br>1.0<br>† | 0.25<br>0.1<br>NS | 20.8<br>5.2<br>* | 6.0<br>2.9<br>§ | 9.3<br>2.2<br>* | 15.3<br>4.1<br>† |

50%, Time to 50% maximal change. * $p < 0.001$; † $p < 0.01$. § $p < 0.05$

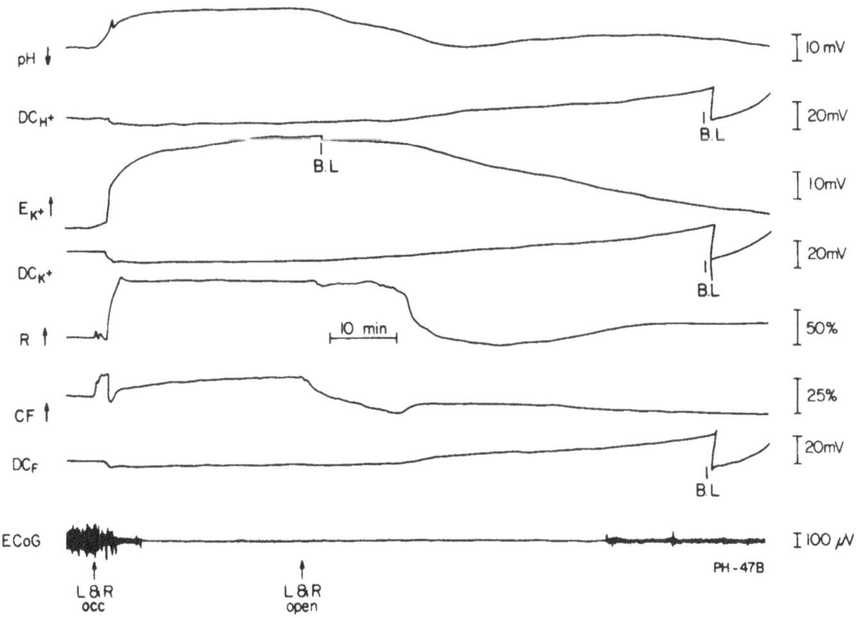


**Fig. 4.49** Effects of 30-min complete cerebral ischemia on various parameters measured from a gerbil brain. BL, a shift of base-line position. Other abbreviations as in Fig. 4.47 [9]

gerbils and 20 min in the remaining three gerbils. In the short-term ischemia group (six gerbils), the occlusion lasted between 3.8 and 5 min.

The records of gerbils 44 and 47 contained movement artifacts during the recovery period and are therefore missing in Table 4.7. We divided the changes in the response of extracellular $K^+$ into two phases according to the rate of $K^+$ accumulation in the extracellular space. The two parameters calculated in the first phase were the amount of $K^+$ accumulated and the time until general depolarization occurred. These two parameters were similar in the two groups of gerbils tested, i.e., 6.4 ± 1.0 mV and 1.7 ± 0.3 min in the 30-min ischemia group compared with 7.3 ± 0.9 mV and 1.5 ± 0.3 min in the 5-min ischemia group. The second phase analysis showed that at 5-min ischemia extracellular $K^+$ did not reach the maximal level, and as a result calculation of the timing was inaccurate. In the long-term ischemic group, extracellular K+ reached its maximal level after 21 min. However, it reached 50% of the maximal increase by 2.5 min, indicating that the depolarization phase is composed of two rates of $K^+$ leakage. The pH response to complete ischemia shows monotonic kinetics. The decrease in pH depends on the duration of the lack of $O_2$ or the accumulation of lactic acid. Due to the large variation between gerbils, the change in pH (mV max), although higher in the 30-min group, was not statistically significant (34.5 vs. 21.3 mV). The third parameter calculated was the DC steady potential near the $K^+$ and pH electrodes. Each brain showed different levels of DC changes during ischemia. These changes did not depend on the site in the brain.

The difference between short-term and long-term ischemia, as well as between the two sites in the brain, was not statistically significant. Three related parameters were analyzed for extracellular $K^+$ and pH recovery: the beginning of the change after reperfusion of the brain, the time to reach baseline levels, and the time to reach 50% of the change. For the two locations of the DC steady potential, we determined the time for complete recovery (Table 4.7). Of the eight parameters calculated, only extracellular $K^+$ and pH did not show a significant difference between long-term and short-term ischemia. Although the extracellular $K^+$ starts to recover later than the pH (4.7 vs. 1.9 min for the 30-min ischemia and 1.9 vs. 0.25 min for the 5-min ischemia), the difference was significant only in the 5-min ischemic group ($p < 0.001$). The recovery time after 30-min ischemia was significantly longer than after 5-min ischemia in the other six parameters calculated. To test the completeness of recovery from the ischemic insult, we activated the brain by inducing SD. The results are summarized in Table 4.8. In gerbils 30 and 41, the SD was induced before and after ischemia, whereas in gerbils 44 and 28 it was only measured after long-term ischemia (20 or 30 min). Several parameters were used to analyze the SD wave and response to it. Accumulation of extracellular $K^+$ was analyzed during the propagation of the wave, and the active pumping back and clearance into the cellular compartment was analyzed during the recovery. The amplitude and duration of the DC steady potential response were also analyzed. The results show that the accumulation of K+ during SD was significantly smaller (26 mV), compared with 5- or 30-min ischemia. Furthermore, the time to reach the maximal level of the extracellular $K^+$ was much shorter in SD (1.1 min) compared with ischemia. The

**Table 4.8** Effects of spreading depression on extracellular $K^+$ and DC potential changes before and after complete ischemia in gerbil brain [9]

| | | | Extracellular $K^+$ | | Recovery Time To | | DC Potential | |
|---|---|---|---|---|---|---|---|---|
| Gerbil No. | Ischemia Duration, min | Treatment | mV max | Time to mV max, min | Baseline, min | mV max/2. Min | mV decrease | Duration of cycle, min |
| 30 | 5 | Control | 33 | 1 | 4.3 | 0.4 | 10 | 1.7 |
| | | Postischemia (60 min) | 37 | 1 | 4.5 | 0.5 | 12 | 2 |
| 41 | 20 | Control | 30 | 1.3 | 2.3 | 0.42 | 7 | 2 |
| | | Postischemia (100 min) | 15 | 1 | 2.3 | 0.75 | 9 | 3.6 |
| | | Control | | | | | | |
| 44 | 20 | Postischemia (100 min) | 13 | 1.3 | 5.2 | 0.33 | 6 | 2.1 |
| 28 | 30 | Postischemia (90 min) | 30 | 0.8 | 3.5 | 0.50 | 10 | 2.5 |
| Mean | | | 26.3 | 1.1 | 3.7 | 0.48 | 9 | 2.3 |
| ±S.E. | | | 4 | 0.08 | 0.5 | 0.06 | 0.9 | 0.28 |

mV max, maximal level of $K^+$ during the wave of spreading depression

time for complete recovery of the extracellular $K^+$ was very short (3.7 min) compared with the 5-min ischemic insult (21 min, $p < 0.01$) or the 30-min ischemic insult (75 min, $p < 0.001$). The results shown in Table 4.8 suggest that even long-term ischemia does not have a dramatic effect on the response to spreading depression measured 60–100 min after recirculation.

#### 4.4.5.7.3 Discussion

The Mongolian gerbil provides a good animal model for studying the effects of complete brain ischemia (bilateral carotid occlusion) on its functional state [6, 32, 38, 69]. This model has been used by several investigators in studying mitochondrial metabolism [70], the concentration of energy metabolites [38], and local cerebral blood flow (CBF) [71]. Recently, CBF and edema formation were also measured after bilateral carotid occlusion in the gerbil [69] as was the appearance of the "no-reflow" phenomenon [72]. In all gerbils studied, either only very few parameters were measured continuously in vivo or the brain was analyzed for metabolite concentrations. In the present study, up to nine parameters were monitored continuously and simultaneously from the brain in vivo. Using this approach, we could detect changes in the metabolic, ionic, and electrical activities before and during the ischemic insult and during the recovery phase [8]. Siesjo [73] has already pointed out that the concept of irreversible damage appearing within 5–10 min of complete ischemia needs reevaluation due to new information accumulated during the past

10–15 years. According to the results published previously [32, 35, 38, 69] and in the present study (Figs. 4.46, 4.47, 4.48, and 4.49), bilateral carotid artery occlusion produces complete cerebral ischemia. The events occurring during long-term ischemia (>10 min) are well known. Due to $O_2$ deprivation, the available energy in the tissue may reach a minimal level, and as a result all energy-dependent functions will be impaired (for details, see Refs. [38, 74]). The remaining main questions involve the recovery process, i.e., how one can define the degree of recovery, what parameters represent the completeness of recovery, how can one test the reversibility of the damage, and so forth. These questions have significant implications to the clinical situations in which complete ischemia occurs.

The main parameters that have been used as criteria for brain recovery are (1) restoration of circulation, (2) tissue metabolite levels and energy balance, (3) intactness of mitochondrial function, and (4) recovery of physiological functions such as EEG or DC steady potential. Using the present approach, we could monitor most of these parameters in vivo.

During bilateral carotid occlusion, complete ischemia was induced as indicated by the maximal redox state of the measured intramitochondrial NADH. This finding was supported by the results obtained when $pO_2$ was also measured and correlated to the NADH redox state (Mayevsky, unpublished results). Kobayashi et al. [38] have shown that the decrease in energy reserves (ATP, phosphocreatine) depends on the duration of the ischemic insult during bilateral carotid occlusion in the gerbil. In our experiments, brain energy production is inhibited due to complete deprivation of $O_2$ in the two groups tested (5- and 30-min ischemia). It is probable that at the end of the ischemic period the energy reserves are different in the two groups tested. This difference must affect the recovery. To analyze the recovery of the brain from the ischemic insult, we used two approaches. We monitored and analyzed various parameters recorded (Table 4.7) and exposed the brain to a short anoxic episode or to an episode of spreading depression (Table 4.8). The response to anoxia (after recovery of all parameters) was unaffected by short- or long-term ischemia (data not shown in Table 4.9). The initial phase in the recovery period depends on restoration of blood flow to the cortex and the delivery of $O_2$ to the mitochondria. It was found that the reoxidation of the mitochondria is a fast process beginning as soon as the two carotid arteries are opened. As shown in Figs. 4.46, 4.47, 4.48, and 4.49, the NADH redox state changed simultaneously with the reopening of the blood vessels, although the kinetics of the oxidation was affected by the duration of ischemic

**Table 4.9** The effects of unilateral carotid artery occlusion on various NADH levels measured from the gerbil brain (means ± S.E.M.) [16]

| | % CF | % CF | % CF | |
|---|---|---|---|---|
| Group | Initial Increase | Steady- State Increase | St. St. max. Inc. | Δ% CF St. St.-Init. |
| +SRI | 41.6 ± 2.6 | 42.6 ± 3.1 | 85.5 ± 4.6 | 1.03 ± 2.0 |
| −SRI | 28.2 ± 2.1 | 17.4 ± 1.8 | 34.4 ± 3.8 | −10.9 ± 1.0 |
| $p <$ (Level of significance) | 0.005 | 0.005 | 0.005 | 0.005 |

insult. As soon as $O_2$ is available to the mitochondria, it is expected that high-energy phosphate compounds will be produced. Ginsberg et al. [35] have shown that after long-term ischemia, mitochondria remain coupled (as measured in vitro after isolation) and appeared to resist the ischemic insult. Kobayashi et al. [38] found that the recovery of high-energy phosphates after complete ischemia began as soon as blood circulation was restored. The completeness of recovery depended on the duration of the ischemic insult. In the present study, we measured both the oxidation of the respiratory chain (NADH fluorescence) and the active pumping of $K^+$ into the intracellular space (ATP utilization). The reason for the delay in the recovery of extracellular $K^+$ compared with the reoxidation of the intramitochondrial NADH is still unknown. Possible explanations are (1) to pump $K^+$ into the cells, a minimal level of ATP is required or (2) the $Na^+$-$K^+$-ATPase activity is inhibited during the ischemic insult and is sometimes needed for the reestablishment of enzymatic activity. The present data provide indirect evidence for $K^+$ removal from the extracellular space by active pumping into the cells requiring $Na^+$-$K^+$-ATPase rather than by passive diffusion into the vascular compartment. In many experiments, we exposed the animal to hypoxia and compared the response with partial or complete ischemia. If blood flow were involved in the removal of extracellular $K^+$, then one would expect to see a decrease in extracellular $K^+$ because flow is increased under these conditions. However, we did not find a difference between hypoxia and ischemia, suggesting that $O_2$ supply is the key factor in regulating extracellular $K^+$. Furthermore, recovery of the pH was very well correlated with the blood flow after the ischemic insult. The delay in pH recovery was shorter than the extracellular $K^+$, and in most animals (either short- or long-term ischemia) it began ~0.5 min after reopening of the carotid arteries. These results suggest that pH recovery is flow- and $O_2$-dependent, whereas $K^+$ recovery is $O_2$- and $Na^+$-$K^+$-ATPase activity-dependent. Our results do not support the recent findings published by Welsh et al. [74] that the magnitude of the NADH increase in response to an anoxic cycle was smaller after the ischemic episodes. In all animals, the response to anoxia was not affected by either short- or long-term ischemia. The response to spreading depression at the end of the recovery period was also unaffected (Table 4.7) by the episodes of ischemia. During the SD cycle, the NADH showed a clear oxidation response, as would be expected if circulation to the tissue and ATP production were normal [6, 7, 65]. We also reveal the different responses to the duration of ischemia. If the insult is very short, as shown in Fig. 4.46, restoration of circulation leads to complete recovery without a concomitant SD wave or general depolarization. With a longer ischemic insult, extracellular $K^+$ will reach a level that can induce a general depolarization (Figs. 4.47, 4.48, and 4.49). If energy is restored soon enough, then the depolarization will develop into a spreading depression wave as shown in Fig. 4.47, where all parameters measured responded as previously described [8, 26, 46]. Under longer periods of ischemia, the general depolarization will persist until reopening of the carotid arteries (Figs. 4.48 and 4.49), and the kinetics of the recovery process will depend on the duration of ischemic insult.

#### 4.4.5.8 The Seventh Study

This study was performed using a slightly modified Protocol 3.6 described in Chap. 3, Table 3.1 [8, 9]. The $H^+$ electrode was replaced by a $Ca^{2+}$ electrode.

##### 4.4.5.8.1 Introduction

The understanding of various pathological events occurring in the ischemic brain of an animal model has significant implications to the treatment of patients undergoing stroke. One of the events that occurs under ischemia is the ischemic depolarization (ID) during which various pathological changes take place in the brain [41, 42, 75]. The mechanism behind the development of ID is not yet clear, although ion homeostasis disturbances are involved [53]. The inhibition of the $Na^+$-$K^+$-ATPase is energy-depletion-dependent, and an immediate response of the brain to ischemia [47] will result in extracellular $K^+$ accumulation. In the Mongolian gerbil, unilateral carotid artery occlusion creates a variable level of ischemia in the ipsilateral hemisphere (in different animals) while the contralateral side is not affected significantly [19, 23, 34, 55].

In the Mongolian gerbil brain, we tested the correlation between energy depletion level (evaluated by intramitochondrial NADH redox state) and the development of ischemic depolarization (ID) developed under ischemia, using the multiprobe assembly (MPA) [8, 9].

Surface fluorometry/reflectometry was used in order to quantitatively evaluate the level of ischemia created under unilateral carotid artery occlusion and to correlate it to the development of the ischemic depolarization (ID) measured by the appearance of the secondary reflectance increase (SRI), a typical optical correlate of the ID [9, 29, 76].

##### 4.4.5.8.2 Methods

**Fluorometer/Reflectometer**

In the present study, a two-channel DC fluorometer/reflectometer was used [19, 30, 45]. The source for the 366-nm excitation light was a 100 W air-cooled mercury lamp passing through the two excitation fiber optic bundles to the brain. The emitted light from the brain that was transmitted through another fiber optic bundle was split into a 90:10 ratio in order to measure the fluorescence light (F: 90%) and the reflected light (R: 10%). The two fluorescence signals and the two reflectance signals, as well as the two corrected fluorescence signals, were recorded on a multichannel Grass instrument. The correction of the fluorescence signals for hemodynamic artifacts was done by subtracting R from the F signal in a 1:1 ratio [30].

**Multiprobe Assembly**

The multiprobe assembly (MPA) used in the present study was described in detail in our previous publications [8, 29]. Hence, only major points will be described herein. The MPA has the following main features: (1) all components have the same type of non-invasive surface contact with the cortex; (2) it can very easily be removed from the skull at the end of the experiment without any damage, so that repetitive experiments can be performed in a short period of time; (3) all signals monitored by the MPA have a very low sensitivity to the movements of the animal, even when exposed to convulsions. In the present study, the pH electrode was substituted by a $Ca^{2+}$ electrode. All the details regarding this new electrode were published elsewhere.

**Animal preparation**

Adult Male Gerbils (Tumblebrook Farms, Mass, USA) were anesthetized by i.p. injection (0.3 ml/100 g) of Equithesin (each ml contains pentobarbital, 9.72 mg; chloral hydrate, 42.51 mg; magnesium sulfate, 21.25 mg; propylene glycol, 44.34% w/v; alcohol, 11.5%, and water). In order to keep the gerbil in the anesthetized state during the experiment and the ischemic period, an additional injection of Equithesin (0.03–0.05 ml) was made every 30 min. After making a midline incision of the skin, the skull was exposed and two holes of 3.5 mm were drilled in the parietal bone area to accommodate the light-guide holders (located epidurally). Two pairs of stainless steel screws (used for the ECoG recording) were also placed in the skull and dental acrylic cement was applied.

In experiments where the MPA was used, a 6-mm hole was drilled in one hemisphere and, after gentle removal of the dura mater, the MPA was placed on the brain using a micromanipulator. Two stainless steel screws were used to connect the MPA and the dental acrylic cement to the skull. The two common carotid arteries were isolated and ligatures of 3-0 silk thread were placed around them.

### 4.4.5.8.3 Results

In the first part of the results, the SRI phenomenon is described. Figure 4.50 shows the responses of the gerbil brain to unilateral carotid artery occlusion repeated twice in the same animal ($R_{occl}$). During the 5 min of ischemia, the NADH became more reduced (increased CF signal) while the reflectance trace showed a small decrease. The percent change of the NADH was about 20% as compared to the pre-ischemic baseline, and as shown in Fig. 4.51, in complete ischemia under unilateral occlusion to be 40% of the maximum (20/50). A small increase in extracellular $K^+$ ($K_e^+$) and a depression in the intensity of the ECoG signal, especially in Part B, were recorded. The extracellular $Ca^{2+}$ ($Ca_e^{2+}$) as well as the DC steady potential around the two ion-sensitive electrodes were not changed. The ID was not developed after 5 min of unilateral occlusion.

When the two carotid arteries were occluded ($L_{occl}$ + $R_{occl}$) in the same animal, as shown in Fig. 4.51, the brain was exposed to complete ischemia as seen in the CF

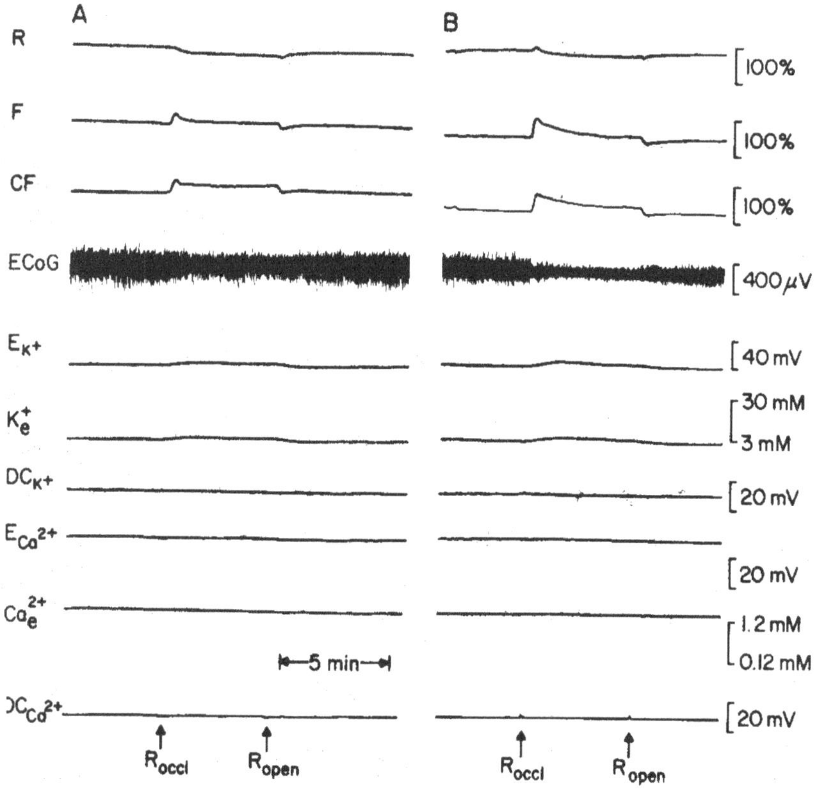


**Fig. 4.50** Effects of two sessions of 5-min unilateral carotid artery occlusion (**a**, **b**) on the metabolic, ionic, and electrical activities in the gerbil brain. The MPA was implanted above the right brain hemisphere. R, F, CF, reflectance, fluorescence, and corrected fluorescence; ECoG, electrocorticogram; $EK^+$, $K_e^+$, $E_{Ca2+}$, $Ca_e^{2+}$, uncorrected and corrected potassium and calcium ion concentrations, respectively; $DC_K+$, $DC_{Ca}^{2+}$, DC steady potential around the $K^+$ and $Ca^{2+}$ electrodes [16]

trace as well as the isoelectric ECoG trace. As expected, the extracellular $K^+$ started to increase immediately after the right carotid occlusion (the MPA was implanted in the right brain hemisphere). After 2–5 min of complete ischemia, the ischemic depolarization was recorded (ID) as seen in all traces recorded. The $K_e^+$ showed a faster increase together with the negativity of the DC potential ($DCK^+$). Also, the extracellular $Ca^{2+}$ ( $Ca_e^{2+}$ ) showed a large decrease together with the negative shift in the DC potential ($DC_{Ca}$). During the ID period, the reflectance trace showed a large increase (SRI) but the corrected fluorescence (CF) only slightly increased. About 15 min after reperfusion, all signals (except ECoG) reached the pre-ischemic level suggesting a complete recovery of the parameters measured.

In the second phase of the study, we used the change in the reflectance trace (SRI) as an indicator of the ID phenomenon. We used 18 gerbils, and NADH was monitored bilaterally together with the 366 nm reflectance in which 65 unilateral carotid occlusions were performed. In 32 cases, the SRI (+SRI) was developed,

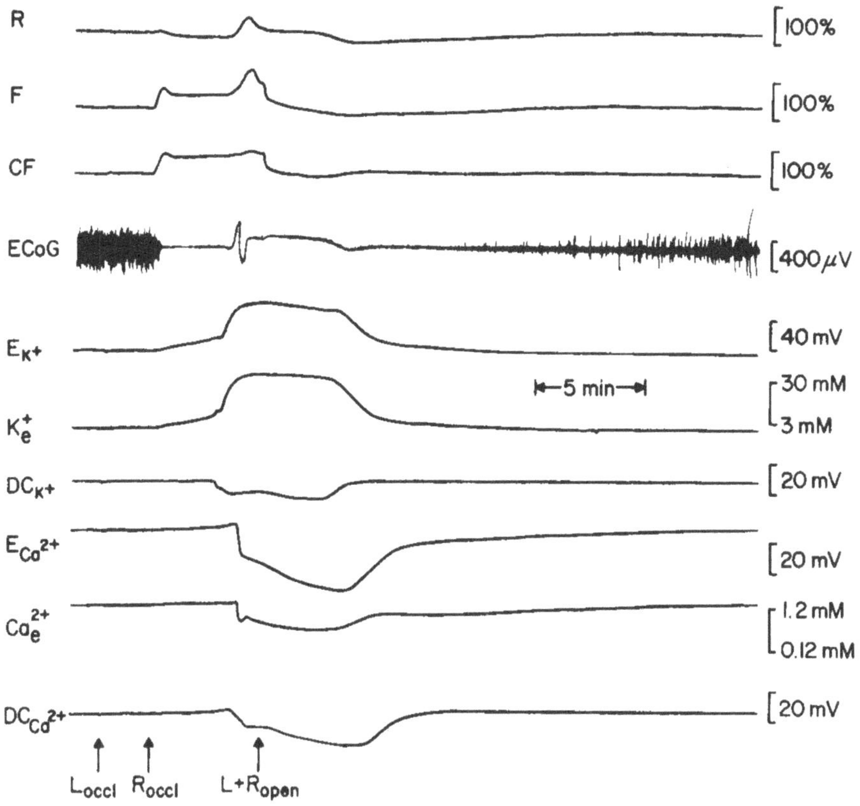


**Fig. 4.51** Metabolic, ionic, and electrical responses to complete ischemia induced by bilateral carotid artery occlusion in the Mongolian gerbil. R, F, CF, reflectance, fluorescence, and corrected fluorescence; ECoG, electrocorticogram; $EK^+$, $K_e^+$, $E_{Ca2+}$, $Ca_e^{2+}$, uncorrected and corrected potassium and calcium ion concentrations, respectively; $DC_K$+, $DC_{Ca}^{2+}$, DC steady potential around the $K^+$ and $Ca^{2+}$ electrodes [16]

while in 33 no secondary change in the reflectance trace was recorded (−SRI). Figure 4.52 shows four typical responses measured from four different gerbils. The two responses shown in A and B were taken from gerbils that developed a mild ischemia (18–20%) under unilateral carotid occlusion and therefore SRI was not recorded. The responses shown in Parts C and D were obtained from gerbils that developed the SRI, and the level of ischemia for C and D was twice that of A and B (32–38%). There are two different kinetics in the responses of NADH (CF). In Parts A and C, the initial increase was followed by a decrease, which later stabilized at a steady-state level. In Parts B and D, the initial increase was followed by an additional increase up to the steady-state level. In each gerbil, NADH was monitored from the two hemispheres, but only the ipsilateral hemisphere of the occlusion side is shown. The contralateral NADH level remained unchanged in all gerbils tested.

**Fig. 4.52** Typical metabolic responses to unilateral carotid artery occlusion in the gerbil brain. The numbers represent the percent increase of NADH level (CF) during the initial step as well as during the steady-state phase. +SRI, −SRI represent two groups of gerbils as explained in the text [16]

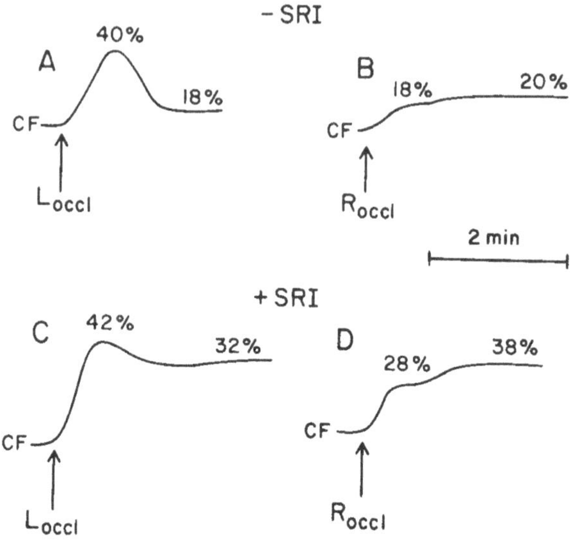


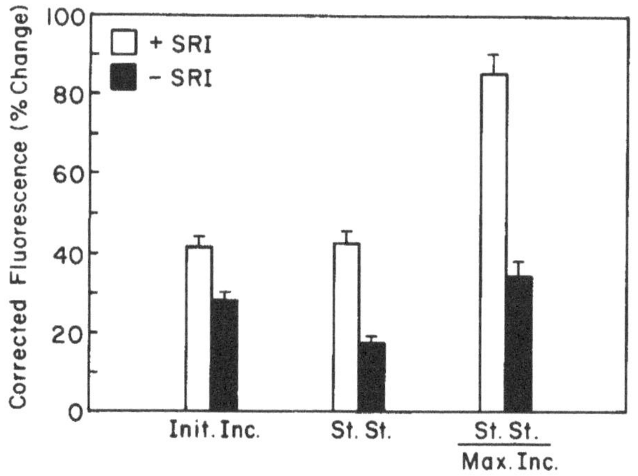


**Fig. 4.53** Effects of unilateral carotid artery occlusion on the NADH level (corrected fluorescence) in two groups of gerbils (means ± SEM). Init. Inc, initial increase after occlusion. st. st., steady-state level during the occlusion. st.st./Max Inc, steady-state level as compared to the maximum increase during anoxia. −SRI, +SRI, Gerbils among which the secondary reflectance increase during the ischemia developed (+) or did not develop (−) [16]

In order to quantitate the results, the 65 responses were analyzed for changes in the NADH level during the initial phase of the occlusion as well as during the steady state (st.st.). We also calculated the steady-state level as compared to the maximal increase of CF during anoxia (100% $N_2$). The results are shown in Table 4.9 and Fig. 4.53, and a large difference was found between the +SRI and the −SRI groups of gerbils. The difference was statistically significant in all three comparisons ($p < 0.005$). Also, when we calculated the ΔCF between the st.st. and initial levels, the difference between the two groups was significant ($p < 0.005$). In the +SRI group the average level of NADH was stable during the occlusion period (41.6 vs. 42.6), while in −SRI group a large decrease was recorded (28.2 vs. 17.4), suggesting that the anterior circulation compensated for part of the ischemia.

#### 4.4.5.8.4 Discussion

One of the main results of brain ischemic event is the development of the so-called ID-ischemic depolarization. This depolarization is very similar to the one developed under severe anoxia—anoxic depolarization [77, 78] and it includes similar ion homeostasis disturbances [39, 40] as well as vascular spasm [76].

The main questions are: what is the triggering mechanism that initiates such a massive pathological state, and whether or not the development of the ID is energy-depletion-level-dependent? To date, there is no published explanation for the ID development although few correlative events were described [75]. In our previous work, we showed the difference between unilateral and bilateral carotid artery ligation in the Mongolian gerbil while monitoring the metabolic ionic and electrical activities [79]. We presented the results only from gerbils that developed the SRI under unilateral carotid occlusion in the development of the ID as evaluated by the appearance of the SRI response.

To determine the level of ischemia during the unilateral occlusion, we monitored the NADH redox state that represents the intracellular oxygen availability under such conditions. The usage of the fiber-optic surface fluorometry/reflectometry enabled us to monitor both the level of ischemia and the vasospasm response to it from the same territory of the cerebral cortex. As shown in Fig. 4.50, one unilateral occlusion did not lead to the SRI development as seen under bilateral occlusion done in the same animal (Fig. 4.51).

The ID phenomenon is of the "all or none" nature; once it is triggered in one spot of the cortex, it will spread to the entire hemisphere. As discussed in the present study, the depletion of energy to a certain low level is a pre-condition to the development of the ID. A clear significant difference between the two groups of gerbils was found (Table 4.9 and Fig. 4.53). The possible mechanism behind this difference is the level of inhibition of the $Na^+$-$K^+$-ATPase activity which determines the rate and level of $K^+$ accumulation in the extracellular space during the ischemic episode. As shown clearly in Fig. 4.50, the initial increase in extracellular $K^+$ was followed by a steady-state level, but under bilateral occlusion (Fig. 4.51) the $K^+$ increased continuously until the appearance of the ID. In our previous work [79], we showed that the ischemic hemisphere had developed the SRI much faster than the control. This also indirectly supports the finding that a different $K^+$ leakage rate was found in the mild ischemic side as compared to the more severe level of ischemia under unilateral occlusion in other animals. The factor that determined the level of ischemia under unilateral occlusion is the structure of the anterior communication system in the circle of Willis [55]. Since the SRI is an "all or none" phenomenon, it is not possible to quantitate this parameter and correlate it with the level of ischemia. We assume

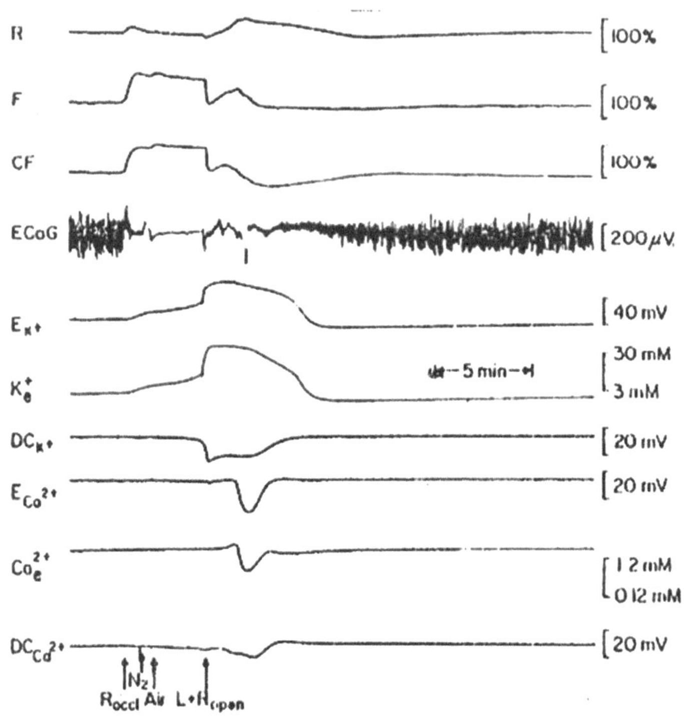


**Fig. 4.54** Effects of ischemia on the metabolic, ionic, and electrical activities in the Mongolian gerbil brain. R, F, CF, reflectance, fluorescence, and corrected fluorescence; ECoG, electrocorticogram; $EK^+$, $K_e^+$, $E_{Ca2+}$, $Ca_e^{2+}$, uncorrected and corrected potassium and calcium ion concentrations, respectively; $DC_K+$, $DC_{Ca}{}^{2+}$, DC steady potential around the $K^+$ and $Ca^{2+}$ electrodes [80]

that determination of the $K^+$ level in parallel will also contribute to the understanding of the ID development.

In summary, the results presented here show very clearly that the development of the ID state under partial ischemia is dependent on the degree of energy depiction. The nature of the development of the ID is of the "all or none" nature, i.e., the depletion of energy has to reach a critical low level in order to initiate the process of ID. It would seem to us that the accumulation of extracellular $K^+$ and its relationship to the changes in $Ca^{2+}$ are involved in it, but further studies are needed to clarify this mechanism.

#### 4.4.5.9 The Eighth Study

In this study, I used the same Protocol 3.6 described in Chap. 3, Table 3.1 [8, 9]. The $H^+$ electrode was replaced by a $Ca^{2+}$ electrode. Figures 4.54 and 4.55 were published in 1990 [80]. Figures 4.56, 4.57, and 4.58 were published in 1992 [81].

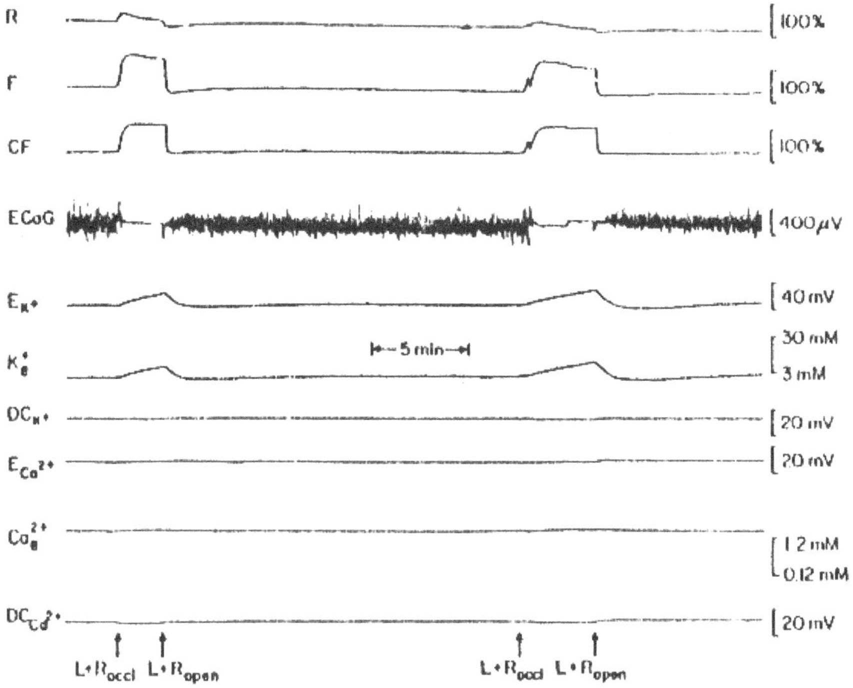


**Fig. 4.55** Responses to bilateral carotid arteries occlusion ($L + R_{occl}$) in the Mongolian gerbil brain. R, F, CF, reflectance, fluorescence, and corrected fluorescence; ECoG, electrocorticogram; $EK^+$, $K_e^+$, $E_{Ca2+}$, $Ca_e^{2+}$, uncorrected and corrected potassium and calcium ion concentrations, respectively; $DC_K+$, $DC_{Ca}^{2+}$, DC steady potential around the $K^+$ and $Ca^{2+}$ electrodes [80]

The effects of ischemia are shown in Fig. 4.54. In this gerbil, which was 27 days old (with 21 g body weight), by the occlusion of the right carotid artery ($R_{occl}$) while the left one was occluded before (data not shown in the figure), the level of ischemia achieved was 80–90% of its maximum. The addition of $N_2$ to the bilateral occluded gerbil led to an additional increase in NADH. Also, the ECoG was not depressed to the minimal level as recorded after the addition of $N_2$. During the ischemic period (more than 4 min), extracellular $K^+$ ( $K_e^+$ ) increased to the range of 10 mM, while a slight increase in extracellular $Ca^{2+}$ was noted ( $Ca_e^{2+}$ ). The reopening of the two carotid arteries was done at the same time the second phase in $K^+$ leakage was recorded. During the reopening of the blood vessels, the NADH became oxidized (decrease in CF signal) to about the pre-ischemic level. After this initial recovery, NADH showed a wave of oxidation (further decrease in CF) simultaneously with the recovery of $K^+$ and $Ca^{2+}$. All signals returned to the control values within 15 min.

When shorter periods of bilateral occlusion were used in the same gerbil, the SD wave or the ischemic depolarization was not developed. As shown in Fig. 4.55, 4 min of bilateral occlusion led to flattening of the ECoG and an increase of $K^+$, but ID was not the end point as expected. Also, the recovery was very fast, including the ECoG.

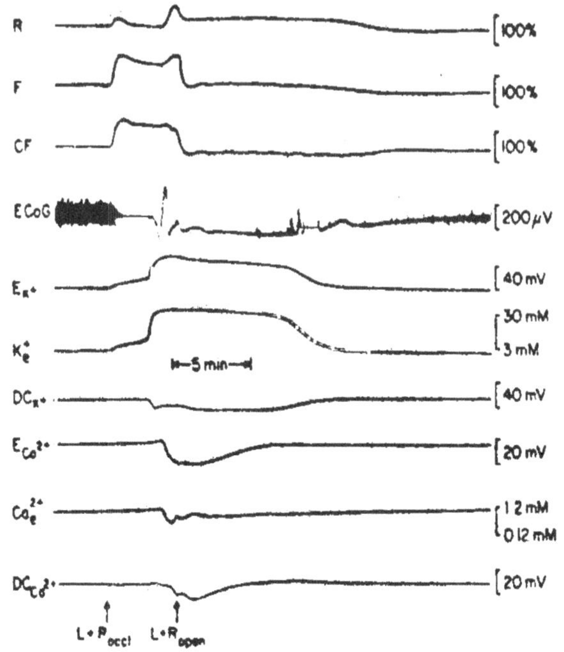


**Fig. 4.56** Responses of the Mongolian gerbil brain to bilateral carotid artery occlusion. The measurement was done using the multiprobe assembly (MPA). $EK^+$, $E_{Ca2+}$, uncorrected potential measured by the two surface electrodes; $K_e^+$, $Ca_e^{2+}$, corrected extracellular potassium and calcium; $DC_K+$, $DC_{Ca}^{2+}$, steady potential measured around the two electrodes used; ECoG, electrocorticogram; R, F, CF, reflectance, fluorescence, and corrected fluorescence, respectively [81]

To evaluate the relationship between the intracellular redox state and ion homeostasis, we used the Mongolian gerbil as an animal model. Occlusion of the two carotid arteries in this animal led to a complete ischemia in most animals [9]. As shown in Fig. 4.56, the decrease in blood flow ($L + R_{occl}$) to the brain led to a large increase (80–90%) in NADH redox state of the mitochondria (CF). Initially, there was very slight change in the reflectance (R) after the occlusion. However, after 2–3 min, a large increase in R was recorded. These changes were probably due to the initial response of a metabolic increase in extracellular $K^+$ ( $K_e^+$ ), followed by the second phase of a large increase due to the massive depolarization. This event of ischemic depolarization led to a large decrease in extracellular $Ca^{2+}$ level ( $Ca_e^{2+}$ ) simultaneously with an increase in the negativity of the DC steady potential near the two electrodes. All these responses represent the failure of the active transport system to operate under the energy-limited state. Upon recirculation, the NADH redox state recovered very quickly to the normoxic level and even showed a small undershoot during the period of $K^+$ pumping back into the cells (a period of about 10 min). $Ca^{2+}$ also recovered within less than 10 min to the pre-ischemic level. The recovery of the ECoG was very slow as compared to other parameters.

Exposure of the same gerbil brain to a short period of anoxia (Fig. 4.57) led to an increase in NADH (CF) as expected, but $K_e^+$ showed only the initial increase which recovered immediately after a change to room air. Extracellular $Ca^{2+}$ did not change during the short anoxic period. In the ischemic episode shown in Fig. 4.57, the left carotid artery was occluded alone for about 2 min and then the right artery was also occluded. The MPA was located on the left cerebral hemisphere and the occlusion

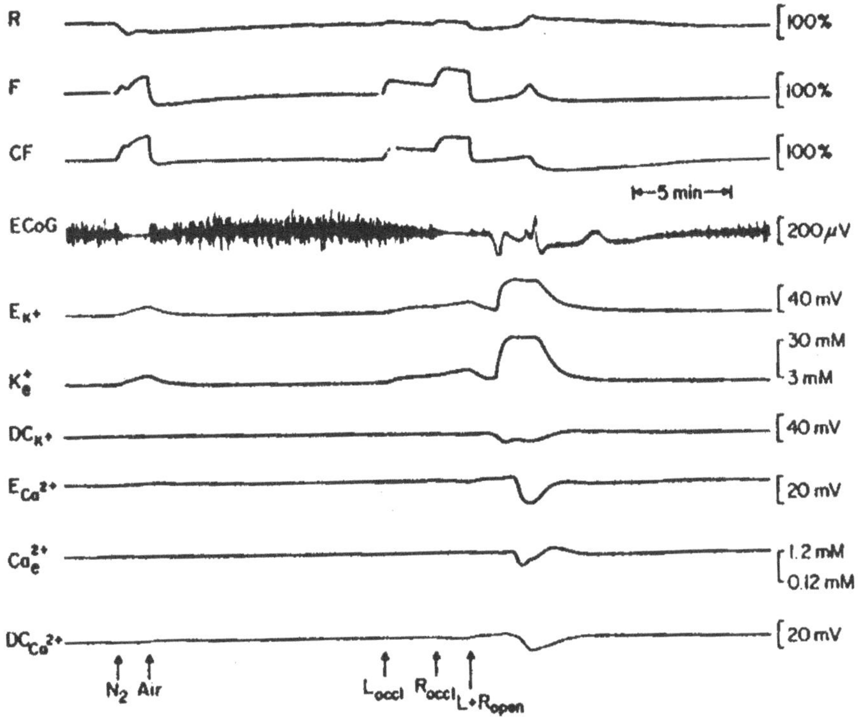


**Fig. 4.57** Effect of anoxia (left) and ischemia (right) on various parameters monitored with the MPA from the gerbil brain. R, F, CF, reflectance, fluorescence, and corrected fluorescence; ECoG, electrocorticogram; $EK^+$, $K_e^+$, $E_{Ca2+}$, $Ca_e^{2+}$, uncorrected and corrected potassium and calcium ion concentrations, respectively; $DC_K+$, $DC_{Ca}{}^{2+}$, DC steady potential around the $K^+$ and $Ca^{2+}$ electrodes [81]

of the left artery led to 50% of the maximum ischemic level. During this phase, $K_e^+$ accumulated as a result of depletion of energy, and this continued after right artery occlusion. The opening of the two carotid arteries was done before the appearance of the ischemic depolarization shown in Fig. 4.56. As a result, NADH recovered as $K_e^+$ started to return to normal values. During the recovery period, a wave of spreading depression (SD) penetrated the monitored area and induced the typical responses to this type of brain stimulation. This wave led to an increase in $K_e^+$ and a decrease in $Ca_e^{2+}$ but the duration of this ionic disturbance was very short as compared to 5 min of ischemia (Fig. 4.56). The recovery time to the baseline levels of all parameters was shorter due to the continuous stimulated activity of the $Na^+$-$K^+$-ATPase and the Ca-ATPase as can be seen by the oxidation wave of the NADH (decreased below baseline). In Fig. 4.58, the bilateral carotid occlusion was followed by anoxia which led to an irreversible state and the gerbil died. The $K_e^+$ level was above 40 mM while $Ca^{2+}$ decreased ten-fold (from 1.2 mM to 0.12 mM).

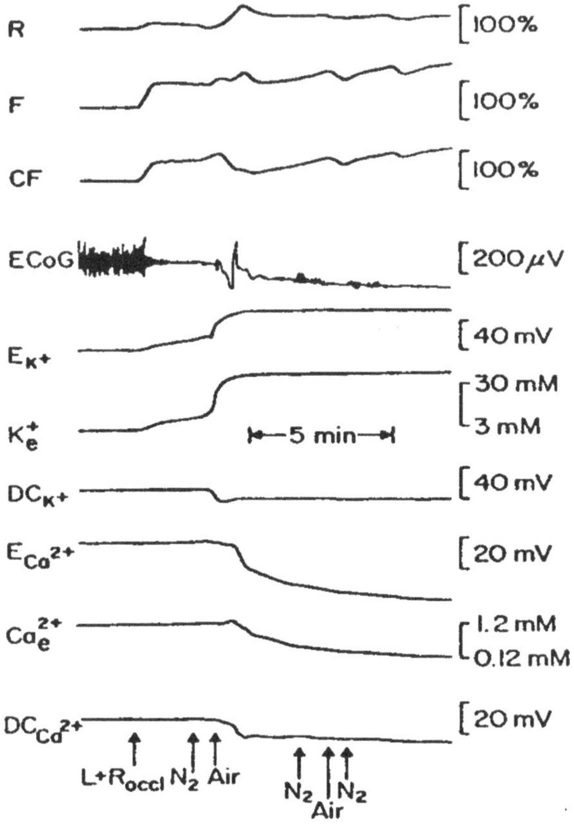


**Fig. 4.58** The responses of the brain to long-term ischemia and anoxia (terminated response). R, F, CF, reflectance, fluorescence, and corrected fluorescence; ECoG, electrocorticogram; EK$^+$, $K_e^+$, $E_{Ca2+}$, $Ca_e^{2+}$, uncorrected and corrected potassium and calcium ion concentrations, respectively; $DC_K$+, $DC_{Ca}{}^{2+}$, DC steady potential around the $K^+$ and $Ca^{2+}$ electrodes [81]

#### 4.4.5.10 The Ninth Study

The next step in the new Protocol 3.10 described in Chap. 3, Table 3.1 [13, 14]. In addition to the $Ca^{2+}$ electrode, I added the measurement of hemoglobin oxygenation described by Frank et al. [82–84]. Figures 4.59 [13] and 4.60 [14] show the effects of unilateral and bilateral carotid occlusions on various parameters, including $HbO_2$ oxygenation, measured in the gerbil brain.

The effects of ischemia on the brain are shown in Fig. 4.59. The ischemia was induced in two steps. After right artery occlusion (MPA was located on the right hemisphere), NADH increased by 10–12% as compared to the normoxic level which represents about 20% of the total signal increase (ischemia + anoxia). The $HbO_2$ also decreased by 20–22% from the normoxic value. This small change in the $O_2$ supply led only to small changes in ions level and ECoG activity. The occlusion of the left carotid artery led to a more severe low flow state but not to a complete one as seen by the further increase of NADH on the addition of $N_2$ to the bilateral carotid occluded brain. This specific gerbil had probably a well-developed posterior communication system in the circle of Willis [55]. The $HbO_2$ shows the same pattern of

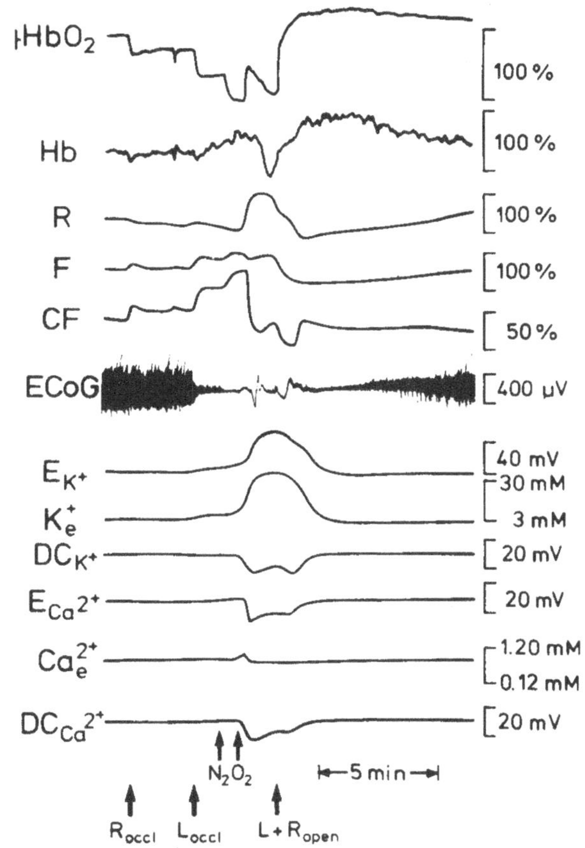


**Fig. 4.59** Effects of unilateral and bilateral carotid occlusion as well as $N_2$ on various parameters monitored from a gerbil brain. R, F, CF, reflectance, fluorescence, and corrected fluorescence; $E_{Ca}^{2+}$, $EK^+$, uncorrected calcium and potassium potentials; $Ca_e^{2+}$ , $K_e^+$ , corrected ion activities; $DC_{Ca}^{2+}$, $DC_K$+, DC steady potential measured around the calcium and potassium electrodes; ECoG, electrocorticogram; $HbCO_2$, hemoglobin oxygenation; Hb, hemoglobin volume or content [13]

changes during the occlusion anoxia combined episode. The $K^+$ response to the ischemia was significant only when the two carotids were occluded. The addition of $O_2$ while the two carotids were occluded led to a 30–40% increase of $HbO_2$ and then the ID was recorded as seen by the ions and DC potential changes. Only when the blood circulation was restored (L + $R_{open}$), $HbO_2$ and NADH became fully recovered to the normoxic level after a hyperoxygenated state. During the ID, a massive vasoconstriction was noted as seen by the large increase in reflectance (SRI) and a large decrease in Hb volume. The recovery from the ID episode includes the typical hyperemic response as indicated by the decrease in the R signal and overshoot of the Hb volume. The development of the ID event started in the $Ca^{2+}$ electrode area and appeared later below the other probes. Due to the coincidence in the ID appearance and the reoxygenation of the brain, the ID response was very similar to a SD wave type of response.

Figure 4.60 shows the response of the gerbil brain to ischemia induced by unilateral and bilateral carotid artery occlusions while MPA was located on the right hemisphere. The 3-min occlusion of the right carotid artery led to partial ischemia

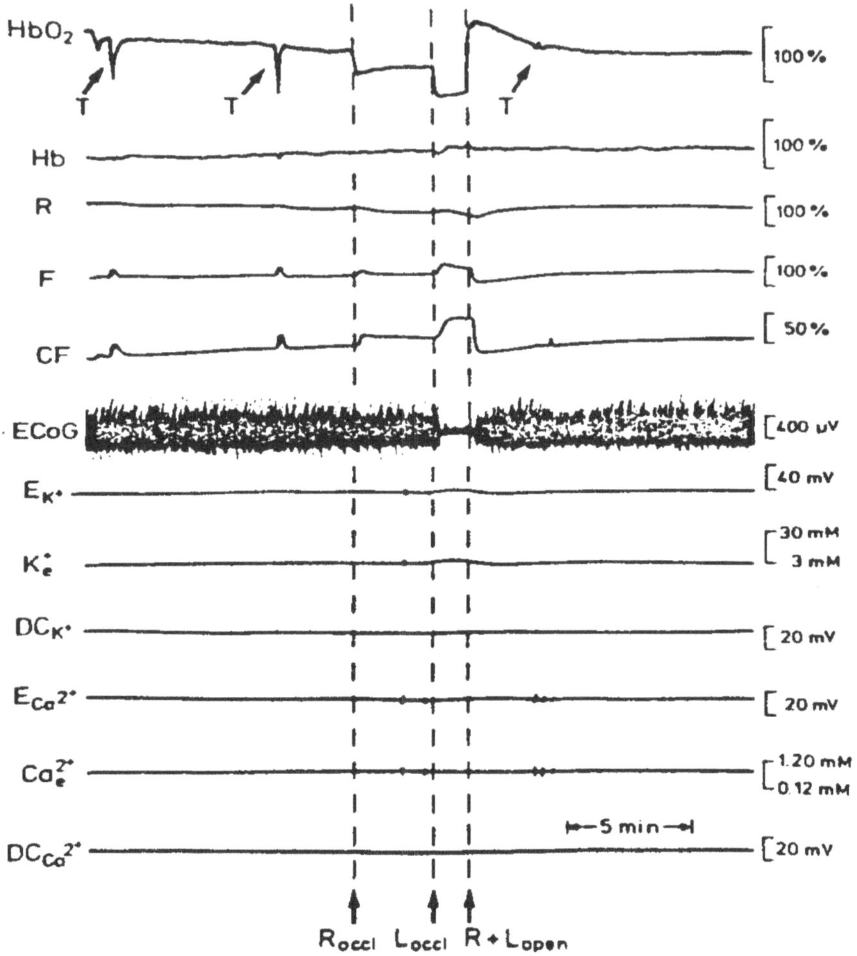


**Fig. 4.60** Effects of unilateral ($R_{occl}$) and bilateral ($L_{occl}$) carotid artery ligation on various parameters monitored from the gerbil brain. $HbO_2$, hemoglobin oxygenation (the pre-ischemic value was taken as 100%); Hb, hemoglobin volume or content; R, F, CF, reflectance, fluorescence, and corrected fluorescence of NADH; ECoG, electrocorticogram; $EK^+$, $E_{Ca}{}^{2+}$, uncorrected potassium and calcium potential; $K_e^+$, $Ca_e^{2+}$, corrected extracellular potassium and calcium levels; $DC_K{}^+$, $DC_{Ca}{}^{2+}$, DC steady potential measured around the potassium and calcium electrodes; T, oxygen consumption test [13]

[55] as seen by the deoxygenation of the $HbO_2$ as well as the increased level of the reduced NADH (CF trace). The addition of the left artery occlusion (while the right one was occluded) created a more severe level of ischemia during which the ECoG became depressed and a small increase in extracellular $K^+$ was noted. No change in extracellular $Ca^{2+}$ level or DC steady potential was noted. When the two carotid arteries were opened, the brain became oxygenated within a few seconds, as seen by the overshoot of the $HbO_2$ trace and the recovery of the NADH redox state to the

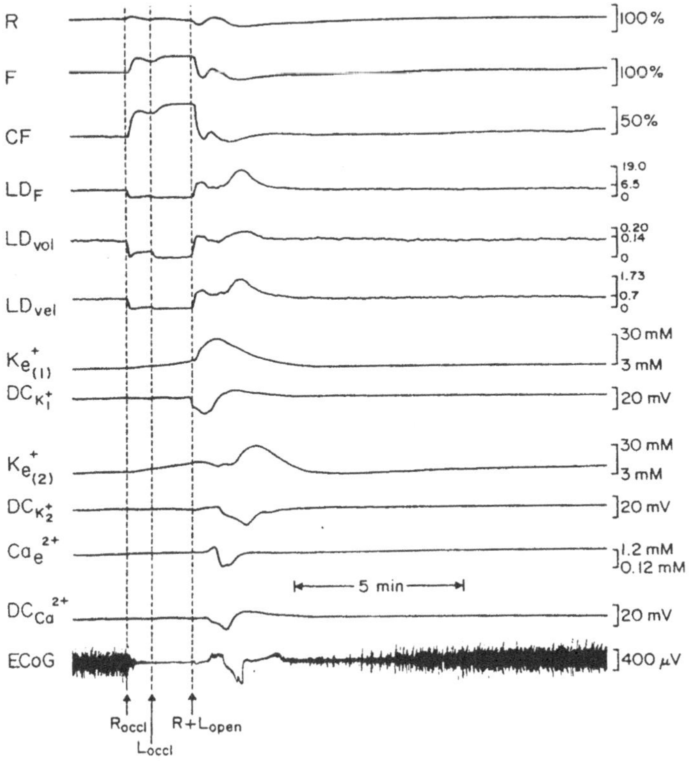


**Fig. 4.61** Effects of unilateral ($R_{occl}$) and bilateral ($L_{occl}$) carotid occlusion on the metabolic, hemodynamic, ionic, and electrical activities in the gerbil brain. R, 366 nm reflectance; F, 450 nm fluorescence; CF, NADH-corrected fluorescence; $LD_F$, $LD_{vol}$, $LD_{vel}$, Laser Doppler flow, volume, and velocity; $K_e^+$ (1), $K_e^+$ (2), $Ca_e^{2+}$, extracellular potassium (two electrodes) and calcium electrodes; $DC_{K+}$ (1), $DC_{K+}$ (2), $DC_{Ca2+}$, DC steady potential measured concentric to the three electrodes; ECoG, electrocorticogram [15]

pre-ischemic level. The letter T in Fig. 4.60 represents. The test for $O_2$ consumption performed before and after the ischemic episode, as will be discussed below.

#### 4.4.5.11 The Tenth Study

In the next 12 figures, I will demonstrate the results obtained in experiments that the measurement of CBF was included, as described in Protocols 3.11 [80] or [15] 3.12 [85] and 3.13 [86]. The details of the protocols are described in Table 3.1.

Figure 4.61 [15] shows typical responses to unilateral ($R_{occl}$) and bilateral ($L_{occl}$) carotid artery occlusion. The two-step reduction in blood supply was detected by the decrease in CBF (LDF) and increase in NADH levels (CF). During the period of ischemia, accumulation of $K^+$ in the extracellular space was recorded ( $K^+_{e(1)}$, $K^+_{e(2)}$ ), but the DC steady potential and the $Ca^{2+}$ levels remained unchanged during the occlusion period. ECoG reached the isoelectric level very soon after the second occlusion. During the reopening of the two carotid arteries, a fast reperfusion was recorded together with the oxidation of NADH. A spontaneous wave of SD was developed during the recovery phase, characterized by the large increase in $K_e^+$ and a decrease in $Ca_e^{2+}$ together with a negative shift in the DC steady potential. During the recovery from the SD wave, a large increase in CBF (300%) was recorded which was accompanied by an oxidation wave of NADH (a decrease in CF).

#### 4.4.5.12 The 11th Study

The aim of this study [87] was to examine the interrelation between hemodynamic events, energy metabolism, extracellular potassium, and electrical activity during the acute phase of transient ischemia in the gerbil brain, in order to enable a better understanding of the pathophysiology during the reperfusion period.

##### 4.4.5.12.1 Methods

The details of the methodology used were described in our publication [87].

**Animal Protocol**
All experimental protocols were approved by the Animal Care and Use Committee of the Bar-Ilan University.

Ninety minutes after surgery, unilateral occlusion was performed for 3–4 min and then released. All parameters were monitored online and recorded on both a Grass polygraph as well as stored in a computer using an A/D converter (Codas System, DATAQ Inc., USA).

**Data Analysis and Statistics**
Since our data acquisition enabled a continuous real-time monitoring, it was necessary to sample signals for quantitative analysis. The leading parameter for sampling of the different parameters and calculations was the CBF. Whenever CBF changed by 25%, data from all other parameters were sampled and calculated. CBF was calculated as percent change of baseline (0% was recorded after death). NADH was calculated as percent change of baseline. Extracellular $K^+$ was converted to mM values according to the Nernst equation and DC was recorded and calculated in mV values. Results are presented as mean ± SEM.

Statistical analysis was carried out using the Student's t-test, Student's paired t-test, and correlation coefficients. Changes were considered significant when $p < 0.05$.

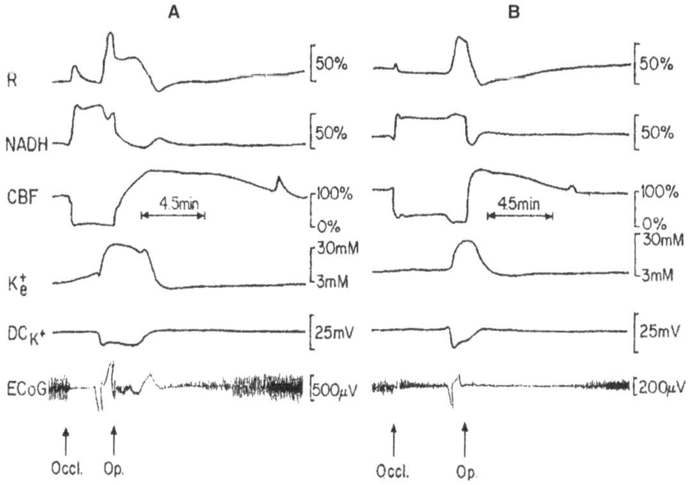


**Fig. 4.62** Changes in various parameters during ischemia and reperfusion. R, reflectance; NADH, mitochondrial NADH redox state; CBF, cerebral blood flow; $K^+$, extracellular potassium; DC, Direct current steady potential surrounding the $K^+$ electrode; ECoG, electrocorticogram; Occl, carotid artery occlusion; Op, carotid artery open [87]

#### 4.4.5.12.2 Results

In total, 20 gerbils used in this study underwent unilateral carotid occlusion for a duration of 3–4 min. Figure 4.62 shows a polygraph tracing, demonstrating the analog changes in various parameters monitored following the ischemia/reperfusion episode in two typical gerbils. A group of 20 animals were divided into two groups according to the level of ischemia obtained after unilateral carotid artery occlusion. As shown in Fig. 4.62a, the level of the CBF remained low and stable until the reperfusion occurred (Op.). This kind of response was monitored in a group of eight gerbils (Group A). In the second group of 12 gerbils (Group B), the initial decrease in CBF was followed by a secondary decrease to the very low flow values compared to those recorded in Group A. The fast initial response of the CBF was followed in the same time scale by the NADH redox state increase and a slow increase in extracellular $K^+$ levels ( $K_e^+$ ). The second major event developed in time after the occlusion is the ischemic depolarization (ID). The typical changes during the ID are an increase in the rate of $K^+$ accumulation, a negative shift in the DC steady potential (also in the bipolar ECoG shift) and a large increase in the reflectance trace. In Fig. 4.62b, the events recorded were similar to those shown in Fig. 4.62a. After the reperfusion, all parameters recovered and reached baseline level within a few minutes. In the two gerbils shown, the CBF was elevated after the reperfusion and a hyperemic response was recorded.

To analyze the responses in a quantitative way, we used the protocol as described in "Materials and Methods." Figure 4.63 describes the three stages of the ischemia procedure:

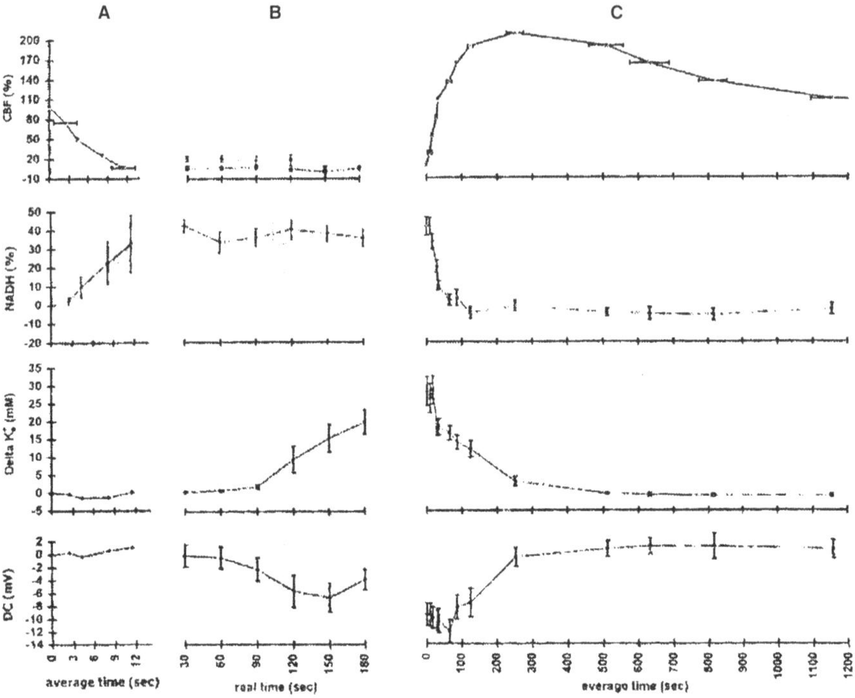


**Fig. 4.63** Changes in CBF (%), NADH (%), extracellular $K^+$ (mM), and DC steady potential (mV). (**a**) immediately after occlusion, (**b**) during 3 min of ischemia; and (**c**) after carotid artery reopened. Time, NADH, $K^+$, and DC were determined every 25% decrease/increase in CBF (a, c). Thus, x-axis represents average time. During ischemia (8), data were calculated every 30 s (x-axis represents real time). 1, $p < 0.05$; 2, $p < 0.01$ [87]

1. The events were recorded immediately after the occlusion of the carotid artery, i.e., time, NADH, $K^+$, and DC were sampled whenever CBF changed by 25% (Fig. 4.63a).
2. After 3 min of ischemia, various parameters were sampled every 30 s (Fig. 4.63b).
3. The events after reopening the carotid artery were sampled whenever CBF changed by 25% (Fig. 4.63c).

The results show that occlusion of the carotid artery resulted in an immediate decrease in CBF which reached a minimum level of 18% (17.9 ± 3.9) in Group A and 5% (4.8 ± 3.0) in Group B (total average of 6.2) within 11.5 ± 1.8 s. The decrease in CBF was accompanied by an increase of approximately 30% in the NADH level. No significant changes in the extracellular potassium ion concentration or in the DC potential were observed at this stage (Fig. 4.63a).

After the initial increase, the NADH levels remained constant during the entire ischemic period. A slow increase in extracellular $K^+$ was observed during the first 90 s of the ischemia, followed by a rapid increase during the phase of ischemic depolarization. Potassium levels before opening the carotid arteries (Fig. 4.63b)

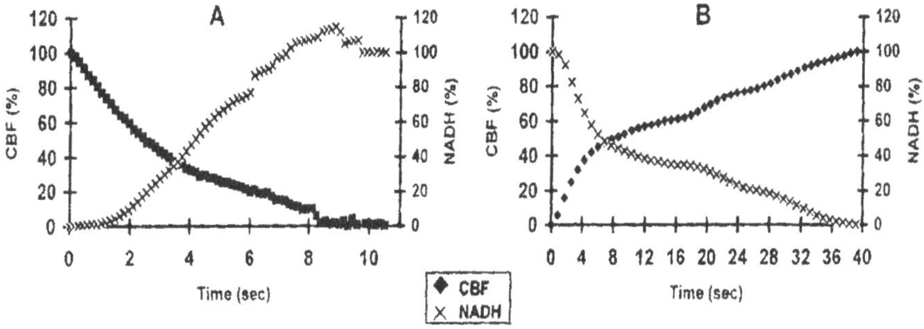


**Fig. 4.64** (**a**) Changes in CBF (%) and NADH (%) during the first 11 s after carotid occlusion. (**b**) Changes in CBF (%) and NADH (%) during the first 34 s after carotid artery was reopened [87]

| | CBF | NADH | $K^+_e$ | DC |
|---|---|---|---|---|
| CBF | | -0.97** | -0.82** | -0.4 |
| NADH | -0.87** | | 0.9** | 0.56** |
| $K^+_e$ | -0.87** | 0.92** | | 0.5* |
| DC | 0.6** | -0.3 | -0.3 | |

**Fig. 4.65** Correlation between various parameters during (gray boxes) the first phase of ischemia (i.e., until minimum flow was recorded) and (white boxes) during reperfusion. *$p < 0.01$, **$p < 0.05$ [87]

were 17 mM (17.0 ± 3.2). These changes in potassium levels were accompanied by a negative shift in the DC potential (−7.5 ± 1.8).

After reopening the carotid artery, the reperfusion was very rapid, and CBF and NADH reached baseline values within 33.5 s (Fig. 4.63c). The CBF continued to increase and a maximal hyperemic response of 193% ±11.5 was recorded 250 s later (251.5 ± 24.0). Full recovery of the CBF occurred only 15 min later. Fifteen seconds after reperfusion (14.5 ± 1.4), extracellular $K^+$ reached a maximum level of 30 mM (30.0 ± 4.1). Extracellular $K^+$ reached baseline levels when the hyperemia was at its maximal level (Fig. 4.63c). The DC potential followed the changes in the $K^+$ levels and reached a minimal level during the first stage of reperfusion (−12.3 mV ± 1.5). DC potential reached the values when CBF was maximal.

In order to clarify the interrelation between NADH and CBF, detailed analysis was performed on the data recorded immediately after occlusion (Fig. 4.64a) and immediately after occlusion (Fig. 4.64b).

CBF decreased immediately after the occlusion while NADH started to increase 1 s (0.75 ± 0.15) after carotid artery occlusion. A significant correlation was found between the decrease in CBF and the increase in NADH ($r = -095$, $p < 0.01$) (Fig. 4.64a).

Although the kinetics of the curve was different, the high correlation between CBF and NADH was kept during the first 30 s (34.0 ± 4.0) of recovery (i.e., until CBF reached the value of 100%, before the hyperemia occurred ($r = 0.99$, $p < 0.01$), Fig. 4.64b). The correlation between CBF and NADH was also calculated during the entire reperfusion episode (i.e., from opening the carotid artery to full recovery of CBF) and was found to be very significant (Fig. 4.65).

#### 4.4.5.12.3 Discussion

Transient global ischemia induced in this study in the Mongolian gerbil resulted in changes that are in agreement with previous studies performed in our laboratory [18]. The immediate decrease in CBF that occurred after carotid artery occlusion was accompanied by an increase in NADH levels. The high correlation between CBF and NADH found in the present study agrees with a large number of reports demonstrating dependence of the brain on continuous supply of oxygen. NADH remained in the reduced state during the entire ischemic episode, due to the lack of oxygen, which resulted in the mitochondria not being able to utilize the NADH. As the ischemia progressed, potassium ions began slowly leaking out of the cell into the extracellular space probably because of ATP store depletion. Significant correlation was found between the NADH and extracellular $K^+$, indicating that as the NADH increased ATP decreased and the sodium/potassium ion pump apparently ceased working. This led to a general depolarization that occurred 1.5 min after occlusion. These findings are in agreement with other studies that observed a decrease in $Na^+$-$K^+$-ATPase activity during ischemia [88]. In a previous study carried out by Mayevsky et al. [86], a clear correlation was found between the NADH redox state and the decrease in the PCr/Pi ratio during ischemia as well as in newborn dogs during hypoxia [89]. This change affects the electrical activity as recorded by a negative DC shift. It can be expected that tissue recovery from ischemia depends on $O_2$ supply. Indeed, after opening the carotid artery, CBF rapidly increased, supplying $O_2$ to the tissue and enabling the rapid oxidation of NADH. Furthermore, NADH returned to its initial level by the time CBF reached 100%. However, CBF continued increasing and hyperemia was observed. This suggests that the mitochondrial need for oxygen necessary for the production of ATP is not the only factor influencing CBF during reperfusion. Other factors, such as an increase in nitric oxide synthesis, may be involved [90–92]. Extracellular potassium ion levels recovered later than NADH, and this raises the question of why NADH was not oxidized beyond the initial level if more ATP was required for the activity of the ion pump. This can be explained by the assumption that ion pump rates are slower than the mitochondrial turnover. This may explain the fact that NADH returned to its baseline level and was not oxidized more. In other words, when the NADH returned to its normal level, ATP was no longer the limiting factor and ion pump activity became the factor determining recovery. The harmful effect of the hyperemia is well known; since brain tissue is "flushed" by unutilized oxygen, secondary damage could be induced by free oxygen radicals [93–95].

**Conclusion**

This study is in agreement with other studies that demonstrate the mutual dependence of tissue oxygen demand for production of ATP and oxygen supply by CBF. It also demonstrates the lack of pools of high-energy compounds such as ATP in the brain. Furthermore, this study suggests that ATP production and oxygen demand are not the only limiting factors in the recovery process from ischemia. Thus, the development of hyperemia after carotid artery opening is probably not due to the extra demand for tissue oxygen but rather to some other *process*, such as an ion pump recovery or nitric oxide synthesis.

#### 4.4.5.13 The 12th Study

**The Functioning Gerbil Brain In Vivo**

*Correlation Between $^{31}P$ NMR Spectroscopy and the Multiparametric Monitoring Approach*

The introduction and methods used in this study were described in details in Chap. 3 according to Protocol 3.13 presented in Table 3.1.

##### 4.4.5.13.1 Results

Each of the gerbils used in the present study was exposed to a short-term (5 min) as well as a long-term (20 min) bilateral carotid artery occlusion (complete ischemia). At least 60 min were allowed for recovery or until the ECoG showed a completely recovered pattern. Figure 4.67 presents typical MPA measurements following 5 min of ischemia (induced by bilateral carotid artery occlusion).

It can be seen that the occlusion of both arteries led to severe ischemia as indicated by the decrease in CBF (LDF) and the increase in the NADH redox state (CF). Two minutes after inducing ischemia, an ischemic depolarization (ID) event was recorded in various parameters monitored. A massive decrease in blood volume was recorded by the laser Doppler flowmeter (LDV) as well as by a large increase recorded in the reflectance trace (R) indicating a decrease in the amount of blood in the area being measured by the probe. Extracellular $K^+$ levels exhibited a two-step increase during the ischemic episode. Initial leakage began simultaneously with the occlusion and a steep slope appeared during the ID event, characterized by a negative shift in the DC potential and a decrease in extracellular calcium. The ECoG became isoelectric immediately after the occlusion, recovering gradually after reperfusion.

Upon the reopening of the carotid arteries, the energy state and blood volume exhibited a rapid return to their pre-ischemic levels. The CBF exhibited a hyperemic response during the initial 5–8 min of recovery. The effects of 5-min ischemia on

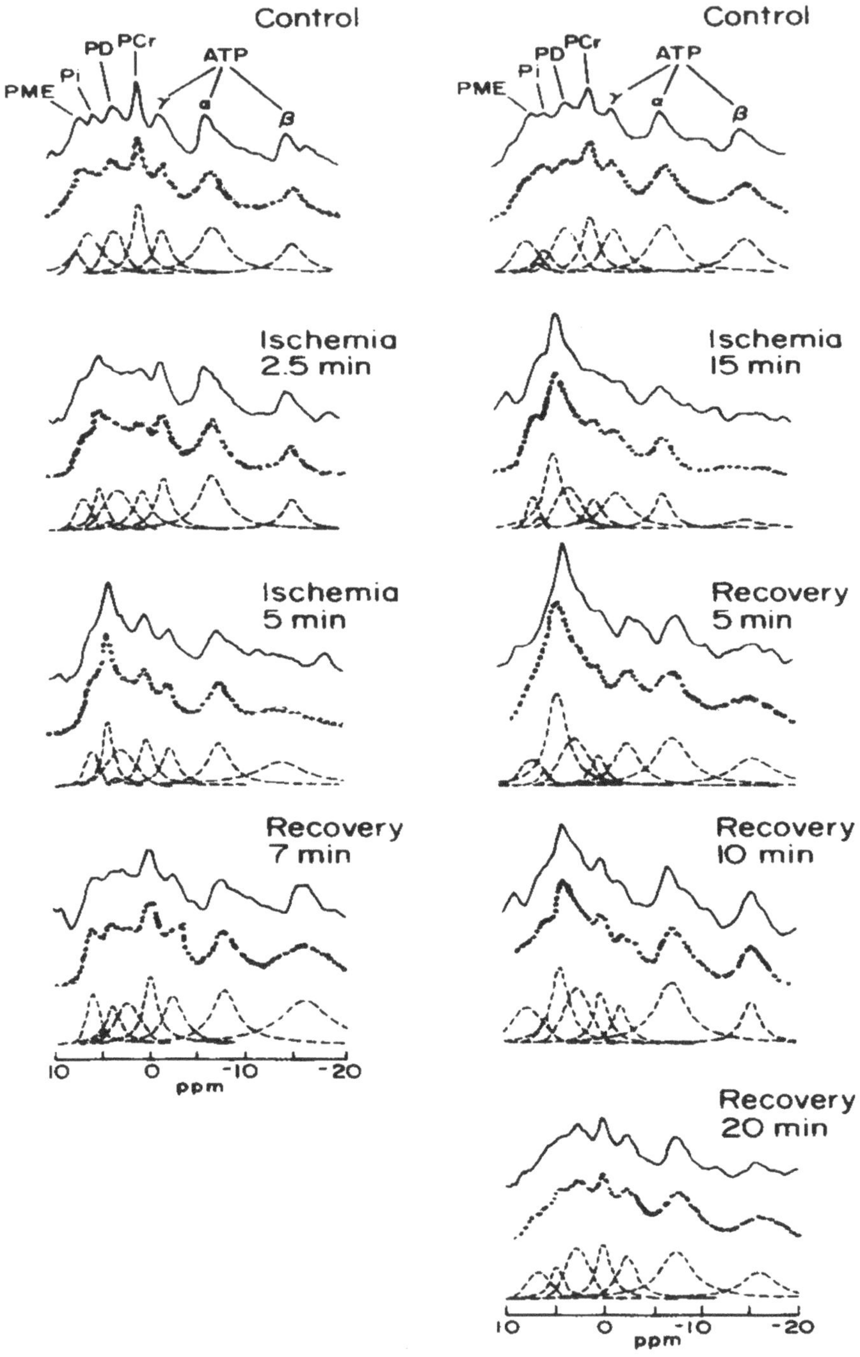


**Fig. 4.66** 31p MRS spectra and Fourier transformations of two gerbils exposed to 5 or 15 min of ischemia. Raw (top trace), fitted (middle trace), and component (bottom trace curves are shown). Peak identifications shown from right to left are three ATP resonance peaks, PCr, phosphodiester (PDE), Pi, and phosphomonoester (PME). The abscissa is frequency in ppm referenced to the PCr resonance at O ppm and the ordinate is relative intensity [86]

**Table 4.10** Relative blood flow. Ions and phosphagens concentration in normal, ischemic, and recovery conditions [86]

| | Flow | K (mM) | Ca (mM) | [Pi] | [PCr] | [ATP] | PCr/Pi |
|---|---|---|---|---|---|---|---|
| Normal | 2.7 ± 2.0 | 2.7 ± 0.5 | 1.07 ± 0.15 | 2.15 ± 0.5 | 2.9 ± 0.3 | 2.4 ± 0.1 | 1.6 ± 0.4 |
| Ischemia | 0.0 ± 0.0* | 62.0 ± 5.3* | 0.11 ± 0.07* | 9.11 ± 1.4* | 0.4 ± 0.3* | 0.5 ± 0.3* | 0.1 ± 0.0* |
| Recovery 1 | 1.5 ± 1.3 | 58.0 ± 5.2* | 0.10 ± 0.05* | 6.0 ± 0.6* | 1.2 ± 0.9* | 1.2 ± 0.8* | 0.5 ± 0.3* |
| Recovery 2 | 3.5 ± 1.5$^{6}$ | 25.8 ± 14.7$^{7,6}$ | 0.30 ± 0.20* | 2.7 ± 12$^{7,6}$ | 3.1 ± 0.8$^{7}$ | 2.2 ± 0.4$^{7,6}$ | 1.4 ± 0.8$^{7,6}$ |

Recovery 1: potassium recovery begins
Recovery 2: potassium recovers at maximum rate
Analysis of variance test: different from normal $^{*}p < 0.05$
Different from ischemia $^{6}p < 0.05$
Recovery 2 is different from Recovery 1 $^{7}p < 0.05$

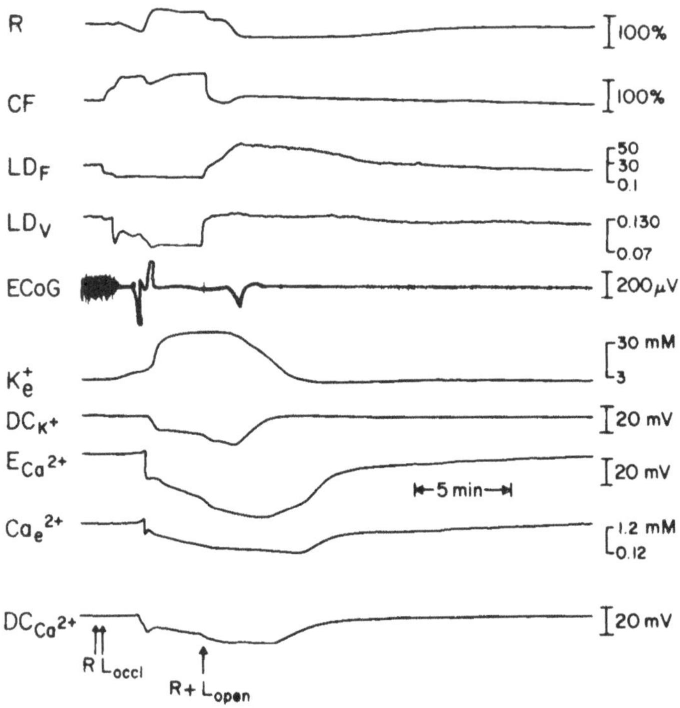


**Fig. 4.67** Effect of bilateral carotid artery occlusion (R. $L_{occl}$) on hemodynamic, metabolic, ionic, and electrical activities in the gerbil brain. R, CF, 366 nm reflectance and 450 nm corrected NADH fluorescence. LDF, LDV, cerebral blood flow and volume measured by the Laser Doppler flowmeter. ECoG, electrocorticogram. $K_e^+$, $Ca_e^{2+}$, $DCK^+$, $DCCa^{2+}$, extracellular levels of potassium, calcium, and DC potential around each of the electrodes. $ECa^{2+}$, electrical potential measured by the calcium electrode [86]

the MRS spectra in two gerbils are shown in Fig. 4.66. The changes in the PCr and ATP peaks during ischemia as well as the recovery are very clear.

A total of nine gerbils were used. However, not all the results were adequate for statistical evaluation. In some of the experiments, only part of the results was recorded due to technical difficulties. All gerbils demonstrated similar qualitative responses to ischemia. The effect of ischemia and two steps of recovery on various parameters measured during four ischemic episodes of 20-min ischemia each was examined by analysis of variance. The results are presented in Table 4.10. It can be seen that during ischemia, the levels of all parameters were significantly different from the control values.

The initial recovery period was divided into two stages according to the rate of K recovery. During the first phase of recovery (the first 5 min), the energy levels as well as the CBF demonstrated a partial return to their pre-ischemic levels. CBF and high-energy phosphate compounds did recover to the pre-ischemic level. CBF at this phase exhibited a hyperemic state as shown in Fig. 3.26b. Extracellular $K^+$ and

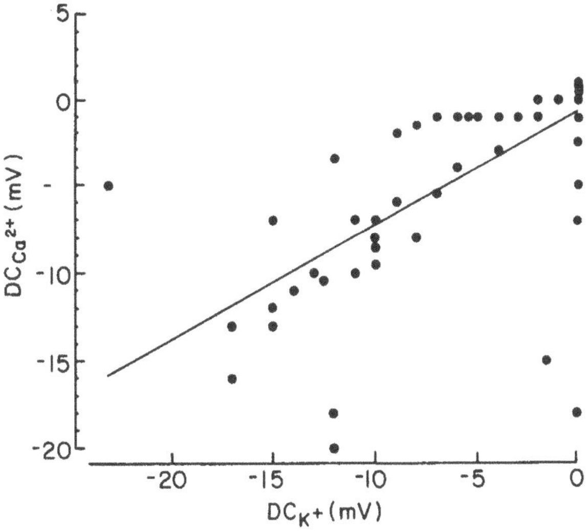


**Fig. 4.68** The correlation between the DC potential measured by the two electrodes located concentric to the $K^+$ and $Ca^{2+}$ electrodes. The baseline values in all four gerbils used were set to 0 [86]

$Ca^{2+}$ at the second recovery phase did not yet reach their pre-ischemic levels and a longer period of time was necessary for regaining normal ionic homeostasis (Fig. 4.67).

In order to test the homogeneity of the responses at various brain locations under MPA, the DC steady potential changes (negative shift during ischemia) measured in the areas of the $K^+$ and $Ca^{2+}$ electrodes were compared. Figure 4.68 demonstrates a significant correlation between the 2 DC measurements, suggesting similar responses at the two measuring sites.

The interrelation between the hemodynamic, metabolic, and ionic homeostasis was evaluated by correlation analyses. The effect of flow on the PCr/Pi or the extracellular $K^+$ levels demonstrated a threshold-like connection although linear regression demonstrated a significant correlation (Fig. 4.69a and b). A more distinct correlation was calculated between the direct measure of energy availability (PCr/Pi) and the extracellular $K^+$ representing ionic homeostasis (Fig. 4.69c).

During the ID event, an uncorrectable artifact was introduced to the NADH signal. Therefore, NADH results from various gerbils were not included in the quantitative analysis. In short, ischemia led to a quick maximal increase in NADH (as shown in Figs. 4.67 and 3.26a) followed by a fast recovery within 5–10 min after the reperfusion.

#### 4.4.5.13.2 Discussion

$^{31}$P-MRS is the sole available tool for noninvasive direct in vivo monitoring of the human brain's energy state. Several investigators have applied this technique to various experimental and clinical situations. For example, Thurlborn et al. (1982) [96] have used this method for discovering a correlation between the $^{31}$P NMR

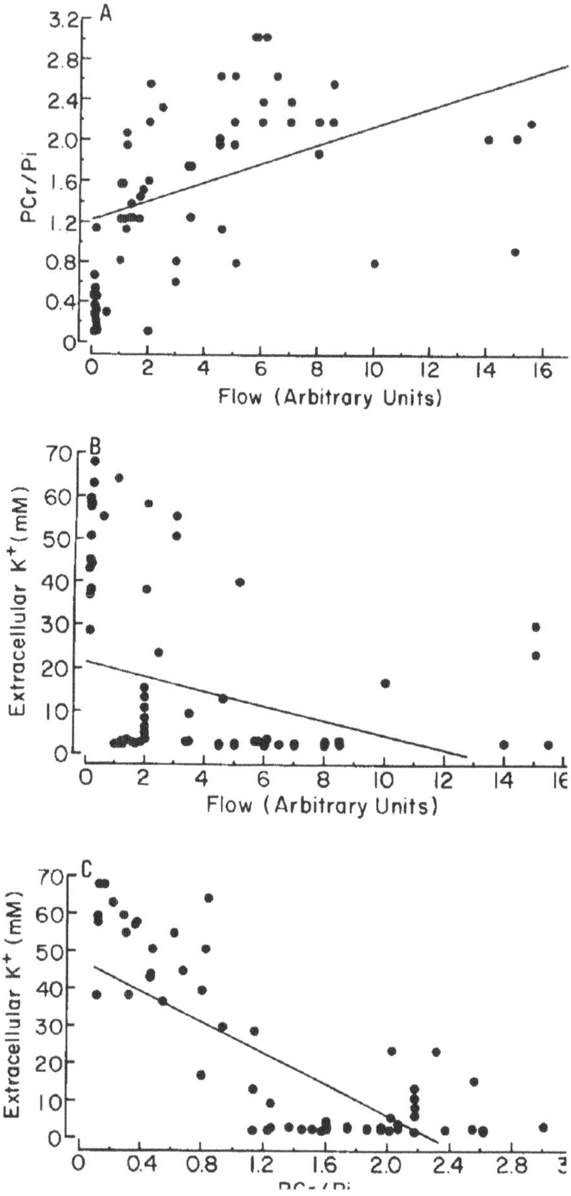


**Fig. 4.69** Correlation between CBF, PCr/Pi, and extracellular K+ measured in four gerbils exposed to ischemia in the NMR magnet. (**a**) CBF vs. PCr/Pi; (**b**) CBF vs. extracellular K+; (**c**) PCr/Pi vs. extracellular K+ [86]

spectra and edema 1 h following carotid ligation in the Mongolian gerbil, while [75] Wyatt et al. (1989) [75] and Azzopardi et al. (1989) [97] used this method for the assessment of perinatal hypoxic–ischemic brain injury. Reviews on the uses of this method have been published by Shulman et al. (1993) [98] and Prichard and Rosen (1994) [99].

However, to enable the complete interpretation of data measured by $^{31}$PMRS, correlation of these data with other physiological and biochemical parameters must be possible [75, 89, 100]. One such attempt had been made by Gyulai et al. (1988) [101] who correlated $^{31}$P-NMR and NADH fluorometry measurements in the in vivo gerbil brain, while [89] studied this correlation in newborn puppies. Since more than 50% of the energy produced in the brain is used by various active transport mechanisms (e.g., $Na^+$-$K^+$-ATPase), it is important to be able to correlate all these activities in the same animal model.

For this study, we adopted the gerbil brain, which is a convenient model for inducing cerebral ischemia, while monitoring MPA and NMR signals in real time. We have also reported the successful use of a multiprobe assembly for measuring various physiological and metabolic parameters in the in vivo brain in the gerbil [80], the rat [102], as well as intraoperatively in patients [103].

#### 4.4.5.14 The 13th Study

In this last study, included in the current chapter, I will describe a new monitoring approach developed in our laboratory and published in 2003 [104].

##### 4.4.5.14.1 Introduction

Monitoring of brain functions during various pathological situations such as neurosurgical procedures, head injury, or even in Parkinson's disease can contribute to the understanding of the clinical state and may improve the patient's final outcome. As of today, the practical monitoring techniques available include EEG monitoring during surgery, [105, 106] jugular bulb oximetry, and brain tissue oxygenation monitoring [107, 108]. Also, various techniques for CBF monitoring such as transit time flow measurements [109], transcranial Doppler ultrasonography, [107, 110] and laser Doppler flowmetry (LDF) [110–115]. As indicated, the use of LDF for CBF monitoring in clinical practice is expanding and used in several animal models as well. This includes local ischemia [116], global ischemia [117], and brain edema [118] under the treatment of various pharmacological reagents such as Flumazenil, Midazolam, and L-arginine [119, 120]. However, in most of the previous reports, CBF was recorded from the surface of the brain. Recently, Kirkeby et al. [121] showed that intracerebral CBF in pigs, using the laser Doppler flowmetry technique, yielded similar results in the CBF monitored from the brain surface and 3, 10, and 20 min inside the cerebral tissue for the investigation of microcirculation fluctuations in the pig brain under the changes in arterial $pCO_2$, arterial pressure, and cerebrospinal fluid pressure [122]. Kurosawa et al. used the LDF for striatal blood flow monitoring in an ischemic model of rats and found to be a reliable index for CBF parameter [123]. In another study, Cao et al. measured local hippocampal CBF in rats under focal stimulation of the septal complex [124]. As pointed out, previous cerebral blood flow in patients in the intensive care unit (ICU), thus its use, may improve the management of comatose patients and also open avenues for a better understanding of the cerebral circulation in pathological states [125, 126]. The use of the LDF for CBF monitoring has the advantage of continuous recording of relative changes in microcirculatory flow in real time. In addition, the technique has shown good correlation with the radioactive microspheres and $H_2$ clearance. However, the LDF has also several limitations and disadvantages. The signal may be unrepresentative due to the small volume of tissue from which the signal is reflected, movement artifacts may rise, the signal is disturbed by external light, as well as if the probe is located close to a large vessel. Therefore, the use of another technique for the evaluation of tissue metabolic activity, like the mitochondrial NADH level, has a significant advantage.

Since the brain has a high $O_2$ consumption and a small reserve of dissolved $O_2$, it depends on continuous oxygen (carried by the blood). More than 50% of energy consumed by the brain is utilized by the active transport mechanisms of ions such as $Na^+$-$K^+$-ATPase [47]. Hence, a direct coupling between the extracellular levels of various ions (such as $K^+$, $Ca^{2+}$, $Na^+$, or $H^+$) and the metabolic activity of the mitochondria reflected in the oxidation reduction state of the respiratory chain enzymes such as NADH [127]. A decrease in $O_2$ supply to the brain will decrease the ATP levels and inhibit active transport mechanisms; the extracellular $K^+$ level will be elevated until the restoration of normal energy supply. As was previously indicated, the redox state of the mitochondria is a sensitive indicator of the intracellular metabolic state and can be used for the evaluation of cellular energy status [30, 128, 129]. Although there are many studies in which NADH was monitored in real time from the surface of the brain in an animal model, there are only a few published studies in which NADH was monitored in intracerebral locations. Mottin et al. monitored NADH from the cortex and the periaqueductal gray matter as well as from the nucleus raphe dorsalis in rats throughout the sleep–wake cycle [130]. In the present study, we used the MPA, a combined probe that enables the monitoring of CBF and mitochondrial NADH in the brain cortex and at intracerebral locations under two models of oxygen deficiency. The aim of this study was to show that intracerebral monitoring of these two parameters is possible and are as reliable as cortically CBF and NADH monitoring. Therefore, the MPA may be used at various situations where intracerebral monitoring is needed, as in Parkinson's disease or during different neurological procedures, providing the physician with important information on the activity of the brain.

#### 4.4.5.14.2 Methods

**The MPA Device**

Monitoring of brain functions during anoxia and complete ischemia was performed by MPA developed in our laboratory [131–133]. MPA includes a bundle of optical fibers for NADH redox state measurement by surface fluorometry as well as for cerebral blood flow (CBF) using laser Doppler flowmetry (LDF). The principle of NADH monitoring from the surface of the brain is that excitation light (366 nm) is passed from the fluorometer to the brain via a bundle of optical fibers made of quartz. The emitted light (450 nm) together with the reflected light at the excitation wavelength (366 nm) is transferred to the fluorometer via another bundle of fibers. To measure the CBF in real time from the same cortical area, we applied the laser Doppler flowmetry [118, 134, 135]. Probe A, used for brain surface monitoring, was 2 mm in diameter, while probe B used for intracortical monitoring was 1 mm in diameter. All of the signals during each experiment were transferred to a multichannel computerized data acquisition and recording system (Labview A/D software, National Instruments Co. USA) in addition to a Grass polygraph for further analysis.

**Animal Preparation**

Adult male gerbils (50–70 g) and Wistar male rats (250–300 g) were anesthetized by i.p. injection of 0.3 ml/100 g Equithesin (each ml contains pentobarbital 9.72 mg; chloral hydrate 42.51 mg; magnesium sulfate 21.25 mg; propylene glycol 44.34% w/v; alcohol 11.5%; and water). In the gerbil model, the common carotid arteries were exposed and prepared for later ligation using special clips. The animal was placed in a head holder and a midline incision of the skin was made exposing the skull. An appropriate hole was drilled in the parietal bone for the fixation of MPA. In the gerbil model, a cannula was fixated to the brain surface and the probe was held in this cannula to enable monitoring from different depths in the brain following complete ischemia. During the experiment, the probe was lowered to different depths and a session of ischemia was induced. In the rat model, a special cannula was prepared in which probe A, which was placed on the brain surface, was fixed and probe B was placed in such a way that enabled brain surface monitoring in the first anoxic episode followed with probe penetration to the brain cortex (5 mm posterior to the bregma and 3 mm in depth posterior to probe A) later on. In both models, the cannula was placed on the brain surface using a micromanipulator and fixation was made using dental acrylic cement.

**Animal Protocols**

All experiment protocols were approved by the Animal Care and Use Committee of the Bar-Ilan University. Thirty minutes after surgery, bilateral occlusion of the common arteries was induced and monitoring was proceeded from the brain surface. In the other experiments, the monitoring probe was inserted to the proper depth and completed ischemia was induced later. In the rat model, two episodes of short anoxia (30 s) were induced. During the first anoxia, both probes monitored the brain cortex. Later on, probe B was lowered into the brain to a depth of 3 mm and another anoxia was induced 1 h later.

**Data Analysis**

Since NADH and CBF measurements are calculated in arbitrary units, we calibrated them as follows. NADH was calculated as a percent change of baseline. CBF was calculated as percent change of baseline where CBF was considered 0% at death and CBF level before treatment was considered 100%.

Statistical analysis was carried out using the Student's paired two-tailed t-test. Changes were considered significant when $p < 0.05$.

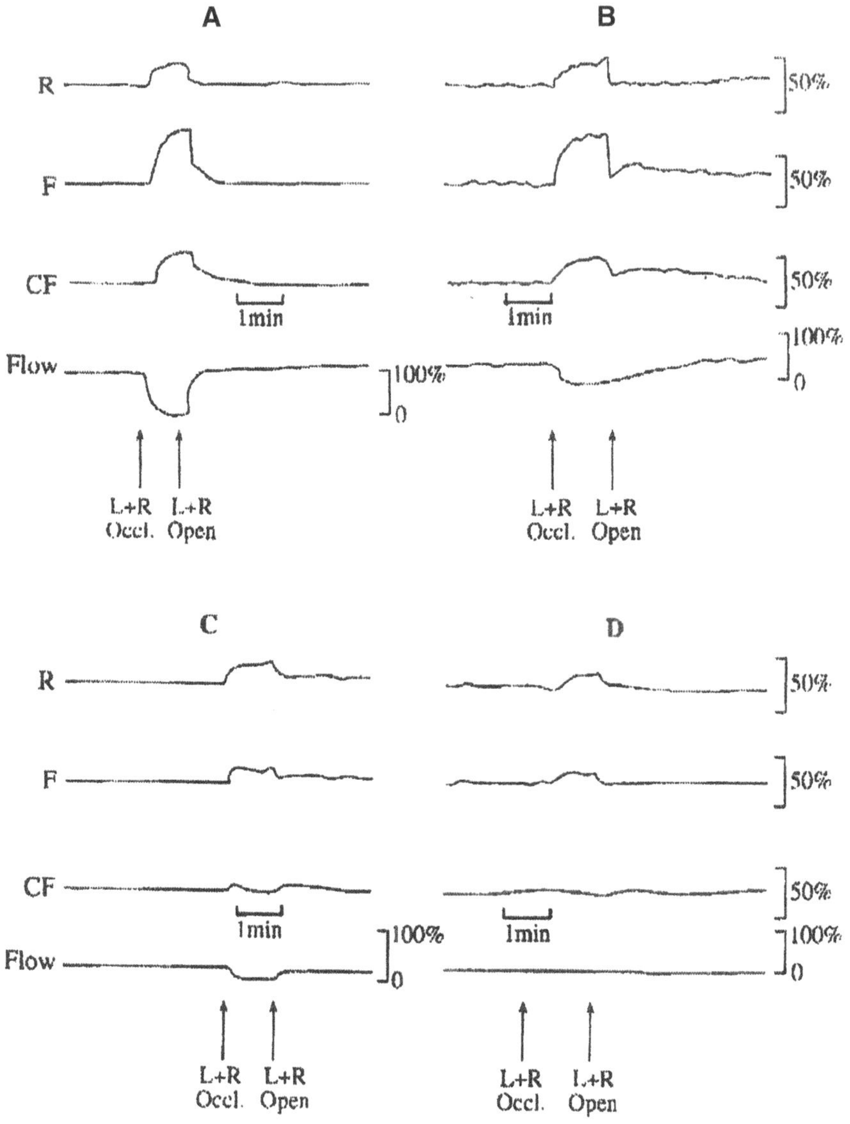


**Fig. 4.70** Typical results of an experiment in which two carotid arteries were occluded in a gerbil model. Monitoring was performed on the brain surface (**a**), and intracortically measurements in the depth of 0.5 mm (**b**), 1 mm (**c**), and 1.5 mm (**d**). Parameters monitored were Reflectance (R), Fluorescence (F), NADH (CF), and CBF (Flow) [104]

#### 4.4.5.14.3 Results

The results of a typical experiment in which both carotid arteries were ligated in a gerbil model are presented in Fig. 4.70. Figure 4.70a–d shows the response of the

**Fig. 4.71** Mean ± SE values of the % of change in CBF and NADH following both carotid arteries occlusion in a gerbil model. Monitoring was performed in three locations: on brain surface (**A**) and intracortically at 0.5 mm (**B**) and at 1 mm (**C**) [104]

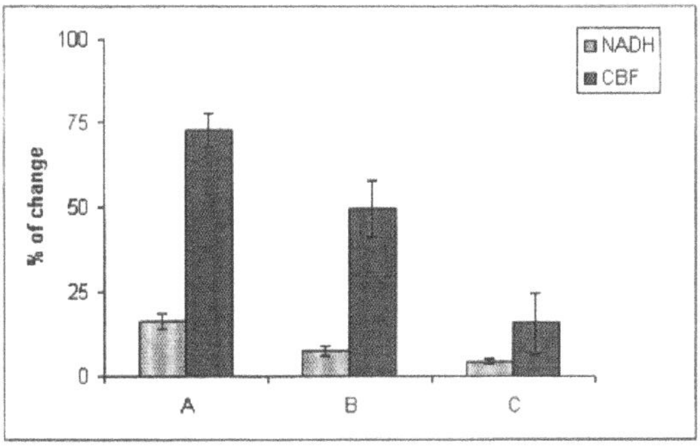


| | A / B | A / C | B / C |
|---|---|---|---|
| NADH | P<0.01 | P<0.001 | Not significant |
| CBF | P<0.05 | P<0.001 | P<0.05 |

reflectance (R), fluorescence (F), NADH (CF), and CBF measured cortically and intracortically at a depth of 0.5–1.5 mm. The occlusion of both carotid arteries caused a dramatic decrease in CBF to very low levels (near 0%), followed by an increase in the reflectance, fluorescence, and NADH levels. When both carotid arteries were reopened, CBF returned to its basal level followed by a full recovery of all NADH parameters. When the probe was inserted 0.5 mm intracortically, the initial CBF dropped to around 50% and CBF response to carotid occlusion was smaller than when it was measured from the brain surface. The changes in NADH parameters were similar to those shown in Fig. 4.70a. In Fig. 4.70c, measurements were from 1 mm in depth. The initial CBF was around 20% and CBF responses to the ischemic episode were minor. The changes in R, F, and NADH were 25%, 20%, and 10% respectively. Figure 4.70d presents measurements in 1.5 mm depth. Initial CBF was very low (near 0%) and no further decrease was observed following carotid occlusion. The changes in NADH parameters were also minor with R and F increase of 30–25% and a change of 5% in CF level.

Figure 4.71 presents mean (±SE) changes in the CBF and NADH levels following carotid arteries occlusion measured cortically (A), intracortically at 0.5 mm (B), and intracortically at 1 mm (C). When measured on the brain surface (A), CBF at the depth of 1 mm (C) decreased by 16 ± 9%, significantly lower than CBF at location A and at location B ($p < 0.001$ and $p < 0.05$, respectively). NADH levels measured at locations A, B, and C were 16 ± 12%, 7 ± 1%, and 4 ± 0.6%, respectively. The t-test analysis showed that the increase in NADH levels measured in location A was significantly lower than in location B ($p < 0.01$) as well as in location C ($p < 0.001$); however, no significant differences were observed between NADH levels at locations B and C.

The changes in CBF in the rat model are presented in Fig. 4.72. When CBF was monitored cortically (A), under a short anoxia, a decrease in CBF developed following an increase in CBF to hyperemic levels (nearly 350% of the basal level) which later decreased and reached the basal level. When the probe was inserted into the

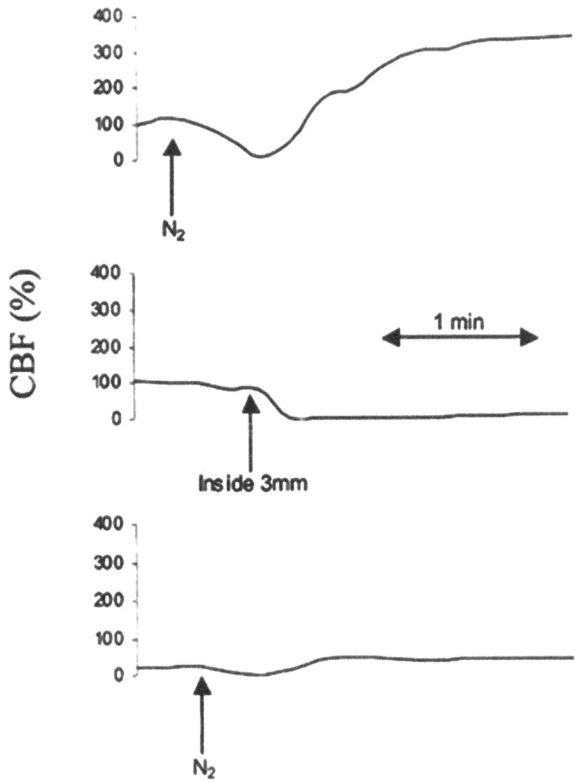


**Fig. 4.72** The changes in CBF in a rat model using probe B following two episodes of short anoxia ($N_2$) and the effect of probe insertion 3 mm inside the rat brain [104]

brain to the depth of 3 mm (30 min after anoxia), CBF decreased dramatically, reaching the level of 2% followed by a recovery to the level of 14%. One hour later, CBF was 24%, and a second anoxia caused a decrease to the level of 2% following a recovery to hyperemic levels of 44%.

Figure 4.73 presents average results of the responses to short anoxia in rats. Anoxia of 30 s was introduced twice. In the first anoxia, both of the probes (A and B) were placed on the surface of the cortex, whereas in the second anoxia, probe A remained on the surface while probe B was introduced 3 mm into the brain; thus, monitoring was intracortical. The left column presents the changes in probe A following two treatments of anoxia and the right column presents the measurements in probe B which were on the surface in the first anoxia ($N_2$–1) and intracortically in the second anoxia ($N_2$–2). Anoxia at probe A caused a significant decrease in CBF to the level of 44 ± 15% and 45 ± 11% at $N_2$–1 and $N_2$–2, respectively, followed by a dramatic increase in CBF to hyperemic levels of 267 ± 43% and 185 ± 41% at $N_2$–1 and $N_2$–2, respectively. These hyperemic levels of CBF returned to their basal levels within an hour (data not shown). In probe B, anoxia also caused a decrease in CBF of 21 ± 9% and 65 ± 32% in $N_2$–1 and $N_2$–2, respectively, followed by hyperemia to the levels of 265 ± 46% and 176 ± 50%, respectively. The reflectance parameter showed no significant changes through the entire experiment in both

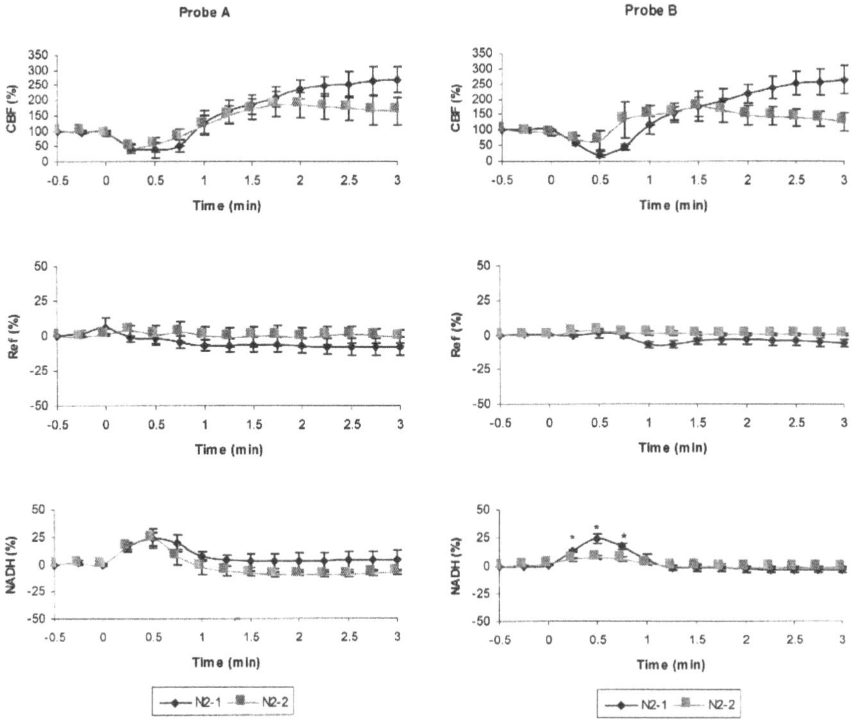


**Fig. 4.73** Mean ± SE values of CBF, Reflectance (Ref), and NADH in a rat model following two episodes of anoxia ($N_2$–1, $N_2$–2-). Probe A was placed on the brain surface during $N_2$–1 and $N_2$–2, while probe B was located on the brain surface in $N_2$–1 and 3 mm inside the brain at $N_2$–2 (*)—$p < 0.05$ [104]

probes and under both treatments. Anoxia caused a significant increase of 24 ± 9% and 23 ± 6% in NADH levels in probe A ($N_2$–1 and $N_2$–2), respectively. A similar increase in NADH was measured in probe B at $N_2$–1 (14 ± 5%); however, a significantly smaller increase was observed in probe B at $N_2$–2 ($p < 0.05$).

#### 4.4.5.14.4 Discussion

The use of the LDF and CBF monitoring in experimental animals as well as in patients showed that this method reflects relative changes in microcirculatory blood flow. Monitoring of mitochondrial function using NADH fluorometry provides information on the balance between oxygen supply and demand in the brain tissue. Combination of these two techniques in a single probe enables real-time monitoring of the hemodynamic and metabolic state of the brain. As discussed in the present study, the MPA was located both on the brain cortex and in intracerebral locations. In the gerbil as well as in the rat model, as the probe was inserted deeper into the brain, the CBF signal was smaller and there were several experiments in which probe penetration yielded a diminished LDF signal. Similar results were presented in another study performed in a gerbil model where probe insertion caused a uniform reduction of flow in the whole hemisphere [136]. In addition, in a pig study, probe insertion caused the loss of CBF measurement [122]. Regarding the question whether probes or electrodes insertion to the brain produce serious damage to the tissue is difficult to evaluate. In Kirkeby's study, no microscopic examination of the tissue around the probes was made; however, other studies that used histology around $H_2$ electrodes have not revealed any significant damage [136, 137]. In another study, Verhaegen et al., who used several electrodes for $H_2$ clearance measurements at different diameters, showed that electrodes >0.25 mm in diameter, in rat brain, causes a low flow reading in the first 60 min post electrode insertion [138]. In our study, the lower CBF signal measured immediately after the probe insertion showed a partial recovery in time. Previously, some possible reasons for a decrease in CBF in intracerebral location were given. First, this includes the direct damage to the tissue and second, probe insertion may affect the local vasoreactivity response [122]. However, we believe that although the insertion of the probe may cause damage to the brain tissue, the tissue monitored below the probe is intact. It may also be that in these locations microcirculation is indeed lower, therefore oxygen supply is smaller, the basal NADH level is elevated, and thus anoxia caused a smaller increase in NADH level. Our results showed high correlation between CBF and NADH in both probes. In probe B, anoxia led to a smaller decrease in CBF as compared to the measurements in probe A since the basal CBF was already reduced. As a result, no significant increase in NADH was observed. Likewise, in the gerbil model, as the probe was inserted deeper, the responses of CBF and NADH were smaller. As demonstrated, the advantage of NADH monitoring concomitantly with CBF monitoring enables better evaluation of the hemodynamic and metabolic state of the brain tissue. Intracortical NADH monitoring has a significant advantage in Parkinson's patients as well as in major depression patients since there are evidence that mitochondrial electron transport abnormalities have been implicated in several degenerative diseases of the nervous system, including Parkinson disease [139–143]. In addition, it was observed that peripherally given NADH in Parkinson's patients and in major depression patients improved their state [144]. Therefore, in his article, Rex emphasized the importance of bedside monitoring of NADH inside the brain in such patients [144]. In addition, a previous study in our laboratory showed stable measurements of NADH in the striatum and the substantia nigra in a rat model for a period of up to 10 days [145]. In the present study, we demonstrate for the first time simultaneous monitoring of CBF and NADH in intracortical location in an animal model. Our results show that the use of the MPA inside the cortex can shed light on the hemodynamic and energetic state in deeper layers of the brain; thus, it could provide significant information in cases like Parkinson's disease and various illness of the central nervous system.

## References


1. Mayevsky A, Barbiro-Michaely E (2013) Shedding light on mitochondrial function by real time monitoring of NADH fluorescence: I. Basic methodology and animal studies. J Clin Monit Comp 27:1–34
2. Chance B, Schoener B (1962) Correlation of oxidation-reduction changes of intracellular reduced pyridine nucleotide and changes in electro-encephalogram of the rat in anoxia. Nature 195:956–958
3. Chance B, Schoener B, Schindler F (1963) In: Dickens F, Neil E (eds) The intracellular oxidation-reduction state. Pergamon Press, London, pp 367–392
4. Chance B, Cohen P, Jobsis F, Schoener B (1962) Intracellular oxidation-reduction states *in vivo*. Science 137:499–508
5. Chance B, Cohen P, Jobsis F, Schoener B (1962) Localized fluorometry of oxidation-reduction states of intracellular pyridine nucleotide in brain and kidney cortex of the anesthetized rat. Science 136:325
6. Mayevsky A, Crowe W, Mela L (1980) The interrelation between brain oxidative metabolism and extracellular potassium in the unanesthetized gerbil. Neurol Res 1:213–226
7. Mayevsky A, Lebourdais S, Chance B (1980) The interrelation between brain $PO_2$ and NADH oxidation- reduction state in the gerbil. J Neurosci Res 5:173–182
8. Friedli CM, Sclarsky DS, Mayevsky A (1982) Multiprobe monitoring of ionic, metabolic, and electrical activities in the awake brain. Am J Phys 243:R462–R469
9. Mayevsky A, Friedli CM, Reivich M (1985) Metabolic, ionic and electrical responses of the gerbil brain to ischemia. Am J Phys 248:R99–R107
10. Mayevsky A, Zarchin N (1988) In: Tomita M, Sawada T, Naritomi H, Heiss WD (eds) Microcirculatory and ionic events during and after incomplete or complete cerebral ischemia in the Mongolian gerbil. Excerpta Medica, pp 157–169
11. Mayevsky A, Zarchin N (1988) Microcirculatory and ionic events during and after incomplete or complete cerebral ischemia in the Mongolian gerbil. In: Tomita M, Sawada T, Naritomi H, Heiss WD (eds) Cerebral hyperemia and ischemia. Excerpta Medica, pp 157–169
12. Mayevsky A (1993) Biochemical and physiological activities of the brain as *in vivo* markers of brain pathology. In: Bernstein EF, Callow AD, Nicolaides AN, Shifrin EG (eds) Cerebral Revascularization. Med-Orion Pub, pp 51–69
13. Mayevsky A, Frank KH, Nioka S, Kessler M, Chance B (1990) Oxygen supply and brain function *in vivo*: a multiparametric monitoring approach in the Mongolian gerbil. In: Piper J, Goldstick TK, Meyer M (eds) Oxygen transport to tissue XII. Plenum Pub, pp 303–313
14. Mayevsky A, Frank K, Muck M, Nioka S, Kessler M, Chance B (1992) Multiparametric evaluation of brain functions in the Mongolian gerbil *in vivo*. J Basic Clin Physiol Pharmacol 3:323–342
15. Mayevsky A (1992) Cerebral blood flow and brain mitochondrial redox state responses to various perturbations in gerbils. Adv Exp Med Biol 317:707–716
16. Mayevsky A, Cohen S (1990) Brain ischemic depolarization and vasospasm in the Mongolian gerbil: the dependence on energy depletion levels. Adv Exp Med Biol 277:315–322
17. Mayevsky A, Zarchin N, Tannenbaum B (1984) Brain responses to experimental oxygen deficiency in the Mongolian gerbil. Adv Exp Med Biol 180:191–202
18. Mayevsky A (1990) Level of ischemia and brain functions in the Mongolian gerbil *in vivo*. Brain Res 524:1–9
19. Mayevsky A (1978) Pyridine nucleotide oxidation reduction state of the cerebral cortex in the awake gerbil. J Neurosci Res 3:369–374
20. Mayevsky A (1978) Ischemia in the brain: the effects of carotid artery ligation and decapitation on the energy state of the awake and anesthetized rat. Brain Res 140:217–230
21. Mayevsky A (1976) Brain energy metabolism of the conscious rat exposed to various physiological and pathological situations. Brain Res 113:327–338

22. Mayevsky A, Chance B (1983) Multisite measurements of NADH redox state from cerebral cortex of the awake animal. Adv Exp Med Biol 159:143–155
23. Levine S, Payan H (1966) Effects of ischemia and other procedures on the brain and retina of the gerbil (*Meriones unguiculatus*). Exp Neurol 16:255–262
24. Levine S, Sohn D (1969) Cerebral ischemia in infant and adult gerbils. Relation to incomplete circle of Willis. Arch Pathol 87:315–317
25. Mayevsky A, Bar-Sagie D, Levy N, Amosi M (1978) The effects of ischemia on the metabolic response of the cerebral cortex of the gerbil. (Abstract). Fed Proc 37:97
26. Crowe W, Mayevsky A, Mela L (1979) The dynamics of $K^+$ leakage and recovery in cerebral ischemia. In: Leter A, Saba T, Mela L (eds) Advances in shock research. Alan R. Liss, New York, pp 221–232
27. Mayevsky A, Chance B (1975) Metabolic responses of the awake cerebral cortex to anoxia hypoxia spreading depression and epileptiform activity. Brain Res 98:149–165
28. Jobsis FF, O'Connor M, Vitale A, Vreman H (1971) Intracellular redox changes in functioning cerebral cortex. I. Metabolic effects of epileptiform activity. J Neurophysiol 3465:735–749
29. Mayevsky A, Zarchin N (1987) Metabolic ionic and electrical activities during and after incomplete or complete cerebral ischemia in the Mongolian gerbil. Adv Exp Med Biol 215:265–273
30. Mayevsky A (1984) Brain NADH redox state monitored *in vivo* by fiber optic surface fluorometry. Brain Res Rev 7:49–68
31. Mayevsky A (1983) Multiparameter monitoring of the awake brain under hyperbaric oxygenation. J Appl Physiol 54:740–748
32. Mayevsky A (1978) Shedding light on the awake brain. In: Dutton PL, Leigh J, Scarpa A (eds) Frontiers in bioenrgetics: from elrctrons to tissues. Academic Press, New York, pp 1467–1476
33. Zarchin N, Mayevsky A (1981) The effects of age on the metabolic and electrical responses to decapitation in the awake and anesthetized rat brain. Mech Ageing Dev 16:285–294
34. Levy DE, Brierley JB (1974) Communications between vertebro-basilar and carotid arterial circulations in the gerbil. Exp Neurol 45:503–508
35. Ginsberg MD, Mela L, Wrobel-Kuhl K, Reivich M (1977) Mitochondrial metabolism following bilateral cerebral ischemia in the gerbil. Ann Neurol 1:519–527
36. Loskota WJ, Lomax P (1975) The Mongolian gerbil (*Meriones unguiculatus*) as a model for the study of the epilepsies: EEG records of seizures. Electroencephalogr Clin Neurophysiol 38:597–604
37. Robinson DG (1980) The gerbil stroke model. Gerbil Digest 7
38. Kobayashi M, Lust WD, Passonneau JV (1977) Concentrations of energy metabolites and cyclic nucleotides during and after bilateral ischemia in the gerbil cerebral cortex. J Neurochem 29:53–59
39. Dora E, Zeuthen T (1976) Brain metabolism and ion movements in the brain cortex of the rat during anoxia. In: Kessler M (ed) Ion and enzyme electrodes in biology and medicine. University.Park Press, Baltimore, pp 294–298
40. Vyskocil F, Kritz N, Bures J (1972) Potassium-selective microelectrodes used for measuring the extracellular brain potassium during spreading depression and anoxic depolarization in rats. Brain Res 39:255–259
41. Hansen AJ (1985) Effect of anoxia on ion distribution in the brain. Physiol Rev 65:101–148
42. Siesjo BK, Bengtsson F (1989) Calcium fluxes, calcium antagonists, and calcium-related pathology in brain ischemia, hypoglycemia, and spreading depression: a unifying hypothesis. J Cereb Blood Flow Metab 9:127–140
43. Mayevsky A, Sclarsky DS (1983) Correlation of brain NADH redox state, $K^+$, $PO_2$ and electrical activity during hypoxia, ischemia and spreading depression. Adv Exp Med Biol 159:129–141

44. Zhu X, Zuo L, Cardounel AJ, Zweier JL, He G (2007) Characterization of in vivo tissue redox status, oxygenation, and formation of reactive oxygen species in postischemic myocardium. Antioxid Redox Signal 9:447–455
45. Mayevsky A, Chance B (1982) Intracellular oxidation-reduction state measured *in situ* by a multicannel fiber-optic surface fluorometer. Science 217:537–540
46. Crowe W, Mayevsky A, Mela L (1977) Application of a solid membrane ion selective electrode to *in vivo* measurements. Am J Phys 233:C56–C60
47. Erecinska M, Silver IA (1989) ATP and brain function. J Cereb Blood Flow Metab 9:2–19
48. Hansen AJ, Zeuthen T (1981) Extracellular ion concentrations during spreading depression and ischemia in the rat brain cortex. Acta Physiol Scand 113:437–445
49. Hossmann KA, Sakaki S, Zimmerman V (1977) Cation activities in reversible ischemia of the cat brain. Stroke 8:77–81
50. Heuser D, Guggenberger H (1985) Ionic changes in brain ischaemia and alterations produced by drugs. Br J Anaesth 57:23–33
51. Cohen S, Mayevsky A (1989) Effects of nimodipine on the responses to cerebral ischemia in the Mongolian gerbil. Adv Exp Med Biol 248:429–438
52. Dierking H, Tegtmeier E, Holler M, Peters TL (1987) Effects of flunarizine, nimodipine, and diphenylhydantoine on K, Ca and DC in the flunarizine, nimodipine, and diphenylhydantoine on hypoxic cortex of the rat brain *in vivo*. J Cereb Blood Flow Metab 7:S 156
53. Mori K, Iwayama K, Kawano T, Kaminogo M (1978) DC potential and extracellular $K^+$ and $Ca^{2+}$ at critical levels of brain ischemia in cats. J Cereb Blood Flow Metab 7:s112
54. Kato H, Nakano S, Kogure K (1989) Repeated innocent ischemia to the brain can injure the selectively vulnerable neurons in the gerbil and the rat. J Cereb Blood Flow Metabol 9:S642
55. Mayevsky A, Breuer Z (1990) In: Schurr A, Rigor BM (eds) The Mongolian gerbil as a model for cerebral ischemia. CRC Press, pp 27–46
56. Dora E, Zeuthen T, Silver IA, Kovach AGB (1979) Effect of arterial hypoxia on the cerebrocortical redox state, vascular volume, oxygen tension, electrical activity and potassium ion concentration. Acta Physiol Acad Sci Hung 54:319–331
57. LaManna JC, Cordingley G, Rosenthal M (1977) Phenobarbital actions *in vivo*: effects on extra cellular potassium activity and oxidative metabolism in cat cerebral cortex. J Pharmacol Exp Ther 200:560–569
58. Leniger-Follert E, Urbanics R, Harbig K, Lubbers DW (1977) In: Ingvar DH, Lassen NA (eds) The behavior of local pH and NADH-fluorescence during and after direct activation of the brain cortex. Acta Neurologica Scandinavica, Munkgaard, Copenhagen, pp 214–215
59. Lewis DV, Schuette WH (1975) NADH fluorescence and $[K^+]_o$ changes during hippocampal electrical stimulation. J Neurophysiol 38:405–417
60. Lothman E, LaManna J, Cordingley G, Rosenthal M, Somjen G (1975) Responses of electrical potential potassium levels, and oxidative metabolic activity of the cerebral neocortex of cats. Brain Res 88:15–36
61. Astrup J, Symon L, Branston NM, Lassen NA (1977) Cortical evoked potential and extracellular K+ and H+ at critical levels of brain ischemia. Stroke 8:51–57
62. Lehmenkuhler A, Speckmann EJ, Garpers H (1976) Cortical spreading depression in relation to potassium activity, oxygen tension, local flow and carbon dioxide tension. In: Kessler M, Clark LC, Lubbers DW, Silver IA, Simon W (eds) Ion and enzyme electrodes in biology and medicine. Urban & Schwarzenberg, Munich, pp 311–315
63. Hansen AJ, Olsen CE (1980) Brain extracellular space during spreading depression and ischemia. Acta Physiol Scand 108:355–365
64. Mayevsky A, Zeuthen T, Chance B (1974) Measurements of extracellular potassium, ECoG and pyridine nucleotide levels during cortical spreading depression in rats. Brain Res 76:347–349
65. Mayevsky A, Zarchin N, Friedli CM (1982) Factors affecting the oxygen balance in the awake cerebral cortex exposed to spreading depression. Brain Res 236:93–105

66. Urbanics R, Leniger-Follert E, Lubbers DW (1978) Time course of changes of extracellular H+ and K+ activities during and after direct electrical stimulation of the brain cortex. Pflugers Arch 378:47–53
67. Somjen GG, Rosenthal M, Cordingley G, LaManna J, Lothman E (1976) Potassium, neuroglia, and oxidative metabolism in central gray matter. Fed Proc 35:1266–1271
68. Tschirgi RD, Inanaga K, Taylor JL, Walker RM, Sonnenschein RR (1957) Changes in cortical pH and blood flow accompanying spreading cortical depression and convulsion. Am J Phys 190:557–562
69. Crockard A, Iannotti F, Hunstock AT, Smith RD, Harris RJ, Symon L (1980) Cerebral blood flow and edema following carotid occlusion in the gerbil. Stroke 11:494–498
70. Thomas S, Tabibnia F, Schuhmann MU, Brinker T, Samii M (2000) ICP and MABP following traumatic subarachnoid hemorrhage in the rat. Acta Neurochir Suppl 76:203–205
71. Kelly RA, Halsey JH (1976) Comparison of local blood flow and oxygen availability at the same locus in the ischemic gerbil brain. Stroke 7:274
72. Ito U, Ohno K, Yamaguchi T, Tomita H, Inaba Y, Kashima M (1980) Transient appearance of "no-reflow" phenomenon in Mongolian gerbils. Stroke 11:517–521
73. Siesjo BK (1978) Ischemia (studies in gerbils). In: Siesjo BK (ed) Brain energy metabolism. Wiley, New York, pp 497–498
74. Welsh FA, O'Connor MJ, Marcy VR, Spatacco AJ, Johns RL (1982) Factors limiting regeneration of ATP following temporary ischemia in cat brain. Stroke 13:234–242
75. Harris RJ, Symon L, Branston NM, Bayhan M (1981) Changes in extracellular calcium activity in cerebral ischaemia. J Cereb Blood Flow Metab 1:203–209
76. Mayevsky A, Chance B (1976) The effect of decapitation on the oxidation-reduction state of NADH and ECoG in the brain of the awake rat. Adv Exp Med Biol 75:307–312
77. Marshall WH (1959) Spreading cortical depression of Leão. Physiol Rev 39:239–279
78. Leão AAP (1947) Further observations on the spreading depression of activity in the cerebral cortex. J Neurophysiol 10:409–414
79. Mayevsky A, Zarchin N (1981) The effects of unilateral carotid occlusion on the responses to decapitation in the gerbil brain. Brain Res 206:155–160
80. Mayevsky A, Yoles E, Zarchin N, Kaushansky D (1990) Brain vascular ionic and metabolic responses to ischemia in the Mongolian gerbil. J Basic Clin Physiol Pharmacol 1:207–220
81. Mayevsky A (1992) Interrelation between intracellular redox state and ion homeostasis in the brain *in vivo*. In: Frank K, Kessler M (eds) Quantitative spectroscopy in tissues. Verlasgruppe GmbH, pp 155–168
82. Frank KH, Schabert A, Friedl A, Brunner M, Hoper J, Kerl G, Kessler M (1984) Correlation between tissue $PO_2$ and intracapillary Hb spectra. Adv Exp Med Biol 169:811–819
83. Frank KH, Rettig V, Friedl A, Brunner M, Ellermann R, Kerl G, Hoper J (1984) Measurements of intracapillary haemoglobia spectra in the beating heart, skeletal muscle and the liver, using the Erlangen micro-lightguide spectrophotometer. Pfluger Archiv 400(Suppl):219
84. Frank KH, Kessler M, Appelbaum K, Dummler W (1989) The Erlangen micro-lightguide spectrophotometer EMPHO I. Phys Med Biol 34:1883–1900
85. Mayevsky A, Zarchin N, Sonn J (1993) Brain redox state and $O_2$ balance in experimental spreading depression and ischemia. In: Lehmenkuhler A, Grotemeyer K-H, Tegtmeier F (eds) Migraine - basic mechanisms and treatment. Urban & Schwarzenberg, Munchen-Wier, Baltimore, pp 379–393
86. Mayevsky A, Nioka S, Wang DJ, Chance B (1997) The functioning gerbil brain *in vivo*. Correlation between 31P NMR spectroscopy and the multiparametric monitoring approach. Adv Exp Med Biol 411:41–53
87. Meilin S, Zarchin N, Mayevsky A (1999) Inter-relation between hemodynamic, metabolic, ionic and electrical activities during ischemia and reperfusion in the gerbil brain. Neurol Res 21:699–704

88. Turkozkan N, Aykol S, Bilgihan A, Yavuz O, Cayci B, Dogulu F (1996) The effect of 2-chloroadenosine on the ATP level Na,K ATPase activity in experimental brain ischemia of gerbil. Gen Pharmacol 27:165–166
89. Mayevsky A, Subramanian VH, Chance B (1988) Brain oxidative metabolism of the newborn dog: correlation between $^{31}$P NMR spectroscopy and pyridine nucleotides redox state. J CBF Metab 8:201–207
90. Strosznajder J, Chalimoniuk M (1996) Biphasic enhancement of nitric oxide synthase activity and cGMP level following brain ischemia in gerbils. Acta Neurobiol Exp (Wars) 56:71–81
91. Nagafuji T, Sugiyama M, Matsui T, Muto A, Naito S (1995) Nitric oxide synthase in cerebral ischemia. Possible contribution of nitric oxide synthase activation in brain microvessels to cerebral ischemic injury. Mol Chem Neuropathol 26:107–157
92. Strasser A, Yasuma Y, McCarron RM, Ishii H, Stanimirovic D, Spatz M (1994) Effect of nitro-L-arginine on cerebral blood flow and monoamine metabolism during ischemia/reperfusion in the mongolian gerbil. Brain Res 664:197–201
93. Truelove D, Shuaib A, Ijaz S, Ishaqzay R, Kalra J (1994) Neuronal protection with superoxide dismutase in repetitive forebrain ischemia in gerbils. Free Radic Biol Med 17:445–450
94. Truelove D, Shuaib A, Ijaz S, Richardson S, Kalra J (1994) Superoxide dismutase, catalase, and U78517F attenuate neuronal damage in gerbils with repeated brief ischemic insults. Neurochem Res 19:665–671
95. Tokuda Y, Uozumi T, Kawasaki T (1993) The superoxide dismutase activities of cerebral tissues, assayed by the chemiluminescence method, in the gerbil focal ischemia/reperfusion and global ischemia models. Neurochem Int 23:107–114
96. Thulborn KR, du Boulay GH, Duchen LW, Radda G (1982) A $^{31}$P nuclear magnetic resonance in vivo study of cerebral ischaemia in the gerbil. J Cereb Blood Flow Metab 2:299–306
97. Heffez DS, Passonneau JV (1985) Effect of nimodipine on cerebral metabolism during ischemia and recirculation in the mongolian gerbil. J Cereb Blood Flow Metab 5:523–528
98. Mies G, Hossmann KA (1983) A method for regional evaluation of potassium depletion during graded focal ischemia in gerbils. J CBF Metab 3:S313–S314
99. Milde LN, Milde JH, Michenfelder JD (1986) Delayed treatment with nimodipine improves cerebral blood flow after complete cerebral ischemia in the dog. J Cereb Blood Flow Metab 6:332–337
100. Nowicki JP, MacKenzie ET, Young AR (1982) Brain ischaemia, calcium and calcium antagonists. Pathol Biol (Paris) 30:282–288
101. Gyulai L, Chance B, Ligeti L, McDonald G, Cone J (1988) Correlated *in vivo* $^{31}$P-NMR and NADH fluorometric studies on gerbil brain in graded hypoxia and hyperoxia. Am J Phys 254:C699–C708
102. Mayevsky A, Duckrow RB, Yoles E, Zarchin N, Kanshansky D (1990) Brain mitochondrial redox state, tissue hemodynamic and extracellular ion responses to four-vessel occlusion and spreading depression in the rat. Neurol Res 12:243–248
103. Mayevsky A, Flamm ES, Pennie W, Chance B (1991) A fiber optic based multiprobes system for intraoperative monitoring of brain functions. SPIE 1431:303–313
104. Barbiro-Michaely E, Zuckerman T, Zarchin N, Rinkevich S, Knoller N, Hadani M, Mayevsky A (2003) *In vivo* simultaneous cortical and intracortical monitoring of cerebral blood flow and mitochondrial NADH redox state in experimental animals. Proc SPIE 4955:22–30
105. Jones TH, Chiappa KH, Young RR, Ojemann RG, Crowell RM (1979) EEG monitoring for induced hypotension for surgery of intracranial aneurysms. Stroke 10:292–294
106. Tempelhoff R, Modica PA, Rich KM, Grubb RL Jr (1989) Use of computerized electroencephalographic monitoring during aneurysm surgery. J Neurosurg 71:24–31
107. Alvarez del Castillo M (2001) Monitoring neurologic patients in intensive care. Curr Opin Crit Care 7:49–60
108. Critchley GR, O'Neill KS, Bell BA (1998) Cerebral blood flow and tissue oxygenation monitoring during aneurysm surgery. Neurol Res 20(Suppl 1):S44–S47

109. Nakayama N, Kuroda S, Houkin K, Takikawa S, Abe H (2001) Intraoperative measurement of arterial blood flow using a transit time flowmeter: monitoring of hemodynamic changes during cerebrovascular surgery. Acta Neurochir 143:17–24
110. Smielewski P, Czosnyka M, Pickard JD, Kirkpatrick P (1998) Assessment of cerebrovascular reactivity in patients with carotid artery disease using near-infrared spectroscopy. Acta Neurochir Suppl 71:263–265
111. Rosenblum BR, Bonner RF, Oldfield EH (1987) Intraoperative measurement of cortical blood flow adjacent to cerebral AVM using laser Doppler velocimetry. J Neurosurg 66:396–399
112. Arbit E, DiResta GR, Bedford RF, Shah NK, Galicich JH (1989) Intraoperative measurement of cerebral and tumor blood flow with laser-Doppler flowmetry. Neurosurgery 24:166–170
113. Lam JM, Hsiang JN, Poon WS (1997) Monitoring of autoregulation using laser Doppler flowmetry in patients with head injury. J Neurosurg 86:438–445
114. Johnson WD, Bolognese P, Miller JI, Heger IM, Liker MA, Milhorat TH (1996) Continuous postoperative ICBF monitoring in aneurysmal SAH patients using a combined ICP-laser Doppler fiberoptic probe. J Neurosurg Anesthesiol 8:199–207
115. Kirkpatrick PJ, Smielewski P, Czosnyka M, Pickard JD (1994) Continuous monitoring of cortical perfusion by laser Doppler flowmetry in ventilated patients with head injury. J Neurol Neurosurg Psychiatry 57:1382–1388
116. Dirnagl U, Kaplan B, Jacewicz M, Pulsinelli W (1989) Continuous measurement of cerebral cortical blood flow by laser-Doppler flowmetry in a rat stroke model. J CBF Metab 9:589–596
117. Dirnagl U, Thoren P, Villringer A, Sixt G, Them A, Einhaupl KM (1993) Global forebrain ischaemia in the rat: controlled reduction of cerebral blood flow by hypobaric hypotension and two-vessel occlusion. Neurol Res 15:128–130
118. Inao S, Kuchiwaki H, Sugita K (1993) Microcirculation in experimental brain oedema assessed by laser-Doppler flowmetry. Neurol Res 15:264–268
119. Kumano H, Shimomura T, Furuya H, Yomosa H, Okuda T, Sakaki T, Kuro M (1993) Effects of flumazenil during administration of midazolam on pial vessel diameter and regional cerebral blood flow in cats. Acta Anaesthesiol Scand 37:567–570
120. Morikawa E, Moskowitz MA, Huang Z, Yoshida T, Irikura K, Dalkara T (1994) L-arginine infusion promotes nitric oxide-dependent vasodilation, increases regional cerebral blood flow, and reduces infarction volume in the rat. Stroke 25:429–435
121. Kirkeby OJ, Rise IR (1999) Intracerebral laser Doppler blood flow measurements compared to blood flow in porcine internal carotid artery. J Clin Neurosci 6:389–394
122. Kirkeby OJ, Rise IR, Nordsletten L, Skjeldal S, Hall C, Risoe C (1995) Cerebral blood flow measured with intracerebral laser-Dopplerflow probes and radioactive microspheres. J Appl Physiol 79:1479–1486
123. Kurosawa M, Fuxe K, Hallstrom A, Goiny M, Cintra A, Ungerstedt U (1991) Responses of blood flow, extracellular lactate, and dopamine in the striatum to intrastriatal injection of endothelin-1 in anesthetized rats. J Cardiovasc Pharmacol 17(Suppl 7):S340–S342
124. Cao WH, Inanami O, Sato A, Sato Y (1989) Stimulation of the septal complex increases local cerebral blood flow in the hippocampus in anesthetized rats. Neurosci Lett 107:135–140
125. Bolognese P, Miller JI, Heger IM, Milhorat TH (1993) Laser-Doppler flowmetry in neurosurgery. J Neurosurg Anesthesiol 5:151–158
126. Meyerson BA, Gunasekera L, Linderoth B, Gazelius B (1991) Bedside monitoring of regional cortical blood flow in comatose patients using laser Doppler flowmetry. Neurosurgery 29:750–755
127. Rosenthal M, Sick TJ (1988) Measurement of metabolic activity associated with ion shifts. In: The neuronal microenvironment, pp 187–246
128. Thom SR, Fisher D, Xu YA, Notarfrancesco K, Ischiropoulos H (2000) Adaptive responses and apoptosis in endothelial cells exposed to carbon monoxide. Proc Natl Acad Sci USA 97:1305–1310
129. Brandes R, Bers DM (1999) Analysis of the mechanisms of mitochondrial NADH regulation in cardiac trabeculae. Biophys J 77:1666–1682

130. Mottin S, Laporte P, Jouvet M, Cespuglio R (1997) Determination of NADH in the rat brain during sleep-wake states with an optic fibre sensor and time-resolved fluorescence procedures. Neuroscience 79:683–693
131. Osbakken M, Mayevsky A (1996) Multiparameter monitoring and analysis of *in vivo* ischemic and hypoxic heart. J Basic Clin Physiol Pharmacol 7:97–113
132. Meilin S, Rogatsky GG, Thom SR, Zarchin N, Guggenheimer-Furman E, Mayevsky A (1996) Effects of carbon monoxide exposure on the brain may be mediated by nitric oxide. J Appl Physiol 81:1078–1083
133. Mayevsky A, Meilin S, Rogatsky GG, Zarchin N, Sonn J (1995) Multiparametric monitoring of the awake brain exposed to carbon monoxide. J Appl Physiol 78:1188–1196
134. Bederson JB, Germano IM, Guarino L (1995) Cortical blood flow and cerebral perfusion pressure in a new noncraniotomy model of subarachnoid hemorrhage in the rat. Stroke 26:1086-1091; discussion 1091-1082
135. Vinas FC, Dujovny M, Hodgkinson D (1995) Early hemodynamic changes at the microcirculatory level and effects of mannitol following focal cryogenic injury. Neurol Res 17:465–468
136. Tomida S, Wagner HG, Klatzo I, Nowak TS (1989) Effect of acute electrode placement on regional CBF in the gerbil: a comparision of blood flow measured by Hydrogen Clearance, {$^{3}$H} Nicotine and [$^{14}$C] Iodoantipyrine techniques. J CBF Metabol 9:79
137. Benveniste H, Drejer J, Schousboe A, Diemer NH (1987) Regional cerebral glucose phosphorylation and blood flow after insertion of a microdialysis fiber through the dorsal hippocampus in the rat. J Neurochem 49:729–734
138. Verhaegen MJ, Todd MM, Warner DS, James B, Weeks JB (1992) The role of electrode size on the incidence of spreading depression and on cortical cerebral blood flow as measured by H2 clearance. J Cereb Blood Flow Metab 12:230–237
139. Schapira AH (1993) Mitochondrial complex I deficiency in Parkinson's disease. Adv Neurol 60:288–291
140. Beal MF, Hyman BT, Koroshetz W (1993) Do defects in mitochondrial energy metabolism underlie the pathology of neurodegenerative diseases? Trends Neurosci 16:125–131
141. Schapira AH, Cooper JM (1992) Mitochondrial function in neurodegeneration and ageing. Mutat Res 275:133–143
142. Villa RF, Arnaboldi R, Ghigini B, Gorini A (1992) Mitochondrial factors involved in Parkinson's disease by MPTP toxicity in Macaca fascicularis and drug effect. Neurochem Res 17:1147–1154
143. Beal MF (1992) Does impairment of energy metabolism result in excitotoxic neuronal death in neurodegenerative illnesses? Ann Neurol 31:119–130
144. Rex A, Hentschke MP, Fink H (2002) Bioavailability of reduced nicotinamide-adenine-dinucleotide (NADH) in the central nervous system of the anaesthetized rat measured by laser-induced fluorescence spectroscopy. Pharmacol Toxicol 90:220–225
145. Elkon C, Mayevsky A, Yadid G (1996) *In vivo* monitoring of mitochondrial NADH in the Nigro-striatal pathway in the rat. Isr J Med Sci 32(Suppl):S37

# Chapter 5
# Brain Activation and Multiparametric Responses in Gerbils

**Abstract** Brain activation could be obtained by three techniques that were used frequently in the rat or cat models. The use of the Mongolian gerbil as an animal model in cortical spreading depression (CSD) studies was performed mostly or only by our research team. In this chapter, I summarize all of our results obtained after using various multiparametric monitoring systems in gerbils exposed to CSD in normoxia as well as under various pathophysiological conditions.

In order to study the interrelation between the energy state of the brain and ionic homeostasis in vivo, we used the multiparametric monitoring approach developed by us a few years ago. The energy state was evaluated by monitoring the intramitochondrial redox state using the surface NADH fluorometry reflectometry. Relative cerebral blood flow (CBF) was measured by a laser Doppler flowmeter. Extracellular $K^+$ and $Ca^{2+}$ levels were monitored by mini-electrodes surrounded by DC steady potential electrodes. Spontaneous electrocortical activity (ECoG) was measured using bipolar electrodes.

We exposed gerbils to the responses to CSD in the normoxic brain as well as after exposing the gerbil to ischemia, hypoxia, and deep anesthesia. It was found that the ability of the brain to recover from the CSD wave is dependent upon the availability of oxygen in the brain, which is a critical factor enabling the maintenance of ionic homeostasis in the brain tissue.

**Keywords** Cortical spreading depression · Ionic homeostasis · NADH redox state · Extracellular potassium · Microcirculation blood flow

## 5.1 Introduction

Brain activation can be induced in at least three ways, namely, direct cortical stimulation, epileptic activity, and cortical spreading depression (CSD). In this chapter, I present and discuss the effects of CSD and epileptic activity induced in the gerbil brain on the redox state of the mitochondria as well as other key physiological



A. Mayevsky, *The Mongolian Gerbil Brain*,
https://doi.org/10.1007/978-3-031-69549-0_5

responses developed in a stimulated brain. As shown by Chance and Williams [1, 2], the activation of the mitochondria by increased ADP is coupled with oxidation of NADH (decreased NADH levels) and is known as the state 4 to state 3 transition in isolated mitochondria. Most of the studies in this field were performed in rats, cats, and mice. In this chapter, I discuss the effects of induced CSD (using a solution of KCl) under normoxia, ischemia, hypoxia, and deep anesthesia on mitochondrial NADH redox state as well as other brain physiological functions.

Figure 5.1 (Mayevsky, unpublished 2024) presents, in a schematic way, the relationship between various physiological parameters monitored in vivo under different experimental perturbations in gerbils. The two perturbations shown are cortical spreading depression and epileptic activity that induced cortical depolarization (purple arrows). The two perturbations affect the same type of processes in the brain, such as cerebral blood flow (CBF), NADH, and extracellular levels of ions. When the brain is activated by CSD or epileptic activity (lower left side), the events will propagate toward the activation of the brain, and energy consumption is increased (purple arrows). The coupling between the two pathological events is clear and manifested by the electrical activity (electrocorticogram (ECoG)) as well as extracellular $K^+$ levels and NADH redox state. During the epileptic stage, extracellular potassium increased and started to recover to the baseline, but then a larger elevation was recorded. There is a clear correlation between the various parameters during epilepsy and CSD. The decrease in NADH was smaller during the first stage, followed by an oxidation cycle typical for CSD.

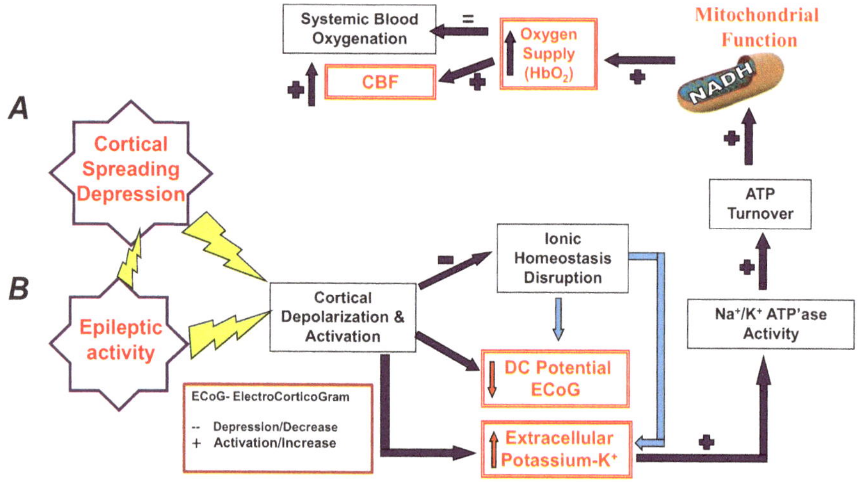


**Fig. 5.1** Schematic representation of the effects of cortical spreading depression on energetic-mitochondrial function, blood oxygenation, ionic homeostasis, and CBF in the Mongolian gerbil

## 5.2 Cortical Spreading Depression (CSD) Responses

### *5.2.1 In the Normoxic Brain*

In order to elicit spreading depression, a 0.5 M KCl solution was washed through the push-pull cannula, and the results (Fig. 5.2) are very similar to those obtained from the brain of an awake rat [4]. The oxidation-reduction cycles were stopped by washing the brain with NaCl solution (W in Fig. 5.2). The oxidation-reduction cycles can be seen in the uncorrected as well as the corrected traces. Owing to the biphasic response of the reflectance, the corrected trace shows cycles that are more symmetrical and have a larger amplitude of oxidation.

#### 5.2.1.1 Propagation of the CSD Wave

In order to monitor NADH redox state from various sites in the awake brain we developed a new 4-channel DC fluorometer reflectometer connected to four points on the brain. The source for the excitation light was a 100-W water-cooled Mercury

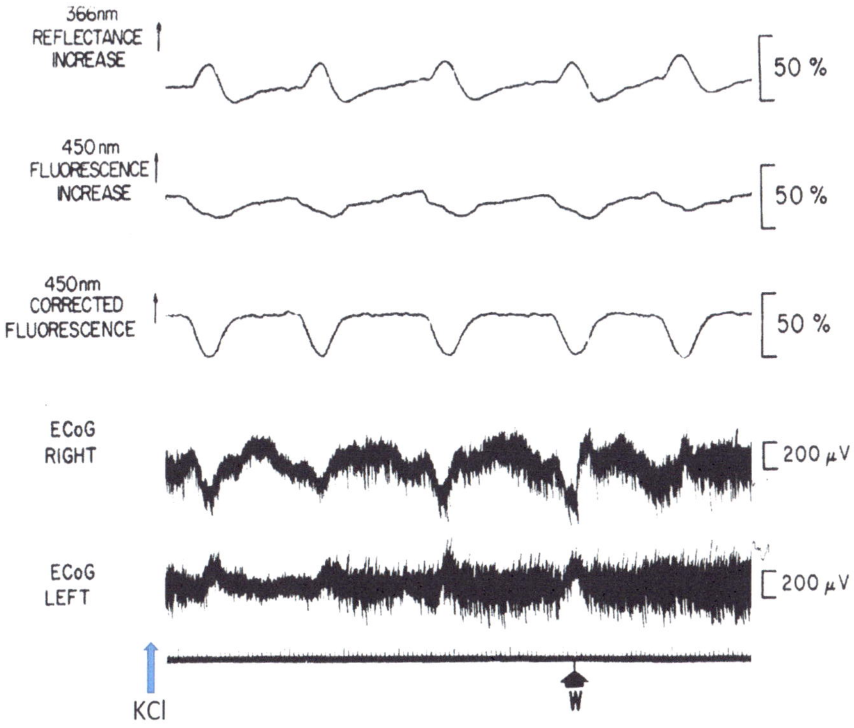


**Fig. 5.2** The first recorded responses, of the gerbil brain, to cortical spreading depression [3]

lamp split into four branches of the light guide after passing the lens and 366-nm filter. The emitted light from the brain, transmitted through the other bundle of fibers, is split in a 90:10 ratio in front of the two photomultipliers (RCA 93IB). The fluorescence and reflectance signals were amplified and calibrated as described previously [5, 6]. The signals obtained from the normoxic brain after the operation served as 100% of the signals, and all changes were calculated as percentage change as compared to the normoxic level. Two types of light guides were used in the present study; a 2 mm active diameter of fibers was used in studies where the light guides were placed as shown in Fig. 3.13. In other experiments, we used the small diameter light guides (0.8 mm active diameter). The rest of the details are presented in Table 3.1 (Chap. 3), Protocol 3.7.

The propagation of a spreading depression (SD) wave is shown in Fig. 5.3. The wave was initiated by washing the dura with a 1 M KCl solution via the push-pull cannula (labeled KCl in the scheme). All four sites measured showed an oxidation cycle as a response to SD (decrease in CF). The reflectance (R) trace also showed a biphasic response, namely, an initial increase and later on a long-lasting decrease until a complete recovery to the baseline levels. The vertical dashed line is a continuation of the time when the wave was under site 1 and could serve as a reference point to calculate the speed of the SD wave propagation. For example, the distance between sites 1 and 3 is about 2.5 mm, and the difference in the response was 50 s, so the speed of the propagation of the response to SD was 3 mm/min, as was calculated previously by monitoring electrical events [8]. The shape of the oxidation cycle was not identical in all four sites, and further investigation is needed in order to clarify the nature of the SD wave shape. In order to study the propagation speed

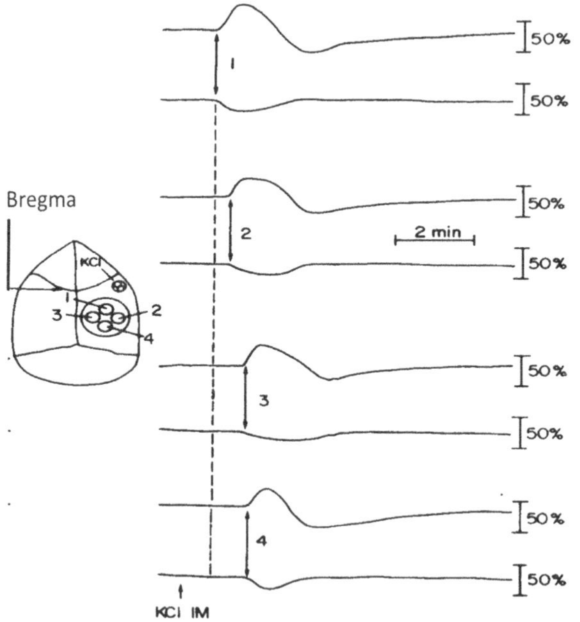


**Fig. 5.3** Effects of CSD on the reflectance and corrected fluorescence measured from four sites in the gerbil brain [7]

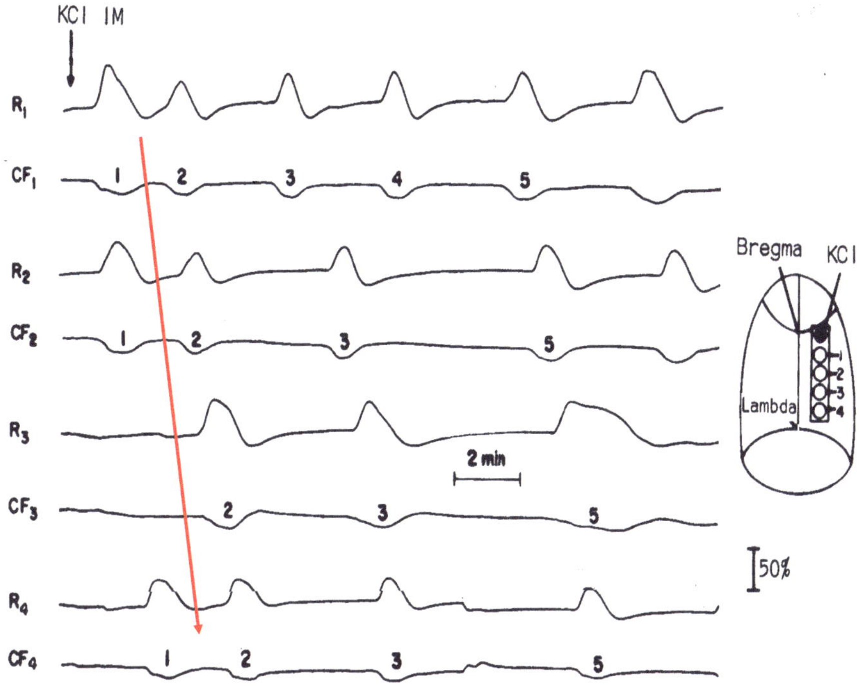


**Fig. 5.4** Repetitive metabolic responses to spreading depression measured from four sites in the awake gerbil brain [7]

of the SD wave, a second type of light guide holder was constructed, and the results are shown in Fig. 5.4. The light guides were organized along the KCl application site. In this specific gerbil, the response in the site 3 wave was not as expected, namely, only three waves were recorded while four waves were measured in the site 4. We showed in the past [9] that not all the metabolic waves started at one point in the brain will propagate through the entire cortex. As can be seen in Fig. 5.4, in site 1, five cycles were recorded, and cycle number 4 did not propagate throughout the cortex. The distance between sites 1 and 4 was 5 mm, and the difference in the responses was about 130 s, so the speed of this gerbil was 2.3 mm/min.

#### 5.2.1.2 Multiparametric Responses

**Protocol 3.3 (Chap. 3, Table 3.1)**

The gerbil brain was used as a model to monitor NADH, which was measured together with extracellular $K^+$, DC potential, and ECoG from one hemisphere. The advantage of the gerbil model is that one can induce ischemia in one hemisphere and have the contralateral one in a normoxic state while exposing the two sides to SD (internal control).

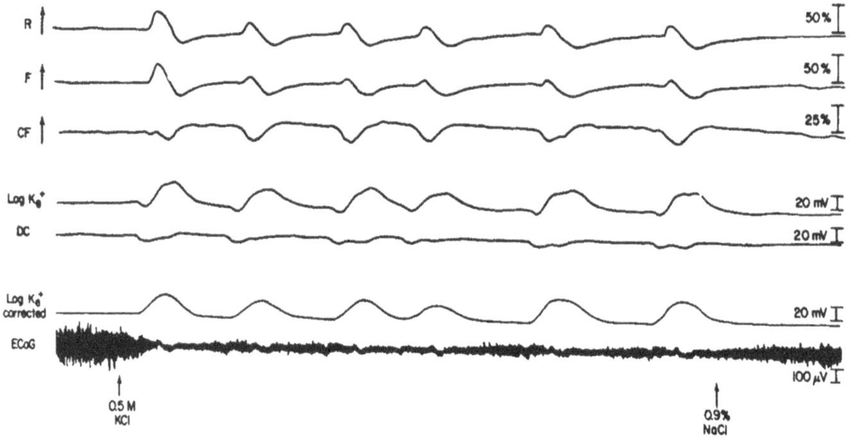


**Fig. 5.5** Repetitive metabolic, ionic, and electrical responses to cortical spreading depression. All parameters were monitored from one hemisphere using a multiprobe cannula [10]. Ref, Flu, and CFlu, reflectance, fluorescence, and corrected fluorescence of NADH; ECoG, electrocorticogram; Log $K_e^+$ and Log $K_e^+$ corrected, extracellular potassium uncorrected and corrected; DC, direct current potential

The exposure of the gerbil brain to SD did not have any damaging effects if the oxygen supply was unlimited. Figure 5.5 shows a repetitive response to SD initiated in the awake gerbil brain. One can see the direct correlation between the pumping back of $K^+$ into the intracellular space and the oxidation cycle of the NADH as was described previously [11]. After washing the brain with NaCl (0.9%), the cycles stopped, and ECoG was also restored to normal.

The metabolic responses to the Leao spreading depression [12] phenomenon (SD) have been studied by various investigators in different animal models. Some investigators showed that SD caused an increase in oxygen utilization as evaluated by measuring $pO_2$ change in the cortex (for details, see Ref. [13]). In order to supply the need for extra $O_2$, hemodynamic changes take place, and indeed, some researchers showed blood flow changes during cortical SD [14–16]. Using fluorometric/reflectometric techniques for monitoring mitochondrial signals, the same conclusion was drawn, namely, that SD caused an increase in energy utilization. In 1973, Rosenthal and Somjen [17] showed an oxidation of NADH due to SD in the cerveau isole cat preparation. Using the awake rat model, we also showed oxidation of NADH due to SD [9]. We were able to correlate the NADH oxidation cycle with the pumping activity of the extracellular $K^+$ in the recovery phase of SD in rats [18].

**Protocol 3.9 (Chap. 3, Table 3.1)**

In order to test the functional state of the brain we recorded, from the same gerbil, responses to SD developed after the ischemic episodes. As can be seen from Fig. 5.6 [19], the brain displayed a normal response to the extra pumping activity of $K^+$ induced during the SD event. The level of $K^+$ increased to about 25 mM and

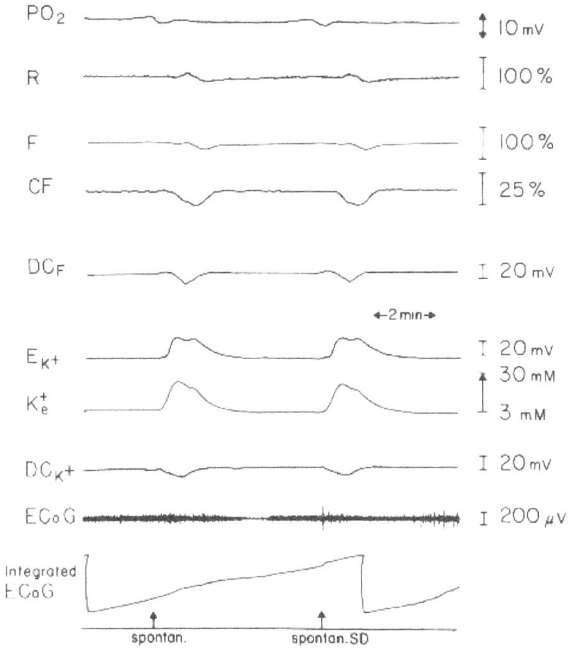


**Fig. 5.6** Metabolic, ionic, and electrical responses to spreading depression (SD) cycles developed spontaneously. $PO_2$, tissue surface $O_2$ levels; R, 366-nm reflectance; F, 450-nm fluorescence; CF, NADH-corrected fluorescence; DCF and DCK, DC steady potential around the optical fibers and the $K^+$ electrode potential and corrected (for DC changes) $K^+$ levels, respectively; ECoG, electrocorticogram [19]

recovered within 2 min to the normal level. This episode led to a clear oxidation cycle of NADH as recorded by a decrease in the CF signal (15–20%).

**Protocol 3.3 Modified (Chap. 3, Table 3.1)**

Another response to SD is shown in Fig. 5.7 [11]. This gerbil makes it possible to assess the responses of the two hemispheres to the same stimulus. In the right hemisphere, three NADH oxidation cycles were measured in response to a 0.5 M KCl stimulus. Only two $K_e^+$ responses were typical of SD, while the middle response showed a small increase of $K_e^+$. $K_e^+$ returned to baseline levels after SD, much slower in the left than in the right hemisphere. The results in Fig. 5.7 demonstrate the characteristic response to SD under normoxic conditions: the oxidation cycles of NADH together with elevated $K_e^+$ and its recovery. This correlation was described previously by us and others [18, 20, 21]. The results shown in Fig. 5.7 are exceptional because of the level of $K_e^+$ in the middle cycle. In all the gerbils used in this study and in others, it has been observed that in a normal brain, SD is an "all or none" phenomenon. If the wave is initiated and spread, the same elevated level of $K_e^+$ can be expected, although this level varies from animal to animal. However, there is a variability in response between two hemispheres of the same brain (Fig. 5.7). In the left hemisphere, the recovery phase for $K_e^+$ level was much longer than that measured in the right hemisphere. $K_e^+$ increase has the same slope in both hemispheres; we may assume that energy state or ATPase activity was not normal in the left hemisphere. Support for this conclusion may be found in the behavior of the same brain during anoxia, namely, after $N_2$ anoxia, $K_e^+$ was elevated in the left

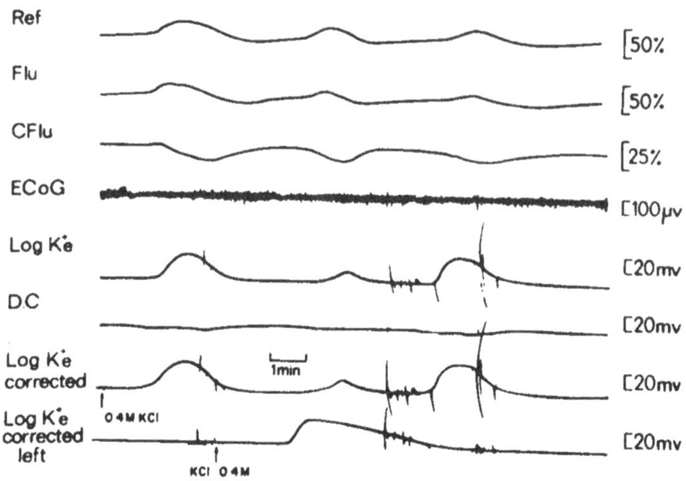


**Fig. 5.7** The effects of spreading depression on the levels of NADH and $K_e^+$ in the gerbil. NADH was monitored only in the right hemisphere, while $K_e^+$ was measured in both the right hemisphere (traces 5–7) and the left hemisphere (trace 8) [11]

hemisphere and reached the same level measured during spreading depression (Fig. not shown).

**Protocol 3.6 (Chap. 3, Table 3.1)**

The effects of cortical spreading depression are shown in Fig. 5.8 [22]. The stimulation for the development of the SD waves was a 0.5 M KCl solution applied topically 2 mm anteriorly to the MPA location. Due to different locations of the various probes in relation to the KCl application site, the time for the appearance of the responses varied. The waves propagated through most of the MPA area (four waves in the $K^+$ electrode site), and the typical changes in ions shift were recorded. In order to pump back the $K^+$ and the $Ca^{2+}$, the energy metabolism was stimulated, as seen by the oxidation cycles of the NADH. This shift in the redox state was accompanied by an increase in blood flow and was detected by the decrease in the reflectance trace.

Another set of responses to SD using the same protocol are shown in Fig. 5.9 [23], recorded from the gerbil brain. A solution of 1 M KCl was applied epidurally, 2 mm anteriorly, to the MPA via a push-pull cannula. As can be seen, the wave reached the $K^+$ electrode first and moved toward the $Ca^{2+}$ electrode and finally to the NADH monitoring area. The shifts of ions between extra- and intracellular spaces led to the stimulation of the pump, and the NADH became more oxidized. The cycles were quite regular and appeared one after the other in the $K^+$ monitored area. Not all cycles were propagated to the light guide area; therefore, only six NADH cycles were recorded out of seven or eight measured in the $K^+$ and $Ca^{2+}$ areas.

Figure 5.10 shows another set of responses to SD using the same protocol [24]. The brain was washed with 0.5 M KCl, which remained on the brain. As a result, six SD cycles were recorded in all parameters measured. The ECoG was depressed

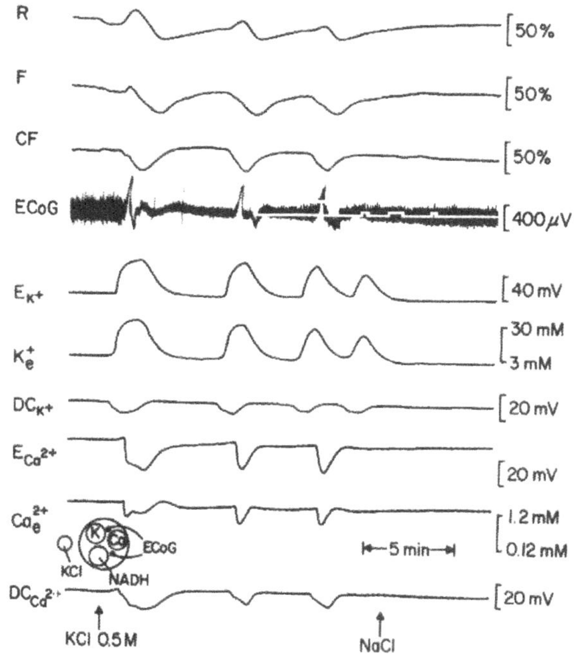


Fig. 5.8 Effects of cortical spreading depression on the metabolic ionic and electrical activities in the gerbil brain [22]

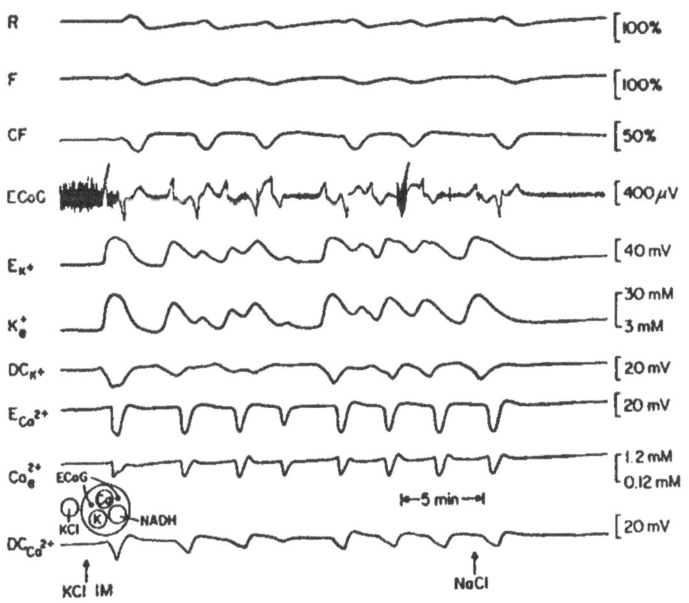


Fig. 5.9 Effects of cortical spreading depression on the metabolic ionic and electrical activities in the gerbil brain [23]

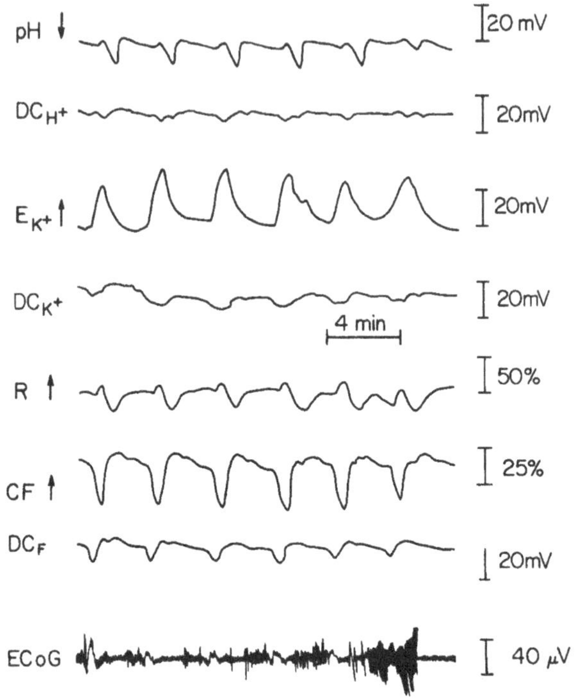


**Fig. 5.10** Repetitive responses of awake brain to spreading depression initiated by washing the dura with KCl solution (not shown). For abbreviations, see text [24]

during the entire period of the SD, and toward the end, the beginning of epileptic activity was recorded. The three DC channels also showed cycles of negative shift. The main change in the pH was an alkalinization response. The extracellular $K^+$ showed the typical increase–decrease cycle, and as a result, the NADH showed the oxidation cycles.

**Protocol 3.10 (Chap. 3, Table 3.1)**

The SD wave was induced by the insertion of a needle into the cortical area next to the MPA, and the propagated wave was monitored (Fig. 5.11) [25]. The $Ca^{2+}$ electrode was the first to detect the wave due to the MPA positioning on the brain in relation to the SD initiation spot. The ionic homeostasis disturbances due to the SD waves, including the increase in $K^+$ and NADH responses, had started about the same time as can be seen also from the probe location scheme. The main new results are the $HbO_2$ responses to the SD wave and its correlation to the NADH kinetic changes. The main change seen in the $HbO_2$ trace is the increase in oxygenation, although a small short initial deoxygenation could be detected. In parallel, the Hb volume was increased, and it correlated well to the decrease in the R signal. Those changes in the $HbO_2$ oxygenation and volume are secondary to the primary oxidation of the NADH, as seen by the decrease in the CF signal. Those metabolic changes recorded in tissue oxygenation as well as the oxidation of the mitochondrial redox state correlated very well to the increase in energy demand by the $Na^+$-$K^+$-ATPase and other active transport systems. The short delay in the $HbO_2$ recorded

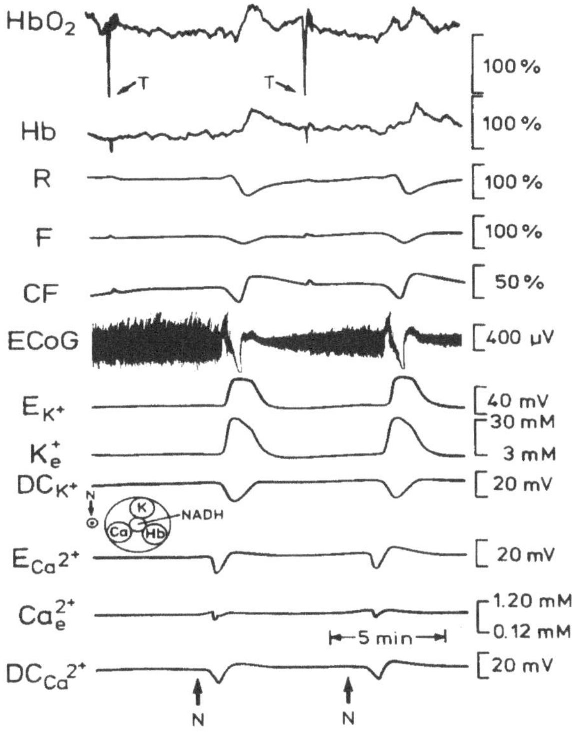


**Fig. 5.11** Metabolic ionic and electrical responses to cortical spreading depression induced by needle insertion (N) 2 mm anterior to the MPA. Two short R + L occlusion episode (T) can be seen and were used for the determination of $HbO_2$ unloading [25]

responses is due to the location of the $HbO_2$ probe in relation to the needle insertion site.

It is very clear from the data presented in Fig. 5.11 that during SD wave passage, the large increase in blood volume is able to provide enough $O_2$ to keep the Hb in a hyperoxygenated state, although $O_2$ consumption was massively stimulated as indicated by the oxidation of the intramitochondrial NADH. This change in $O_2$ consumption is associated mainly with the increase in $Na^+$-$K^+$-ATPase activity, which led to a state 4 to state 3 transition [26] and thus NADH became more oxidized. The same type of results were described by other groups (e.g. [27, 28],). This oxidation response of the NADH to SD wave is typical of the normoxic brain and may change toward an increase of NADH according to the $O_2$ balance created in the brain [10].

**Protocol 3.11 (Chap. 3, Table 3.1)**

The effects of repetitive cycles of spreading depression (SD), induced by 1 M KCl, on the hemodynamic, metabolic, and ionic activities are shown in Fig. 5.12 [29]. In this protocol, the measurement of cerebral blood flow was added to the multiparametric monitoring system. The application site of KCl is seen in the scheme shown inside the figure. The front of the propagating SD wave affected the various electrodes at slightly different timing.

As seen in line A, the sequence was that the $Ca^{2+}$ and $K^+_{(1)}$ were the first sites to be affected, followed by the light guide area, and the $K^+_{(2)}$ was the last one to

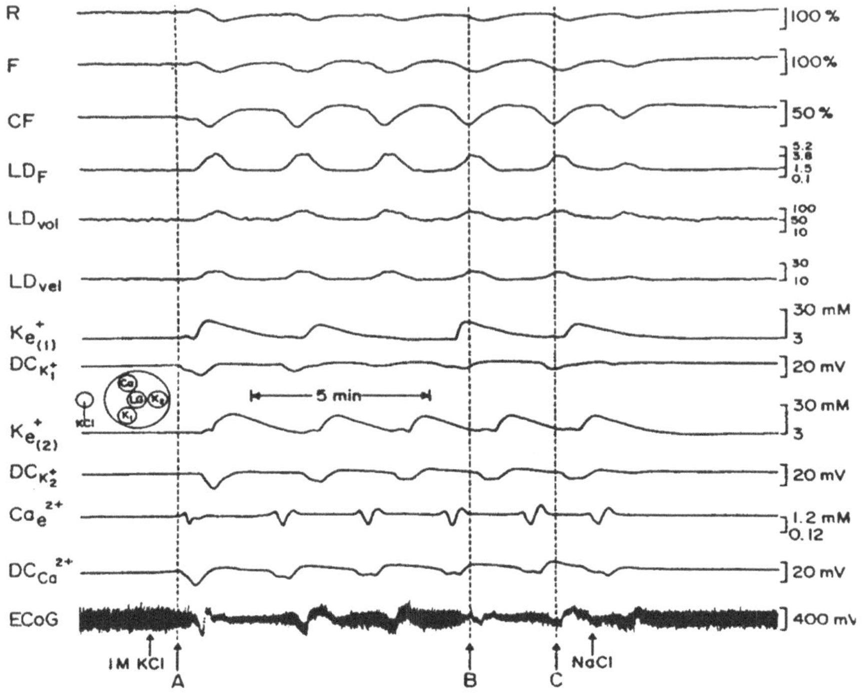


**Fig. 5.12** Metabolic, hemodynamic, ionic, and electrical responses to cortical spreading depression induced by epidural application of KCl solution [29]

respond. Lines B and C show that the peak of the NADH oxidation precedes the peak of the maximal CBF recorded (laser Doppler flowmeter (LDF)). Since the 1M KCl was left in the washing cannula, repetitive cycles were recorded. Due to the depolarization wave developed as seen by the negative shift in the DC potential and the increase in the extracellular $K^+$, the energy metabolism was activated. Intramitochondrial NADH became oxidized (decrease in CF signal), and a large increase in CBF was recorded in order to compensate for the extra $O_2$ needed [12, 14, 16, 30].

### 5.2.2 *Effects of Hyperoxia*

The effects of repetitive cycles of SD on the various parameters are shown in Fig. 5.13 [31]. This gerbil was operated on as described in the Methods section. A special push-pull surface cannula was located epidurally 2–3 mm anterior to the MPA and was cemented to the skull. The SD wave was initiated by an 0.5 M KCl epidural application and was stopped by washing the dura with physiological saline.

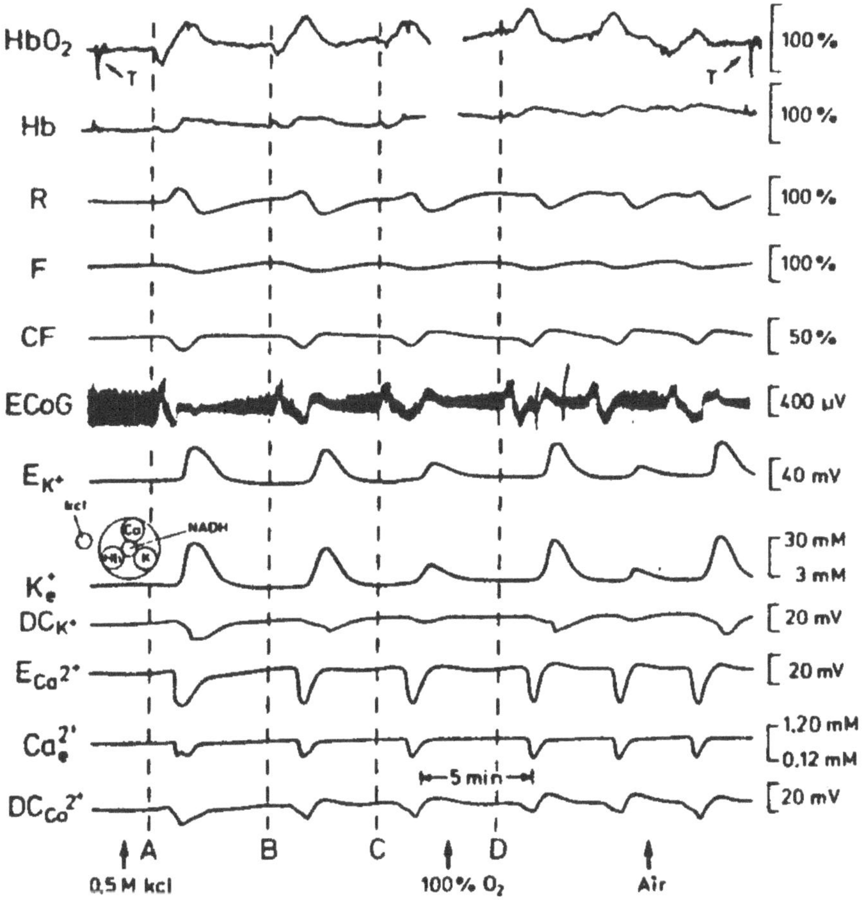


**Fig. 5.13** The effects of cortical spreading depression on the metabolic, ionic, and electrical activities of the gerbil brain. The relationship between the probes' location and the KCl (0.5 M) application site is shown inside the figure (center, left side). The gerbil was breathing air and then was switched to 100% $O_2$ as well as returned to air during the repeated cycles of SD [31]. R,F, and CF, reflectance, fluorescence, and corrected fluorescence; $ECa^{2+}$, $EK^+$, uncorrected calcium and potassium potentials; $Ca_e^{2+}$ and $K_e^+$, corrected ion activities; $DCCa^{2+}$ and $DCK^+$, DC steady potential measured around the calcium and potassium electrodes; ECoG, electrocorticogram; $HbO_2$, hemoglobin oxygenation; HB, hemoglobin volume or content [31]

Since the light guide of the EMPHO was the closest probe to the KCl application site, it shows the earliest response to the SD wave. As seen in line A, the $HbO_2$ trace shows an initial decrease in oxygenation followed by a larger and longer increase. The same biphasic response can be seen in the reflectance trace (R). The NADH redox state trace (CF) shows the typical symmetrical oxidation wave, well correlated to the high activity of the $Na^+$-$K^+$-ATPase keeping the ion gradient recovery after the large depolarization induced by the spreading depression. The same responses were recorded when the second and third waves passed below the MPA, as seen in lines B and C (Fig. 5.13). The main difference between the three cycles developed under normoxia was the diminution of the initial decrease of the $HbO_2$ levels, which could also be seen when the waves were induced under hyperoxic conditions (D).

As seen, the potassium electrode did not record a uniform change during the repetitive waves due to being far away from the KCl application site. Also, it seems that the first cycle led to a slightly different response as compared to the repetitive one, but this topic will need more statistical evaluation to be done using more animals. In order to verify the responses to the SD event, we correlated the changes in CBF to those recorded with the MPA. To do so, we replaced the EMPHO light guide with a laser Doppler flowmeter 1.4 mm in diameter light guide and used it with the same animal model. As seen in Fig. 5.14 [31], two sets of responses to SD were recorded after 0.1 M KCl application to the surface of the brain 2 mm anterior to the

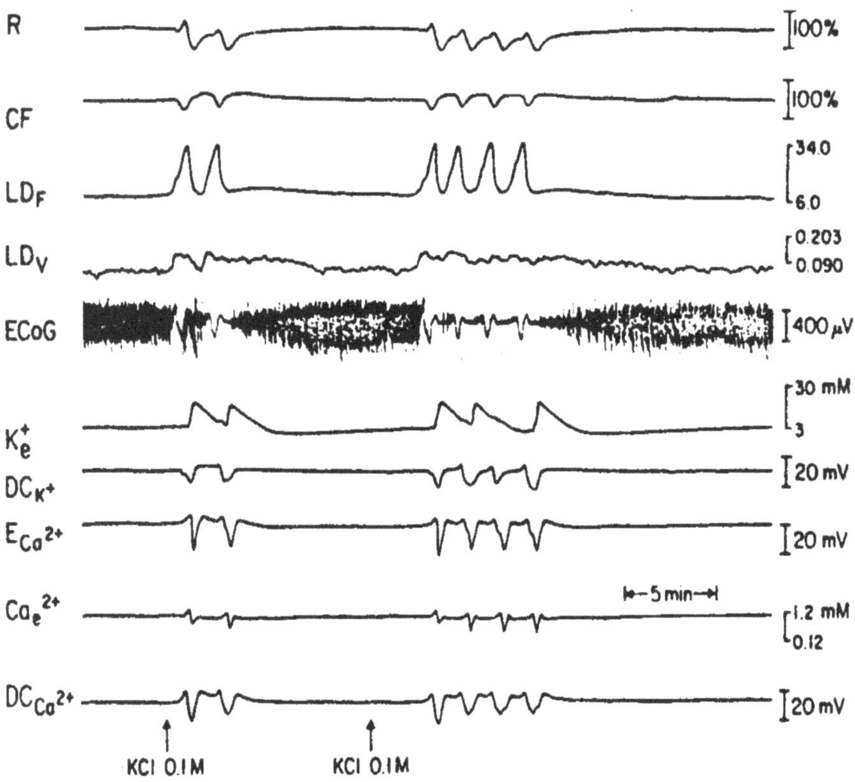


**Fig. 5.14** Responses of the gerbil brain to cortical spreading depression induced by KCl (0.1 M) application, 2 mm anterior to the MPA. LDF and LDV, laser Doppler flowmeter and volume monitored by a light guide inserted into the MPA. R, F, and CF, reflectance, fluorescence, and corrected fluorescence; $ECa^{2+}$, $EK^{+}$, uncorrected calcium and potassium potentials; $Ca_e^{2+}$, $K_e^{+}$, corrected ion activities; $DCCa^{2+}$, $DCK^{+}$, DC steady potential measured around the calcium and potassium electrodes; ECoG, electrocorticogram [31]

MPA. A clear correlation between the oxidation of NADH (CF) and the large increase in relative CBF can be seen during the six cycles recorded. The changes in relative CBF are calculated in relation to the control values before the waves were initiated, and in this animal, it was four to five times higher. The cycles of SD were not so clear on the volume trace, and we suspect it was a reading problem since in other studies a clear connection between flow and volume was found using the same laser Doppler flowmeter. The changes in the other parameters were very similar to those recorded in Fig. 5.13.

The effect of hyperoxia was tested in another gerbil as presented in Fig. 5.15 [32]. In order to understand the nature of the metabolic responses to SD, we incorporated a Laser Doppler flow meter probe inside the MPA. As seen in Fig. 5.15, repetitive increases in CBF were recorded during each wave propagated through the monitored area. The changes in blood flow were in the range of five to ten times increased as compared with the resting flow levels. The changes in blood volume (LDV) were less dramatic and were in the range of 30–40%, and were more pronounced when 100% $O_2$ was given to the gerbil. A very clear correlation was found between the increase in blood volume recorded by the LD flow meter and the change in blood volume measured by the changes in the reflectance trace (R). In order to check the state of the brain during the SD waves, the gerbil was exposed to a short $N_2$ cycle. As seen, the NADH became more reduced (increased CF), while the autoregulation mechanism led to an increase in blood flow (LDF). Here again, the coupling between ionic pumping activity and the metabolic consequences is very clear.

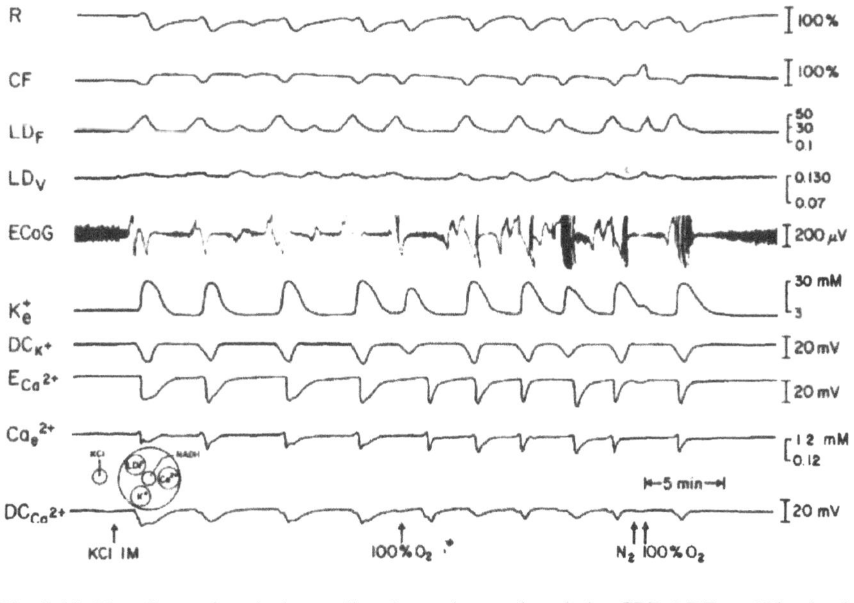


**Fig. 5.15** The effects of cortical spreading depression on the relative CBF (LDF) and blood volume as well as the other parameters monitored from the gerbil brain. LDF and LDV, laser Doppler flowmeter and volume monitored from another optic fiber located in the MPA [22]

In this gerbil, some indication of abnormal electrical activity was recorded after seven or eight cycles of SD (high amplitude and frequency in the ECoG trace).

Since the interrelationship between cortical SD and migraine attacks became of greater interest during the last few years [30, 33], it is important to analyze the basic mechanism underlying brain ischemia and SD. The original definition of the SD was dependent upon the measurements of DC potential and EEG or ECoG activity [12, 34]. The rate of propagation of the wave (3 mm/min) as well as the recovery from it was defined for the normal normoxic brain. Under those conditions, the supply of $O_2$ and substrate to the brain is normal, and the main event to occur is the wave of depolarization propagating from the stimulated site to the rest of the same brain hemisphere. During this phase, the ionic homeostasis is disturbed, leading to an increase or a decrease in extracellular $K^+$ or $Ca^{2+}$ correspondingly [18, 35, 36]. In order to restore the normal ionic distribution, the $Na^+$-$K^+$-ATPase is stimulated, leading to an increase in energy consumption, as can also be detected in real time by monitoring the NADH redox state [17, 18, 37]. As a result, an increase in blood supply was noticed during the wave of SD in order to compensate for the extra $O_2$ needed for the stimulated ATPase activity (for details, see Bures et al. [34]). When $O_2$ supply becomes inadequate (i.e., by partial ischemia), the rate of $O_2$ consumption is higher than the limited supply, and therefore the redox state of the mitochondria becomes more reduced. This was described for SD under ischemia [10, 30] or other activated states of the brain [28, 38, 39]. When ischemia occurs without the SD wave, the energy depletion is the primary event leading to an inhibition of $Na^+$-$K^+$-ATPase, and as a result, the ionic homeostasis is disturbed, leading to an increase or a decrease in extracellular $K^+$ and $Ca^{2+}$, respectively. The only way one can correlate the ionic activities to the energy state is by monitoring in real time most of the parameters related to it. Since it is preferable to measure NADH from the surface of the brain, we developed the surface mini-electrode to measure the extracellular ion activities. In our original description of the ion-selective mini $K^+$ electrode, we showed that only small differences between microelectrodes and surface electrodes were recorded under SD [40]. The advantage of the surface electrode technique is that the integrity of the tissue and microcirculation remain intact as compared to the "artificial" microenvironment created by the insertion of the microelectrode into the brain.

## 5.2.3 CSD in the Ischemic Brain

**Protocol 3.5 (Chap. 3, Table 3.1)**

In the normoxic brain, it was expected and also found that an increase in energy utilization may lead to an oxidation of NADH toward the more active state—State 3 [17, 18]. In order to obtain this oxidation of NADH, oxygen must be available in an unlimited amount [26], which in the normal brain can be achieved via increased blood flow to the active site in the tissue. In 1978, we found [41] that in the partially ischemic awake rat brain, the metabolic response to spreading depression (SD) was

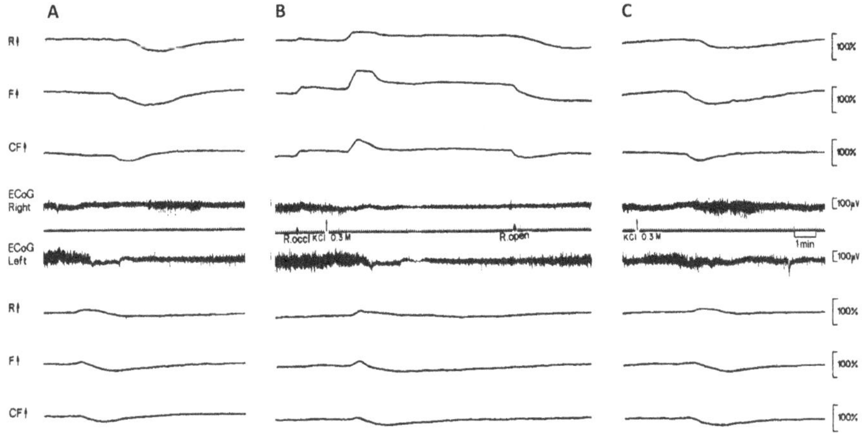


**Fig. 5.16** Intramitochondrial metabolic responses to spreading depression in the normoxic brain (**a**, **c**) and spreading depression in the partial, unilateral ischemic brain (**b**) [10]

a "reduction cycle" rather than the normal "oxidation cycle." We later showed that the same "reduction cycle" as a response to spreading depression was recorded also from the anesthetized gerbil brain [13]. In those studies, NADH was monitored from the cerebral cortex and correlated to the ECoG depression episode during the SD. Only recently [11], we developed the system by which NADH can be correlated with extracellular $K^+$ activity in the brain of an awake gerbil.

Figure 5.16 shows the effects of unilateral carotid occlusion on the metabolic responses to SD in the awake gerbil [10]. The same type of response was described previously [9, 11]. In Fig. 5.16a, c, which were recorded before and after the unilateral partial ischemia application (the time of application of KCl is not shown in Fig. 5.16a). In Fig. 5.16b, after the induction of partial ischemia (25–30% increase of NADH in the right hemisphere), the two hemispheres were exposed to SD, and the occluded side showed a reduction cycle as compared to the typical oxidation cycle measured from the contralateral left normoxic hemisphere. After reopening the right carotid artery, the NADH recovered to the pre-occlusion normoxic level. Partial ischemia had a uniform effect on the metabolic response to SD, namely, a reduction of NADH observed as compared to the normoxic state, due to the large demand for extra energy needed.

**Protocol 3.3 (Chap. 3, Table 3.1)**

In order to evaluate the functional metabolic state of the ischemic brain, we used the spreading depression cycle as a standard· stimulus. Figure 5.17 [11] shows the effects of a contralateral carotid occlusion on metabolic response to SD. The effects of exposure of the unanesthetized brain to KCl solution are shown in Fig. 5.17a, in which a typical response to SD is recorded. This response was a large increase in $K_e^+$, which was accompanied by an NADH oxidation cycle. Occlusion of the left carotid artery did not affect the $K_e^+$ level or NADH level, but the response differed from that seen in the control cycle. The time $K_e^+$ returned to its baseline level was

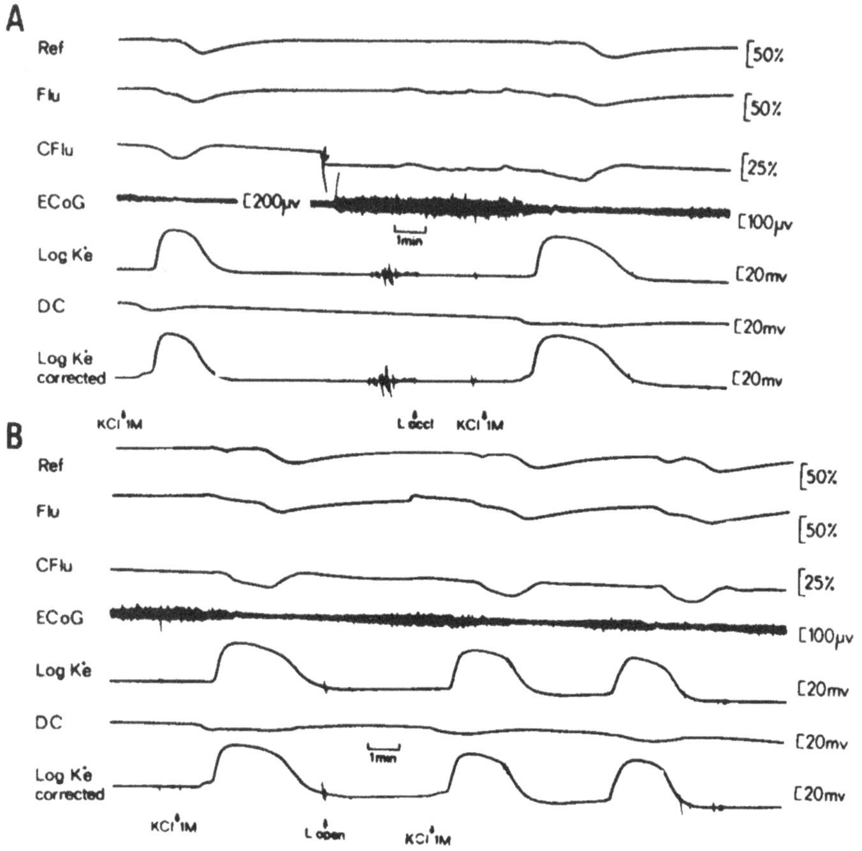


**Fig. 5.17** The effects of unilateral carotid occlusion ($L_{\cdot occl}$) on the response of the unanesthetized gerbil brain to spreading depression. $L_{\cdot open}$ indicates reflow in the left carotid artery [11]. Part (**b**) is a continuation of part (**a**)

over 4 min, as compared to a recovery time of 2.5 min during the control cycle. The same result was found after a second response (Fig. 5.17b). Following the return of flow in the left carotid artery, the response to SD became similar to the type of response seen during control cycles. The effect of occlusion can be seen also in the shape of the oxidation cycle in response to SD. The duration of the cycle is longer in partially ischemic SD cycles. Using another gerbil, the effects of a short bilateral occlusion on the response to SD cycles were tested. When the right carotid was occluded, the NADH level increased by 10–15%, and $K_e^+$ showed a very slow gradual increase (Fig. 5.18a) [11]. [ID: 323] When the left carotid was occluded, while the right carotid was also occluded, a large increase in NADH, which reached a level 40% above the control level, was observed. At the same time, $K_e^+$ continued to increase until the reopening of the two carotids. After recirculation, a spreading depression cycle developed spontaneously, and typical $K_e^+$ and metabolic responses were shown in the last section of Fig. 5.18a. The effects of partial ischemia ($R_{occl}$)

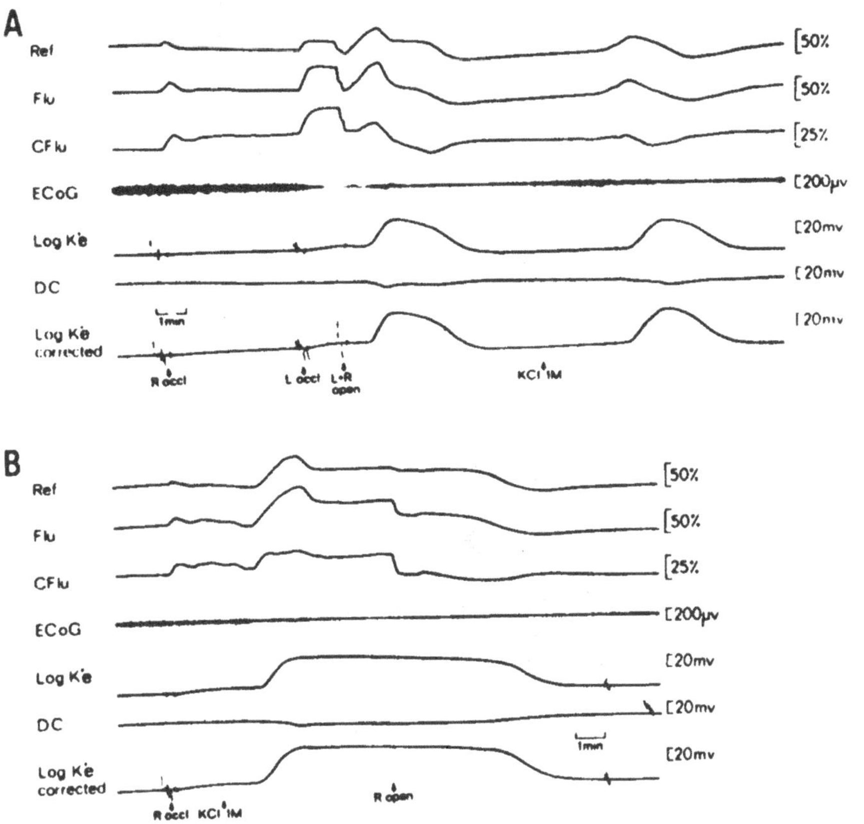


**Fig. 5.18** (**a**) Effect of partial and complete ischemia on the metabolic and $K_e^+$ responses of the brain. (**b**) The response to spreading depression was tested during a unilateral occlusion [11]

on the response to SD are shown in Fig. 5.18b. In the first stage, occlusion of the right carotid induced metabolic and ionic responses similar to those shown in Fig. 5.18a. Exposing the ischemic brain to SD (KCl 1 M) caused a typical increase in $K_e^+$, but the recovery phase was almost completely inhibited. At the same time, NADH increased in a cyclic manner, a finding also seen in the reflected light changes. Full recovery of $K_e^+$ was completed only after recirculation in the right carotid artery and then after a delay of a few minutes. The undershoot of NADH level (CFlu) after recirculation is probably due to the increased pumping activity of $K^+$ ions. The shape of the NADH trace is that of a small oxidation cycle.

**Protocol 3.11 (Chap. 3, Table 3.1)**

In order to monitor the flow-metabolism coupling and correlations under those conditions, we have developed a combined fiber optic probe for the simultaneous measurement of the relative CBF and mitochondrial redox state (NADH fluorescence). Incorporation of this metaflow probe (metabolism-flow) into the multiprobe assembly [29] provides us with information regarding metabolic, ionic, and electrical activities. In the present study, we monitored the extracellular levels of $K^+$ and $Ca^{2+}$ using ion-selective mini-electrodes as well as the DC steady potential and the ECoG.

The interrelation between energy supply and the ability of the brain to recover from spreading depression is shown in Fig. 5.19 [29]. Line A shows the initial effect on the $K^+$ electrode, followed by the other sensors. Due to the occlusion ($R_{occl}$), a small decrease in CBF was recorded (LDF) simultaneously with an increase in NADH redox state (CF). The propagation of the SD wave was blocked by the ischemic episode, and recovery processes started only after the reopening of the occluded artery ($R_{open}$).

In one of the gerbils monitored, an abnormal response to SD was recorded, as shown in Fig. 5.20 [29]. Due to an unclear reason, the delivery of $O_2$ under the SD wave was not sufficient, and the NADH became mainly reduced (increased CF) concomitantly with the initial decrease in CBF (line A). In the second cycle shown

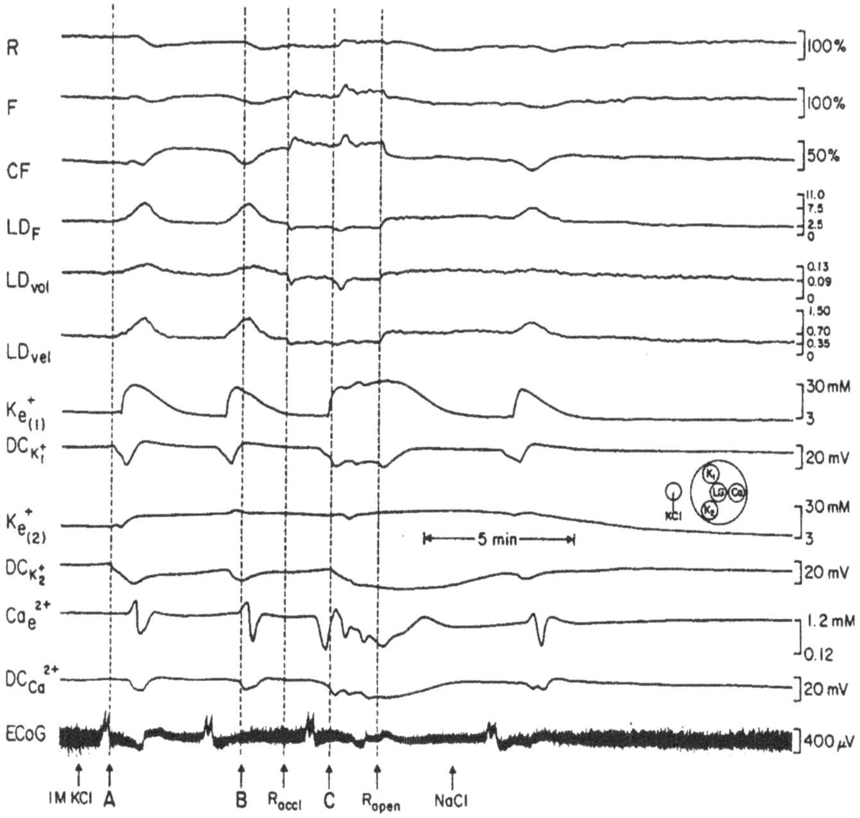


**Fig. 5.19** The effects of partial ischemia induced by unilateral carotid artery occlusion ($R_{occl}$) on the responses to spreading depression [29]

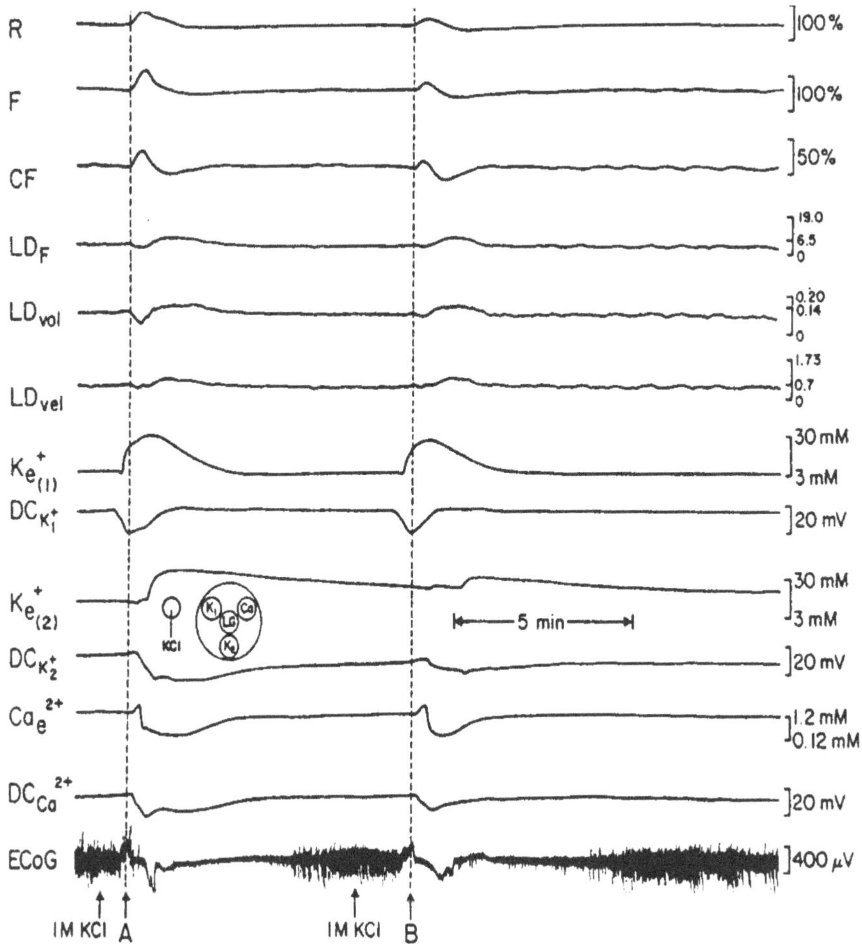


**Fig. 5.20** A nontypical response to spreading depression waves measured in one of the gerbils. All abbreviations are as in Fig. 5.14 [29]

(line B), the response did improve in terms of $O_2$ supply. This kind of response was found in rats exposed to SD under ischemic conditions [10].

The results presented in the last two figures (Figs. 5.19 and 5.20) suggest that the combination of the two optical techniques, namely, the NADH redox state and CBF, created a new powerful tool. It has been shown that the metaflow probe-MFP provides information that is more meaningful as compared to each one of the individual parameters. Also, the criticism about the lack of absolute calibration of the two techniques is diminished by using the MFP, and on the other hand, the information is continuous and in real-time mode, which cannot be achieved using the quantitated techniques.

### 5.2.4 CSD Responses in the Hypoxic Gerbil Brain

**Protocol 3.3 (Chap. 3, Table 3.1)**

Figure 5.21 shows the effect of hypoxia on $K_e^+$ and NADH responses to cortical SD [11]. In the first stage, a control response to SD was measured in the subject. $K_e^+$ showed a typical response, while NADH showed an oxidation cycle. A small increase in NADH was recorded in the initial phase of this cycle. When the gerbil was made hypoxic by administration of 5% $O_2$ in 95% $N_2$, an increased NADH level of about 10% was recorded. At the same time, $K_e^+$ began to increase gradually above its baseline level, finally reaching a point at which a fast, large increase was measured. During this phase, the NADH level remained high while the gerbil breathed from the 5% $O_2$ mixture. When the gerbil was switched back to a room-air breathing mixture, a small transient oxidation of NADH occurred, but in this animal, only 100% $O_2$ supplied enough oxygen to oxidize NADH to its original level. $K_e^+$ also started to decrease only after 100% $O_2$ was introduced.

The effects of hypoxia on the gerbil brain shown in Fig. 5.21 [11] have been described previously for the rat brain [9]. As expected under hypoxic conditions, NADH is reduced, and available energy production decreases. In the present case, it is expected that extracellular levels of $K_e^+$ will accumulate. As shown in Fig. 5.21, after 5% $O_2$ respirations, $K_e^+$ went up concomitantly with the increase of NADH. After reaching a critical level, a large increase in $K_e^+$ was recorded under hypoxic conditions. The recovery phase took about 10 min in this case, and pure oxygen was required to achieve full recovery.

Using the same monitoring system (Protocol 3.3), I exposed another gerbil to 10% oxygen, as seen in Fig. 5.22 [10]. Due to the lower $pO_2$ in the inspired air (10% $O_2$), the NADH increased by 5–10% and returned to the normoxic level after air

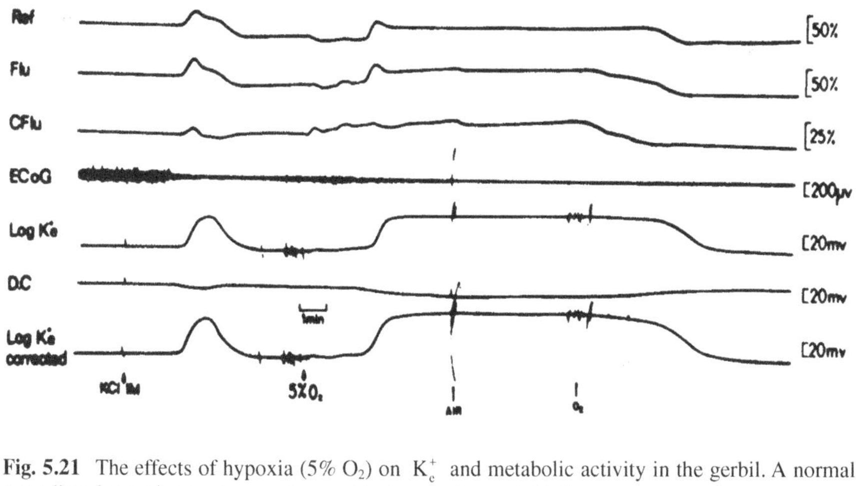


**Fig. 5.21** The effects of hypoxia (5% $O_2$) on $K_e^+$ and metabolic activity in the gerbil. A normal spreading depression cycle was observed before the hypoxia [11]

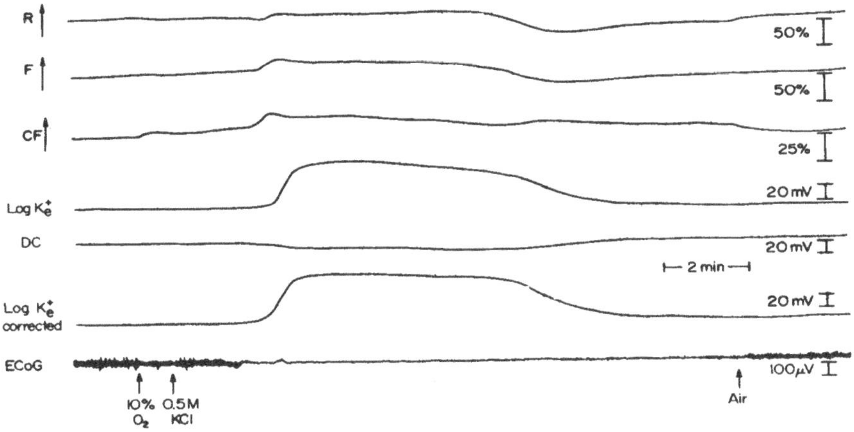


**Fig. 5.22** The effects of hypoxia on the metabolic, ionic, and electrical responses to spreading depression in the gerbil brain. The records do not show the complete recovery of the ECoG due to its later time [10]

breathing was introduced. The response of the NADH to SD under hypoxia was a long reduction cycle concomitant with the slower recovery of $K_e^+$ and the DC potential as compared to the normoxic response measured in the same gerbil.

**Protocol 3.4 (Chap. 3, Table 3.1)**
When $O_2$ consumption was evaluated, Leao and Morrison (1945) [8] found a decreased $O_2$ concentration in blood samples taken from the sagittal sinus during SD. Van Harreveld and Stamm (1952) [42] measured a decreased $pO_2$ in the depressed region. The same fall in $pO_2$ during SD was also described by Marshall (1959) [43] and Lukyanova and Bures (1967) [44]. The described hemodynamic and $pO_2$ responses to SD are indicators of increased metabolic activity during the recovery phase of SD. In 1973, Rosenthal and Somjen [17] and Mayevsky and Chance [37] showed that the increased metabolic activity during SD in the cat can be detected by monitoring the oxidation-reduction state of NADH. Mayevsky and Chance (1974, 1975) [4, 9] described in detail the metabolic response to SD in the awake rat. Using the fluorometer/reflectometer, we showed that during SD a hemodynamic response was measured in the reflectance trace changes. In most of the animals, we found a biphasic change in the reflectance trace, namely, a short increase followed by a large decrease. The large decrease was interpreted as an increase in blood volume in the tissue, which led to more absorption of light so that the amount of reflected light decreased. In the present study, we correlated for the first time the changes in NADH fluorescence, changes in the reflected light, and the changes in $pO_2$ measured with a surface electrode. The interrelation between the three parameters measured was tested under various conditions, such as anoxia, hypoxia, SD, anesthesia, and ischemia. In Fig. 5.23 [13], the responses to SD under normoxic (A) and hypoxic (B) conditions are presented. The NADH shows its typical oxidation cycle under both conditions, although the oxidation cycle has a lower amplitude

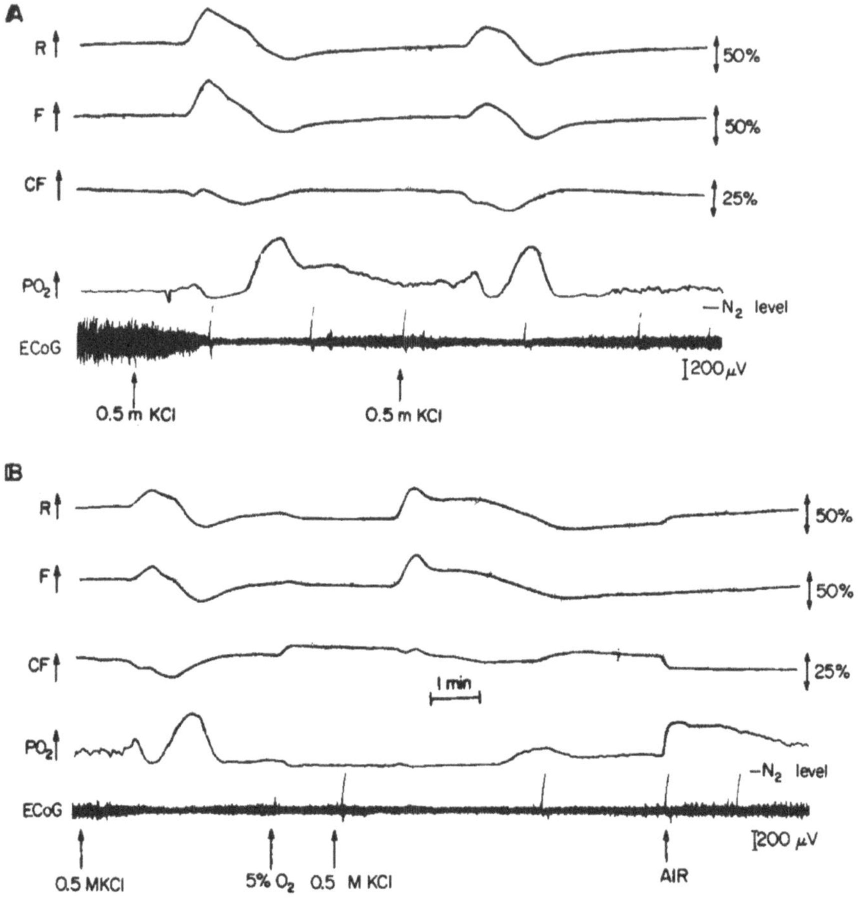


**Fig. 5.23** The responses of the normoxic (**a**) and hypoxic (**b**) brain to spreading depression [13]

with a longer duration during hypoxia. When the gerbil was exposed to 5% $O_2$, an increase in NADH (10%) was recorded, which was returned to the normoxic level after breathing air. Here again, the correlation between the large increase in $pO_2$ after the SD cycle and the reflectance decrease is clear. This overshoot of the $pO_2$ was blocked almost completely under 5% $O_2$. When air was given after the 5% $O_2$, a large overshoot in $pO_2$ was recorded concomitantly with the oxidation of NADH. After the first application of KCl, the ECoG showed depression, but the recovery was not achieved due to the following SD cycles. The effects of graded hypoxia on the responses to SD are shown in Fig. 5.24 [13]. A typical SD response was measured under 95% $O_2$, showing the oxidation cycle and initial decrease in $pO_2$, with its large overshoot in parallel to the decrease in reflectance. Under air respiration, the same pattern of changes was recorded, but the overshoot in the $pO_2$ was smaller. Under 10% $O_2$, the duration of the oxidation cycle was longer, as well as the interval between the initial decrease in $pO_2$ and its overshoot. Here again, the

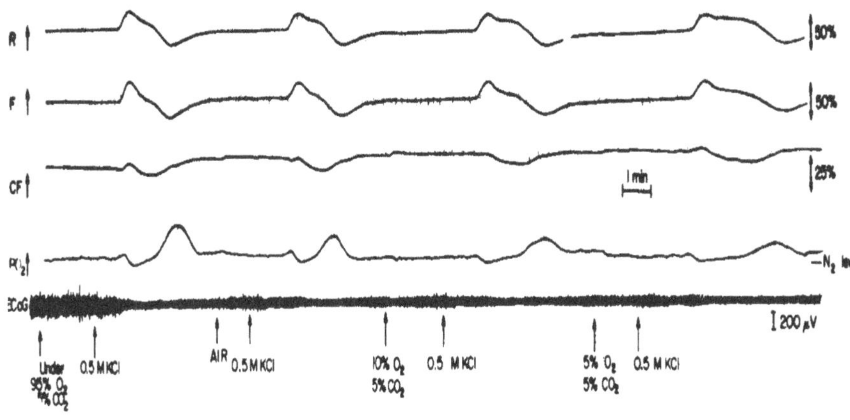


**Fig. 5.24** The interrelation between oxygen level in the breathing mixture and the response to spreading depression in the awake gerbil [13]

$pO_2$ level during the overshoot was lower. The effects of 5% $O_2$ were more drastic, as can be seen from the duration of the SD cycle. By careful analysis of the reflectance trace, one can see there is an interval of time between the initial increase and the dip increase later on. This interval was correlated to the one found in the $pO_2$ trace.

### 5.2.5 CSD Under Deep Anesthesia

**Protocol 3.4 (Chap. 3, Table 3.1)**
Finally, the effects of anesthesia on the responses to SD are shown in Fig. 5.25 [13]. The three cycles shown in Fig. 5.25 were measured after the induction of anesthesia. The qualitative response of the reflectance is the same in all three cycles, although the interval between the increase and decrease was longer in Part (c). This was in correlation to the longer interval in the $pO_2$ response. The NADH showed that anesthetics induced a qualitative change, namely, that the oxidation cycle became biphasic (Fig. 5.25a, b) and finally showed a long reduction cycle (Fig. 5.25c).

In a second study, we obtained the results presented in Fig. 5.26 [13]. In Part (a) of Fig. 5.26, two responses to SD are shown after elicitation by topical application of KCl solution. The reflectance trace shows a typical biphasic response, and the corrected fluorescence (CF) shows the oxidation–reduction cycle. The $pO_2$ trace presents a decreased level of oxygen during the SD cycles. The ECoG was depressed during the cycles, although the recovery to the original amplitude was very slow and not shown in the figure. The $pO_2$ response to SD in the anesthetized state (Part b) is similar to that of the awake brain (Part a). On the other hand, the NADH-corrected trace shows a reduction cycle rather than the typical oxidation cycle.

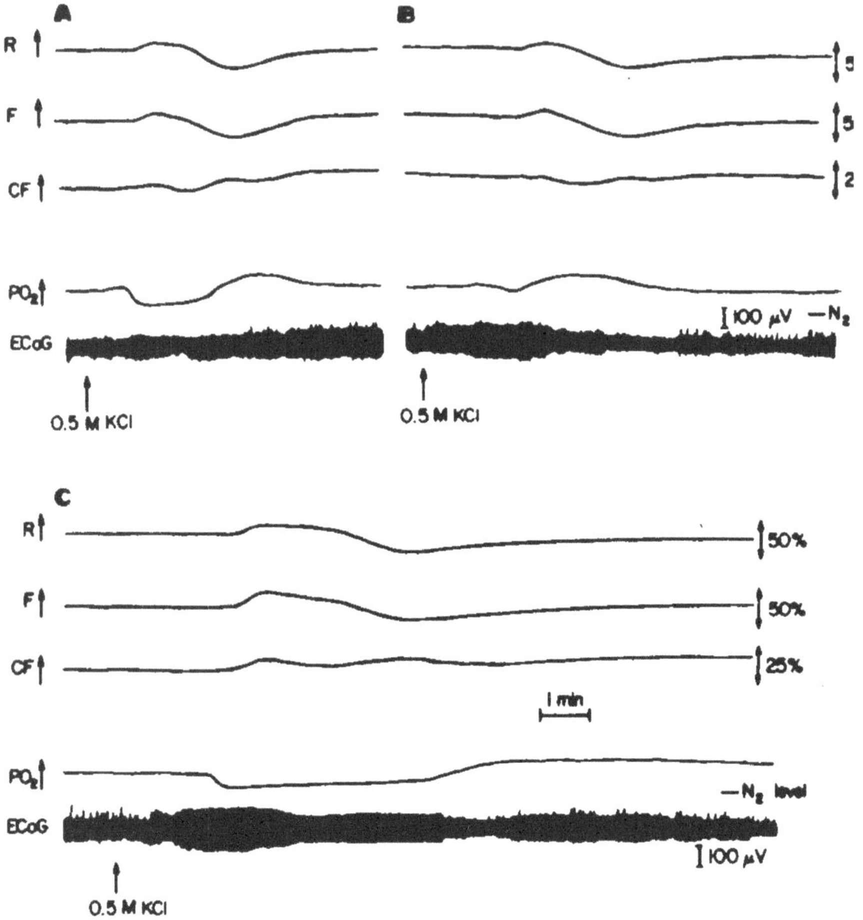


Fig. 5.25 The responses of the brain to spreading depression under three levels of anesthetics-A, B and C [13]

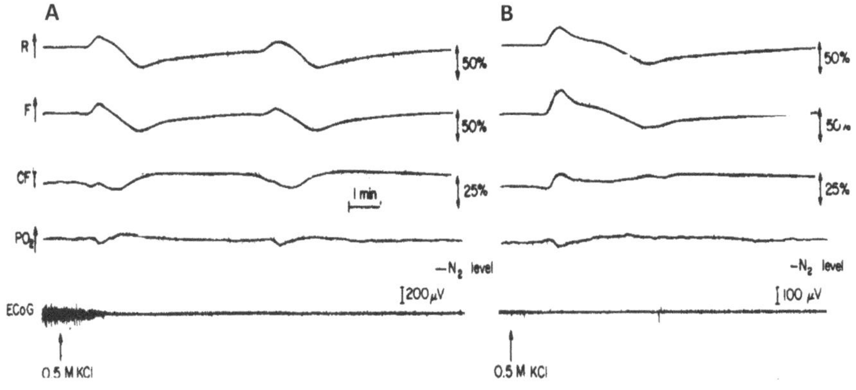


**Fig. 5.26** The effects of spreading depression on NADH, $pO_2$ in the awake (**a**), and deeply anesthetized (**b**) gerbil brain [13]

In the present study, the effects of anesthesia on the response to SD were tested. As was described previously [45], pentobarbital anesthesia did not abolish the electrical response to SD. Figure 5.26b, as compared to Fig. 5.26a, shows that in the awake brain, the typical oxidation cycle was found but, after giving anesthetics, a clear reduction cycle was recorded. Under both conditions mentioned, the $pO_2$ response was the same in the two figures, namely, the typical decrease in $pO_2$ was measured. We may conclude that, under both conditions, the high extracellular $K^+$ level stimulates the ATPase system, and as a result, electron flow in the respiratory chain is accelerated in order to produce more ATP (State 4 to 3 transition) [11, 21]. The response of the NADH fluorescence under those conditions may reflect the oxygen availability in the tissue and the ability of the autoregulatory system to respond to an increase in oxygen consumption. In the awake gerbil, the compensation in $O_2$ delivery was adequate to allow an oxidation cycle, but when the brain was anesthetized, the compensation in $O_2$ supply to the increased demand during SD was inadequate, so a temporary local "hypoxia" was produced and the reduction cycle was recorded.

**Protocol 3.3 (Chap. 3, Table 3.1)**

Figure 5.27 represents a response to SD in the deeply anesthetized gerbil [10]. Here again, after the application of KCI, NADH showed a reduction cycle. The ECoG was depressed during the entire period due to the anesthetics given.

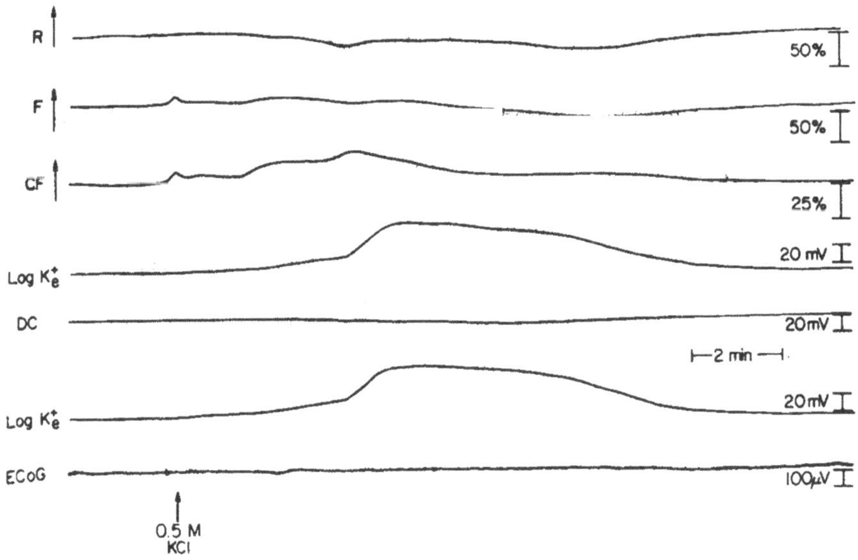


**Fig. 5.27** The effects of deep anesthesia on the metabolic, ionic, and electrical responses to spreading depression in the gerbil brain. The records do not show the complete recovery of the ECoG due to its later time [10]

## 5.3 Spontaneous Epileptic Activity

**Protocol 3.9 (Chap. 3, Table 3.1)**

Figure 5.28 [47] presents the progression of seizure activity in a special strain of gerbils, developing epileptic activity as a result of monotonic noise or other factors [48, 49]. As seen, the exposure of the awake gerbil to noise resulted in the development of epileptic activity followed by a CSD wave. The coupling between the two pathological events is clear and manifested by the electrical activity (ECoG) as well as extracellular $K^+$ levels and NADH redox state. During the epileptic stage, extracellular potassium increased and started to recover to the baseline, but then a larger elevation was recorded due to the developed CSD. There is a clear correlation between the various parameters during the epilepsy and the CSD. The decrease in NADH was smaller during the first stage, followed by an oxidation cycle typical for CSD.

Figure 5.29 [47] shows the continuation record of the same epileptic gerbil that developed the two interest responses. Here again, the two processes, namely, epileptic activity and CSD, are very clear. Figure 5.30 presents schematically the effects of six experimental perturbations on the hemodynamic, metabolic, ionic, and electrical activities in the gerbil brain.

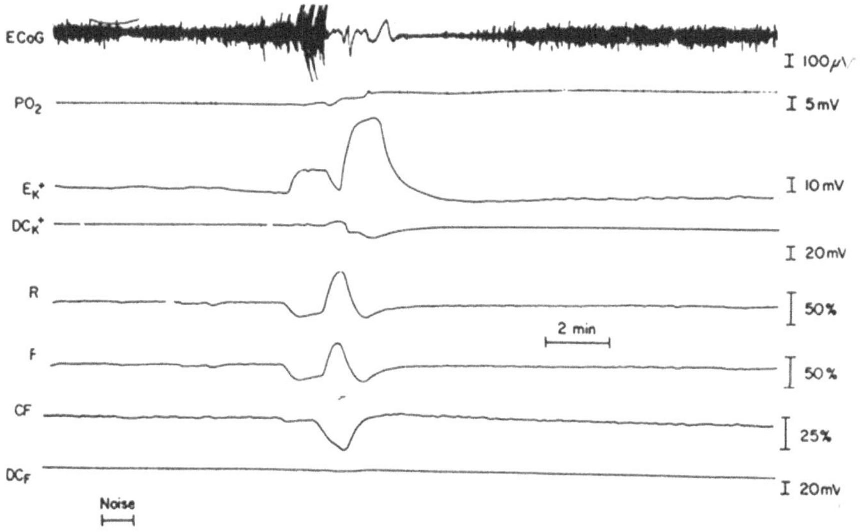


**Fig. 5.28** The development of epileptic activity followed by CSD in the brain of a seizure-prone gerbil. ECoG, electrocorticogram; $PO_2$, partial pressure of $O_2$ in the tissue; $EK^+$ and $DCK^+$, extracellular potassium and DC steady potential measured around the $K^+$ electrode; R, F, and CF, reflectance, NADH fluorescence, and corrected NADH fluorescence [46]

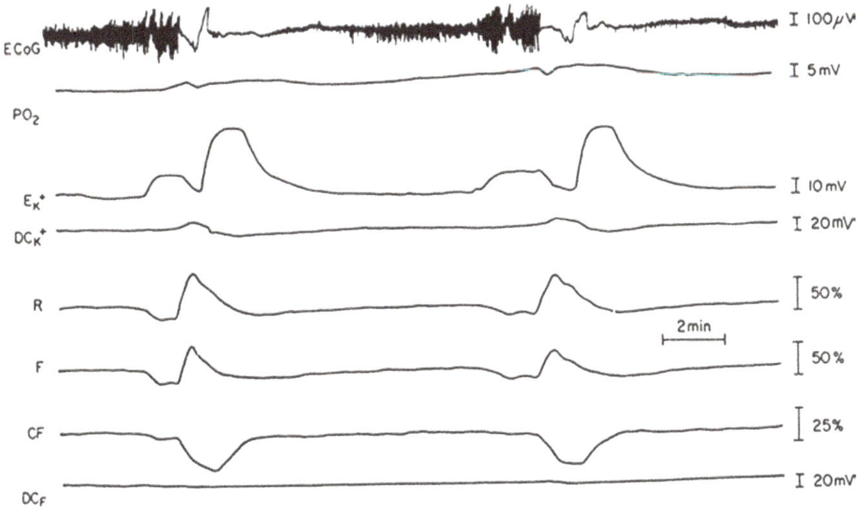


**Fig. 5.29** The development of two episodes of epileptic activity followed by CSD in the brain of a seizure-prone gerbil. This figure is a continuation of Fig. 5.28. ECoG, electrocorticogram; $PO_2$, partial pressure of $O_2$ in the tissue; $EK^+$ and $DCK^+$, extracellular potassium and DC steady potential measured around the $K^+$ electrode; R, F, and CF, reflectance, NADH fluorescence, and corrected NADH fluorescence [47]

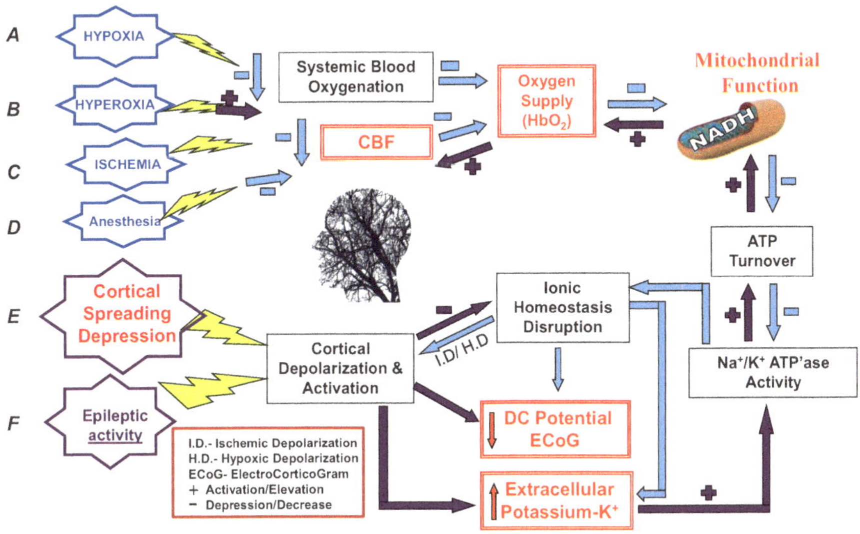


**Fig. 5.30** Schematic representation of the effects of four pathophysiological conditions (hypoxia, hyperoxia, ischemia, and anesthesia) and two brain activation events (CSD and epilepsia), on energetic-mitochondrial function, blood oxygenation, ionic homeostasis, and CBF in the Mongolian gerbil

## 5.4 Discussion

Partial ischemia had a uniform effect on the metabolic response to SD, namely, a reduction of NADH observed as compared to the normoxic state, due to the large demand for extra energy needed. Deep anesthesia and hypoxia have different effects on the three models used. The adult gerbil showed the same change in NADH due to SD (increase). We can conclude that the brain compensation mechanisms activated in the adult gerbil by SD under hypoxia and deep anesthesia cannot deliver the amount of $O_2$ needed for the complete recovery of the SD so a reduction (increase) of NADH was recorded. As shown previously, anesthesia cannot abolish the responses to SD. In the present study, under conditions where NADH showed an increase due to SD, the rate of $K^+$ influx into the cells was slower but was returned to the baseline level of extracellular potassium measured before the SD.

## References

1. Chance B, Williams GR (1956) The respiratory chain and oxidative phosphorylation. In: Nord FF (ed) Advances in enzymology. Interscience Publisher, Inc., New York, pp 65–134
2. Chance B, Williams GR (1955) Respiratory enzymes in oxidative phosphorylation (III-The steady state). J Biol Chem 217:409–427
3. Mayevsky A (1978) Pyridine nucleotide oxidation reduction state of the cerebral cortex in the awake gerbil. J Neurosci Res 3:369–374
4. Mayevsky A, Chance B (1974) Repetitive patterns of metabolic changes during cortical spreading depression of the awake rat. Brain Res 65:529–533
5. Chance B, Mayevsky A, Goodwin C, Mela L (1974) Factors in oxygen delivery to tissue. Microvasc Res 8:276–282
6. Mayevsky A (1976) Brain energy metabolism of the conscious rat exposed to various physiological and pathological situations. Brain Res 113:327–338
7. Mayevsky A, Chance B (1983) Multisite measurements of NADH redox state from cerebral cortex of the awake animal. Adv Exp Med Biol 159:143–155
8. Leão AAP, Morison R (1945) Propagation of spreading cortical depression. J Neurophysiol 8:33–45
9. Mayevsky A, Chance B (1975) Metabolic responses of the awake cerebral cortex to anoxia hypoxia spreading depression and epileptiform activity. Brain Res 98:149–165
10. Mayevsky A, Zarchin N, Friedli CM (1982) Factors affecting the oxygen balance in the awake cerebral cortex exposed to spreading depression. Brain Res 236:93–105
11. Mayevsky A, Crowe W, Mela L (1980) The interrelation between brain oxidative metabolism and extracellular potassium in the unanesthetized gerbil. Neurol Res 1:213–226
12. Leão AAP (1944) Spreading depression of activity in cerebral cortex. J Neurophysiol 7:359–390
13. Mayevsky A, Lebourdais S, Chance B (1980) The interrelation between brain $PO_2$ and NADH oxidation-reduction state in the gerbil. J Neurosci Res 5:173–182
14. Buresova O (1957) Changes in cerebral circulationin rats during spreading EEG depression. Physiol Bohemoslov 6:1–11
15. Van Harreveld A, Ochs S (1957) Electrical and vascular concomitants of spreading depression. Am J Phys 189:159–166

16. Leão AAP (1944) Pial circulation and spreading depression of activity in the cerebral cortex. J Neurophysiol 7:391–396
17. Rosenthal M, Somjen G (1973) Spreading depression, sustained potential shifts, and metabolic activity of cerebral cortex of cats. J Neurophysiol 36:739–749
18. Mayevsky A, Zeuthen T, Chance B (1974) Measurements of extracellular potassium, ECoG and pyridine nucleotide levels during cortical spreading depression in rats. Brain Res 76:347–349
19. Mayevsky A, Zarchin N (1988) Microcirculatory and ionic events during and after incomplete or complete cerebral ischemia in the Mongolian gerbil. In: Tomita M, Sawada T, Naritomi H, Heiss WD (eds) Cerebral hyperemia and ischemia. Excerpta Medica, pp 157–169
20. Lothman E, LaManna J, Cordingley G, Rosenthal M, Somjen G (1975) Responses of electrical potential potassium levels, and oxidative metabolic activity of the cerebral neocortex of cats. Brain Res 88:15–36
21. Crowe W, Mayevsky A, Mela L (1979) The dynamics of $K^+$ leakage and recovery in cerebral ischemia. In: Leter A, Saba T, Mela L (eds) Advances in shock research. Alan R. Liss, New York, pp 221–232
22. Mayevsky A, Yoles E, Zarchin N, Kaushansky D (1990) Brain vascular ionic and metabolic responses to ischemia in the Mongolian gerbil. J Basic Clin Physiol Pharmacol 1:207–220
23. Mayevsky A (1992) Interrelation between intracellular redox state and ion homeostasis in the brain *in vivo*. In: Frank K, Kessler M (eds) Quantitative spectroscopy in tissues. Verlasgruppe GmbH, pp 155–168
24. Friedli CM, Sclarsky DS, Mayevsky A (1982) Multiprobe monitoring of ionic, metabolic, and electrical activities in the awake brain. Am J Phys 243:R462–R469
25. Mayevsky A, Frank KH, Nioka S, Kessler M, Chance B (1990) Oxygen supply and brain function *in vivo*: a multiparametric monitoring approach in the Mongolian gerbil. In: Piper J, Goldstick TK, Meyer M (eds) Oxygen transport to tissue XII. Plenum Pub, pp 303–313
26. Chance B, Williams GR (1955) Respiratory enzymes in oxidative phosphorylation (I-Kinetics of oxygen utilization). J Biol Chem 217:383–393
27. Rosenthal M, Sick TJ (1988) Measurement of metabolic activity associated with ion shifts. In: Neuronal microenvironment. Humana Press, pp 187–246
28. Somjen GG, Rosenthal M, Cordingley G, LaManna J, Lothman E (1976) Potassium, neuroglia, and oxidative metabolism in central gray matter. Fed Proc 35:1266–1271
29. Mayevsky A (1992) Cerebral blood flow and brain mitochondrial redox state responses to various perturbations in gerbils. Adv Exp Med Biol 317:707–716
30. Lauritzen M (1987) Cerebral blood flow in migraine and cortical spreading depression. Acta Neurol Scand Suppl 113:1–40
31. Mayevsky A, Frank K, Muck M, Nioka S, Kessler M, Chance B (1992) Multiparametric evaluation of brain functions in the Mongolian gerbil *in vivo*. J Basic Clin Physiol Pharmacol 3:323–342
32. Lewis DV, O'Connor MJ, Schuette WH (1974) Oxidative metabolism during recurrent seizures in the penicillin treated hippocampus. Electroencephalogr Clin Neurophysiol 36:347–356
33. Lauritzen M (2001) Cortical spreading depression in migraine. Cephalalgia 21:757–760
34. Bures J, Buresova O, Krivanek J (1974) The mechanism and applications of Leão's spreading depression of electroencephalographic activity. Academic, New York
35. Vyskocil F, Kritz N, Bures J (1972) Potassium-selective microelectrodes used for measuring the extracellular brain potassium during spreading depression and anoxic depolarization in rats. Brain Res 39:255–259
36. Kraig RP, Nicholson C (1978) Extracellular ionic variations during spreading depression. Neuroscience 3:1045–1059
37. Mayevsky A, Chance B (1973) In: Bicher HI, Bruley DF (eds) A new long-term method for the measurement of NADH fluorescence in intact rat brain with implanted cannula. Plenum Press, New York, pp 239–244
38. Jobsis FF, O'Connor M, Vitale A, Vreman H (1971) Intracellular redox changes in functioning cerebral cortex. I. Metabolic effects of epileptiform activity. J Neurophysiol 3465:735–749

39. Leniger-Follert E, Urbanics R, Harbig K, Lubbers DW (1977) The behavior of local pH and NADH-fluorescence during and after direct activation of the brain cortex. In: Ingvar DH, Lassen NA (eds) . Acta Neurologica Scandinavica, Munkgaard, Copenhagen, pp 214–215
40. Crowe W, Mayevsky A, Mela L (1977) Application of a solid membrane ion selective electrode to *in vivo* measurements. Am J Phys 233:C56–C60
41. Mayevsky A (1978) Ischemia in the brain: the effects of carotid artery ligation and decapitation on the energy state of the awake and anesthetized rat. Brain Res 140:217–230
42. Van Harreveld A, Stamm JS (1952) Vascular concomitants of spreading cortical depression. J Neurophysiol 15:487–496
43. Marshall WH (1959) Spreading cortical depression of Leão. Physiol Rev 39:239–279
44. Lukyanova LD, Bures J (1967) Changes in $pO_2$ due to spreading depression in cortex and nucleus caudatus of rat. Physiol Bohemoslov 16:449–455
45. Van Harreveld A, Stamm JS (1953) Effect of pentobarbital and ether on the spreading cortical depression. Am J Phys 173:164–170
46. Mayevsky A, Rogatsky G (2007) Mitochondrial function *in vivo* evaluated by NADH fluorescence: from animal models to human studies. Am J Physiol Cell Physiol 292:C615–C640
47. Mayevsky A, Sonn J (2022) Cortical spreading depression of Leao, "from mitochondrial function to brain metabolic score (BMS)". Springer, Cham
48. Mayevsky A, Zarchin N, Yoles E, Tannenbaum B (1986) Oxygen supply to the brain in hypoxic and hyperoxic conditions. In: Nicolau C (ed) Advances in the biosciences. "$O_2$-transport in red blood cells". Pergamon Press, pp 119–132
49. Mayevsky A, Kaplan H, Haveri J, Haselgrove J, Chance B (1986) Three-dimensional metabolic mapping of the freeze-trapped brain: effects of ischemia on the Mongolian gerbil. Brain Res 367:63–72

# Chapter 6
# Neuroprotectants Effects on the Gerbil Brain

**Abstract** Temporary interruption or reduction of cerebral blood flow (CBF) during cerebrovascular surgery may rapidly result in ischemia or cerebral infarction. Nimodipine and thiopental have been shown to have cerebroprotective effects. The studies in this chapter were performed in order to examine whether the multiprobe assembly (MPA), which measures energy metabolism by CBF and mitochondrial NADH as well as extracellular ion concentrations ($K^+$), can shed light on the mechanism of the cerebroprotective effects of Nimodipine or thiopental. The MPA was placed on the brain of Mongolian gerbils, and cerebral ischemia was induced by occlusion of carotid arteries. Burst suppression of the electrocorticogram (ECoG) after thiopental was accompanied by a significant decrease in cerebral blood flow. In animals that received thiopental prior to ischemia, NADH increased to a lesser degree and extracellular potassium ion concentration increased to a lesser degree than in the control animals, indicating that thiopental affords protection of the brain under ischemic conditions due to improved energy metabolism. This study also demonstrates that the MPA can monitor changes occurring in the cerebral cortex even after the ECoG can no longer be used.

**Keywords** EEG burst suppression · Ischemic depolarization · Mitochondrial NADH · Ionic homeostasis · Brain microcirculation

## 6.1 Effects of Nimodipine in Gerbil's Brain Ischemia

### *6.1.1 Introduction*

Brain ischemia has significant implications for the development of damage in patients undergoing stroke treatment. In order to understand the mechanisms involved, a few animal models have been adopted. The Mongolian gerbil, having an incomplete circle of Willis, is widely used as a model for partial or complete ischemia [1–6]. The disturbances of ion homeostasis in the brain under ischemia were



A. Mayevsky, *The Mongolian Gerbil Brain*,
https://doi.org/10.1007/978-3-031-69549-0_6

recorded by a few groups using an extracellular $K^+$ electrode and, more recently, $Ca^{2+}$ electrode [6–11].

The involvement of calcium ions in the development of irreversible brain damage is a well-documented phenomenon [12–14]. During the last years, the effects of a dihydropyridine calcium antagonist, Nimodipine, were studied in experimental as well as clinical conditions [15–18]. The evidence for the beneficial effect of Nimodipine on the responses to brain ischemia is not solid as yet.

The purpose of the present investigation was to develop an animal model in which the effects of drugs such as Nimodipine with a potential protection property against cerebral ischemia could be tested. Using the surface fiber optic fluorometry/reflectometry, the intramitochondrial NADH redox state was monitored simultaneously with the hemodynamic properties of the observed tissue. This technique provides two signals, namely, the 450-nm fluorescence (NADH redox state) and the 366-nm reflectance trace (R), correlated mainly to the changes in tissue blood volume. We showed that under ischemia in the gerbil brain [19], a large increase in the 366 nm reflectance signal secondary reflectance increase (SRI) was recorded about 2 min after decapitation. This change in the R trace was correlated to the large ions shift as recorded by surface electrodes [6]. I am reporting here the effects of Nimodipine on the responses to complete ischemia-induced in the Mongolian gerbil by bilateral carotid artery occlusion.

### 6.1.2 Methods

*In the current study, the following protocol was used—Protocol 3.5 (Chap. 3, Table 3.1)*

**Animal Preparation**
Male gerbils (50–70 g) obtained from Tumblebrook Farms (MA, USA) were anesthetized by IP injection of Equithesin (0.3 mL/100 g) for the duration of the operation. Later on, a small volume of Equithesin was added (0.03–0.05 mL) every 30 min. The skull was exposed, and two holes were drilled in the parietal bone to hold the light guide holder. Two pairs of stainless steel screws (used also as electrocorticogram (ECoG) electrodes) were held in the parietal bone, and dental acrylic cement was used. The two common carotid arteries were isolated, and ligatures of 3-0 silk thread were placed around them. Nimodipine (0.5 mg/mL) was dissolved in the following solution composed of 969 g of polyethylene-glycol 400, 60 g of glycerin, and 100 g of water.

**Experimental Protocol**
The gerbil was connected to the monitoring system 10 min after the operation. In order to test the redox state of the brain, a 1-min $N_2$ exposure was used. Only gerbils showing a typical increase response were used further on. Each animal was exposed to 4–6 bilateral carotid arteries occlusions. The release of the aneurysm clips was

done after recording the typical SRI change in the reflectance trace, and the range was 2–6 min. The interval between successive occlusions was about 60 min.

Injection of Nimodipine, placebo, or saline solution was done 15–20 min before the third and fifth carotid arteries occlusions. Each gerbil was injected with 0.1 mL of Nimodipine, placebo or saline solution. Also, a short $N_2$ exposure was done 5–10 min before the occlusions for maximum response recording. Body temperature was measured and adjusted to be in the physiological range. The ECoG measured bilaterally was used for the evaluation of the recovery after each ischemic episode.

### 6.1.3 Results

A total of 31 gerbils were used, 8 of which were controls, 11 treated with a placebo, and 12 treated with Nimodipine. Each animal was exposed to two ischemic control episodes so that baseline responses were obtained before injections were done. Figure 6.1 shows two typical responses to bilateral carotid artery occlusion in a gerbil before and after placebo injection. Immediately after the occlusion, the redox state of the brain became reduced as seen on both hemispheres (CFR and CF) and reached a plateau level within 1 min. The reflectance trace shows a transient increase at the occlusion point, which later on showed a large secondary increase (SRI). This large increase in the reflectance led to an artifact in the corrected fluorescence trace (an increase or a decrease). Both the R and CF signals returned to the preischemic level after reperfusion was done ($R + L_{open}$). As can be seen, the differences between the responses to ischemia before and after placebo injection were minimal as expected.

In Fig. 6.2, the effects of Nimodipine injection are shown in another gerbil. The same pattern of responses to ischemia was recorded as shown in Fig. 6.1. During the second ischemia episode (after Nimodipine), the amplitude of the SRI was much smaller in the two hemispheres due to the drug injection. Although the NADH response to the occlusion was different in the two hemispheres, the consistency of the response before and after the treatment was very good in each hemisphere. In this specific gerbil, the recovery from the ischemia was not spontaneous, so we applied a short $N_2$ cycle to stimulate the recovery (empirical information).

In order to test the effects of Nimodipine in a statistical way, the analog signals were analyzed and quantitated for each ischemic episode.

The following three parameters were calculated:

1. Time to the appearance of the SRI
2. Maximal amplitude of the SRI
3. Maximal level of NADH during the ischemia (CF)

The results obtained are presented in Figs. 6.3, 6.4, and 6.5, and the general explanation of the way of calculation is as follows: In each gerbil, the responses obtained in the second occlusion were used as a reference to the other four

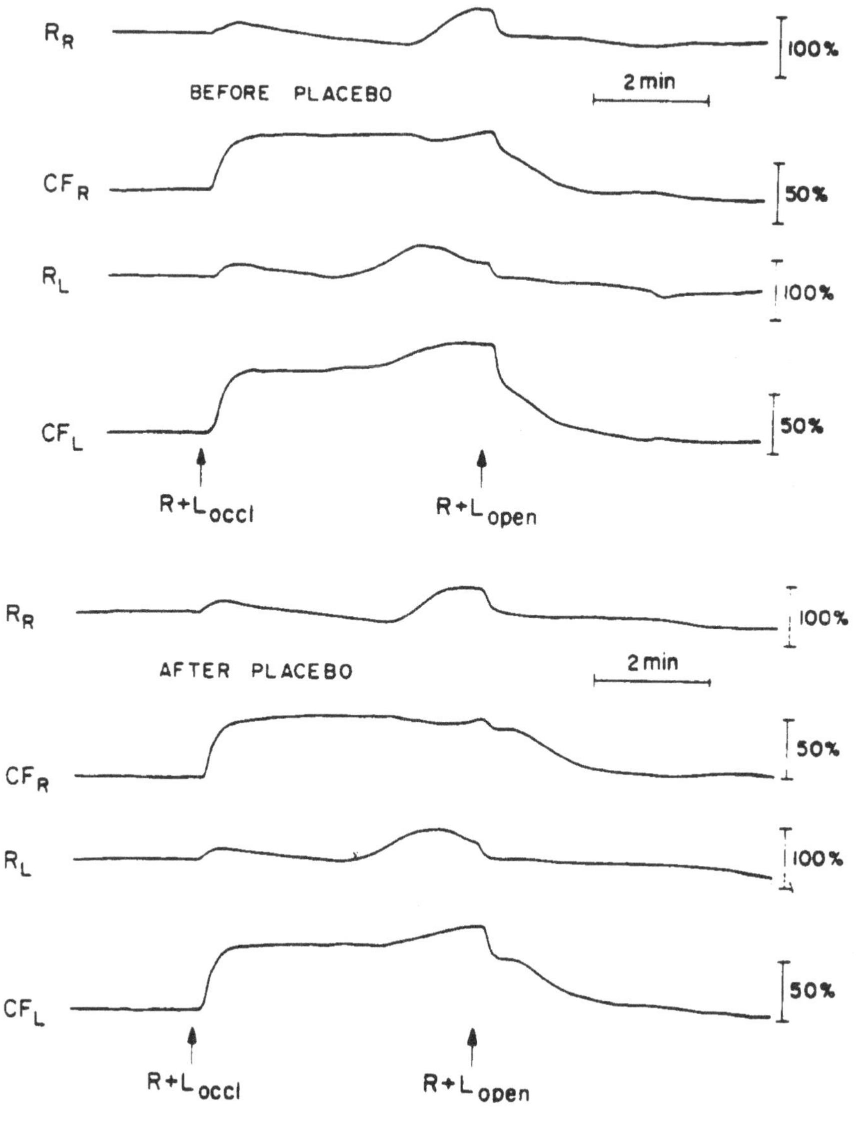


**Fig. 6.1** The effects of bilateral carotid artery occlusion on the metabolic and hemodynamic signals measured from the two hemispheres of the gerbil brain. Two ischemic episodes were induced before and after injection of placebo solution. Each point represents the mean and the standard error of it. $R_R$ and $R_L$—366-nm reflectance measured from the right and left hemispheres, respectively. $CF_R$ and $CF_L$—450-nm corrected fluorescence measured from the right and left hemispheres, respectively [20]

occlusions and got a value of 100%. All other responses are presented in percent change relative to the second occlusion. In all three parameters and all occlusions, the differences calculated between the control and placebo groups were very small

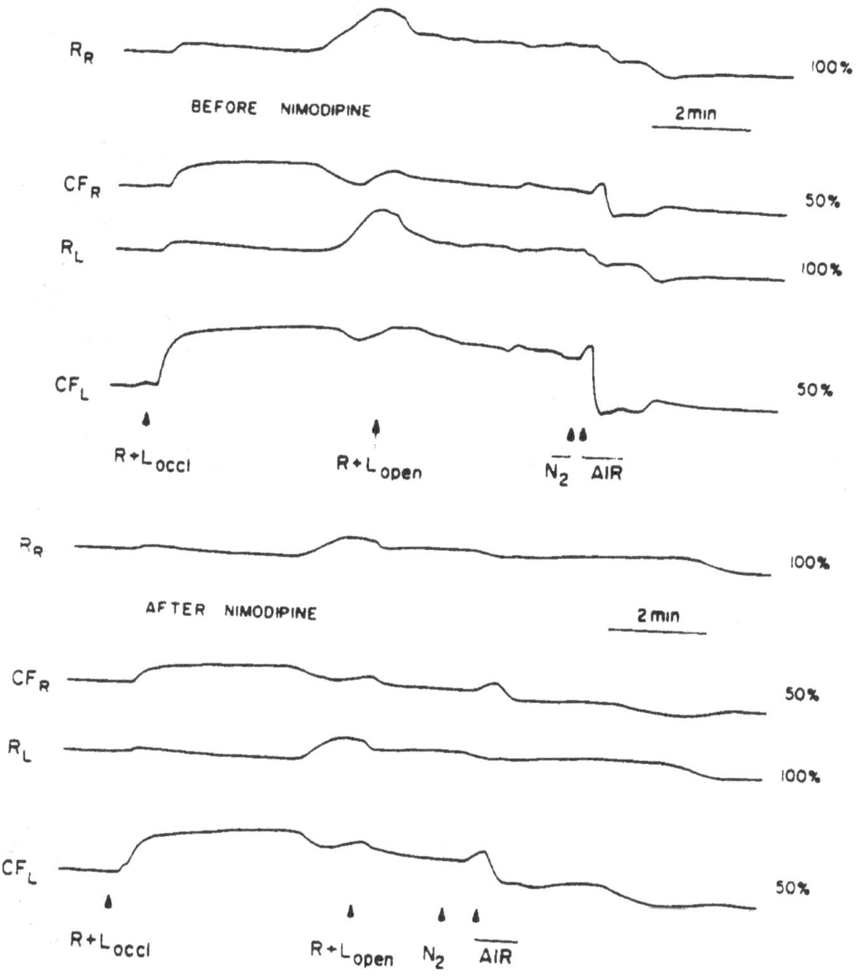


**Fig. 6.2** The effects of Nimodipine injection on the hemodynamic and redox state responses to bilateral carotid artery occlusion in the Mongolian gerbil. All symbols are as in Fig. 6.1 [20]

and statistically insignificant. The main effect was of Nimodipine as expected. As seen in Fig. 6.3, the Nimodipine delayed the appearance of the SRI by 10–40% as compared to the second ischemic episode, which served as a control. A slight decrease in time was recorded in the control and placebo animals.

The amplitude of the change during the SRI was decreased dramatically by the Nimodipine, as shown in Fig. 6.4. Here, a small difference was also recorded between the control and placebo groups. While in the SRI response, the effect of Nimodipine was clear, the NADH response was not so clear. As shown in Fig. 6.5, the maximal increase in NADH (CF) in the ischemic episodes decreased from the second to the last occlusion in the control and placebo animals. In the

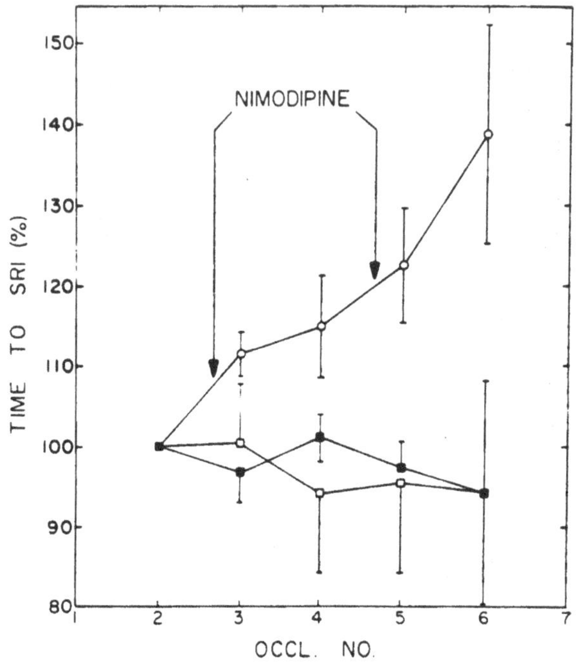


**Fig. 6.3** The effects of calcium channel blocker—Nimodipine on the time in which SRI appeared under bilateral carotid artery occlusion in the gerbils. SRI, secondary reflectance increase. Open and filled squares represent control and placebo-injected animals, respectively. Open circles represent Nimodipine-injected gerbils; the arrows are indicators of the two injection points during the experiment [20]

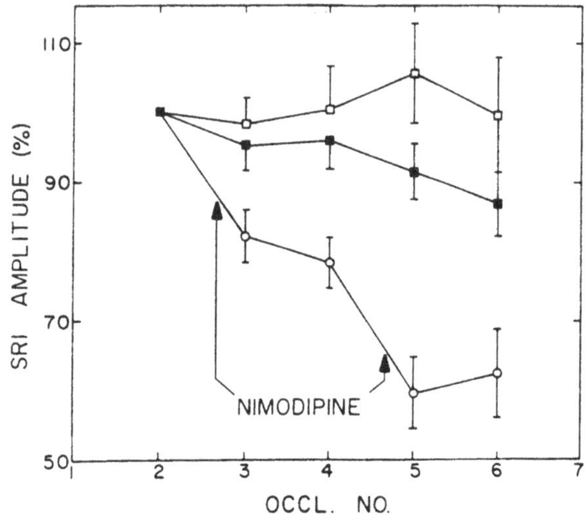


**Fig. 6.4** The responses of the SRI amplitude to Nimodipine under bilateral carotid artery occlusion in the Mongolian gerbil. Symbols are as in Fig. 6.3 [20]

Nimodipine-treated group, an initial increase in the change was recorded, which later on in the occlusion returned to the 100% level area. If one considers the decrease in response in the control and placebo groups, then the Nimodipine group remains high as compared to the control and placebo groups through the last four

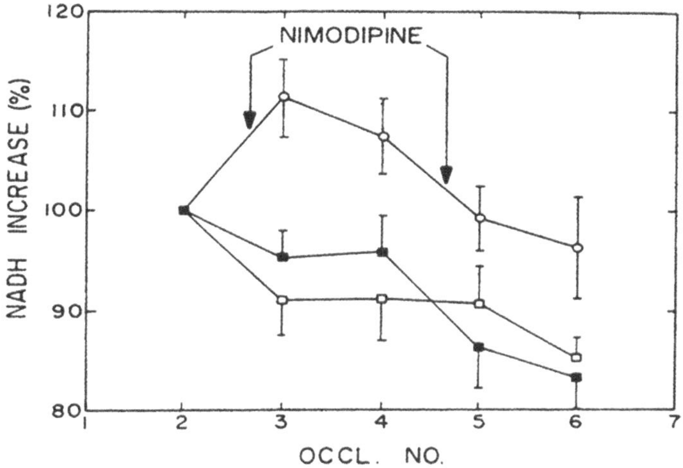


**Fig. 6.5** The effects of Nimodipine on the NADH redox state responses under cerebral ischemia in the gerbil. Symbols are as in Fig. 6.3 [20]

occlusions. The differences between the Nimodipine group and the control or placebo groups were significant through all the occlusions ($p < 0.05$; $p < 0.01$).

### 6.1.4 Discussion

The purpose of the present study was to develop a simple animal model in which the effects of calcium channel blockers could be analyzed on the level of microcirculation as well as the mitochondrial function. Using the surface fluorometry technique, two main parameters were measured, namely, the redox state of the intramitochondrial space and the changes in blood volume of the same measurement site [21].

In the gerbil model (*Meriones unguiculatus*), occlusion of the two carotid arteries produced in most animals is complete or close to a complete ischemia [4, 6]. The bilateral monitoring provides information on the two hemispheres, increasing the validity of the results, and also enables one to occlude one hemisphere and compare it to the unoccluded side in the same animal [19]. Two main phenomena were recorded under ischemic conditions, namely, the decrease in oxygen availability (increase in NADH levels) and the large increase in the reflectance trace (SRI). As suggested and shown previously, the SRI appeared as a result of a massive vasoconstriction that occurred when the large increase in extracellular $K^+$ recorded [6, 11, 22]. More recent studies showed that the Ca influx occurred at the same time as the large increase in $K^+$ and the SRI (Mayevsky, Yoles, and Zarchin, Unpublished results).

Therefore, the SRI was chosen as an indirect parameter to easily evaluate the appearance of the general depolarization event that occurred under ischemic conditions. Recently, two groups suggested the usage *of* ions shift and DC potential negativity as good parameters to evaluate the effects of calcium channel blockers [23, 24]. Therefore, our approach is a complimentary one and could be done more easily in many animals. As shown in Figs. 6.3 and 6.4, the two parameters calculated for the SRI event were affected significantly by the Nimodipine injection. The time for the appearance of the SRI was delayed, and the amplitude of the change was much smaller, suggesting the partial inhibition of the $Ca^{2+}$ influx to the vascular muscle cells, thus decreasing the level of vasoconstriction under ischemia. Those changes in the SRI parameters continued to appear after the first and second injections of Nimodipine and also through all the occlusions.

The responses of the mitochondrial redox state to the injection of Nimodipine were more complicated. As seen in Fig. 6.5, the increased change during the ischemia suggested that Nimodipine led to a more oxidized state of the mitochondria before the ischemia, and therefore the capacity of reduction was larger later on. Another possibility is that the resting oxygen supply was not affected by the Nimodipine and the increased response during ischemia was due to an uncoupling effect.

In summary, the results presented here suggest the beneficial effects of Nimodipine on the pathophysiology of ischemic insult. Nimodipine appears to affect cerebral blood flow (CBF) and NADH levels in the cerebral cortex, as it was round from reflectance and fluorescence changes during and after bilateral occlusions of the carotids in the gerbil. These effects may be attributed to the calcium antagonism effect of the Nimodipine. After 2–5 min of cerebral ischemia an increase in the amplitude of reflectance signal (SRI—secondary reflectance increase) is generally observed. This SRI reflects vasoconstriction of cortex blood vessels, and it is attributed to membrane depolarization and K leakage from cells. Maximum SRI during R + L ischemia was significantly lower ($p < 0.01$) in gerbils after 50 μg Nimodipine injection, relative to control or placebo-treated animals. In Nimodipine-treated animals, the time needed to the beginning of SRI seems to be longer relative to control or placebo-treated animals. According to these preliminary results, it seems that Nimodipine delayed the appearance of the massive depolarization and decreased its influence, as reflected by the delay and the decrease of the SRI amplitude.

## 6.2 Thiopental and Gerbil's Brain Ischemia

*In the current study, the following protocol was used—Protocol 3.5 (Chap. 3, Table 3.1)*

### 6.2.1 Introduction

Temporary interruption or reduction of cerebral blood flow during cerebrovascular surgery may rapidly result in ischemia or cerebral infarction. This is due to the fact that the brain is an obligate aerobe, utilizing approximately 20% of the total oxygen consumption of the body. Anaerobic metabolism via glycolysis cannot supply enough energy to support cerebral metabolism. Barbiturate anesthesia results in a dose-dependent reduction of the cerebral blood flow and the cerebral metabolic rate (CMR), and with the onset of anesthesia, CBF and the cerebral metabolic rate for oxygen ($CMRO_2$) are reduced by 30–50% [25–28]. Various hypoxic and ischemic models have shown that barbiturates protect the brain [29, 30]. In particular, thiopental has been shown to have cerebroprotective effects [31–34]. Thus, during temporary clipping of intracranial vessels during aneurysm surgery, thiopental is frequently used to provide cerebral protection in case of inadequate collateral perfusion. The dose of thiopental demonstrated to provide cerebral protection results in burst suppression of the EEG [35–37]. Since the EEG is the most commonly used monitor of cerebral viability and is used for the detection of cerebral ischemia, burst suppression may render this parameter useless for detecting changes in cerebral metabolism under thiopental surgery. In order to increase the margin of safety during neurosurgical procedures, in which burst suppression of the EEG is induced by thiopental anesthesia, it would be useful to have a reliable monitor of cerebral ischemia and to be able to reduce the metabolic rate of the brain, thus improving the oxygen balance. This would, in effect, prolong the period during which feeding vessels can be occluded and would allow more time in which to perform the necessary surgery without fear of causing irreversible brain damage.

In our laboratory, we have developed a multiprobe assembly (MPA), which enables the continuous, simultaneous, online, in vivo measurement of several electrical, ionic, hemodynamic, and metabolic parameters of the cerebral cortex. These parameters include electrocorticogram, laser Doppler measurement of cerebral blood flow and volume, mitochondrial NADH redox level, extracellular concentration of potassium and calcium ions, brain temperature, and DC steady potential. This assembly has been used successfully in experimental animals [38, 39] and also in patients hospitalized in the neurosurgical unit of a hospital [40]. In previous studies, we have shown that NADH is an indicator for energy metabolism (oxygen supply) and, when coupled to the measurement of extracellular $K^+$ (energy utilization from ion pumps), may supply information on the mechanism of brain function under different conditions, such as during ischemia in the brain. We have used this assembly successfully on the gerbil brain during decapitation [19] and unilateral and bilateral ischemia [39]. In this study, we attempted to determine the relationship between these parameters under thiopental anesthesia in order to shed light on the mechanism of thiopental protective effect against ischemia by the measurement of the supply and the demand for energy under conditions of EEG burst suppression.

## 6.2.2 Methods

### 6.2.2.1 Multiprobe Assembly (MPA)

This unit enables the simultaneous and continuous measurement of numerous parameters from the same area of the cerebral cortex. Hemodynamic changes were measured using laser Doppler flowmetry. The metabolic state was monitored using the light guide fluorometer for measuring changes in the intramitochondrial NADH redox state. Changes in the extracellular concentration of potassium, calcium, and hydrogen ions were measured with the aid of specific mini-electrodes. This system also included electrodes suitable for measuring the spontaneous electrical activity of the brain ECoG, electrodes for measuring the DC potential, and a probe for measuring brain temperature. All these components were mounted in a special design holder that was positioned on the cerebral cortex. Body temperature was monitored in all animals using a rectal thermistor. Details for the preparation of the probes and the MPA have been published previously [38, 41].

### 6.2.2.2 Animal Preparation

All experimental protocols were approved by the Institutional Animal Care and Use Committee of the Bar-Ilan University. Mongolian gerbils weighing 50–70 g were anesthetized by an intraperitoneal injection of 0.3 mL/100 g. Each ml of Equithesin contains 9.72 mg pentobarbital, 42.51 mg chloral hydrate, 21.25 mg magnesium sulfate, 44.34% (w/v) propylene glycol, 11.5% alcohol, and water. A polyethylene tube was inserted into the trachea for artificial ventilation. Both carotid arteries were isolated, and a silk suture was placed under each for future occlusion. The gerbil was then placed on its stomach on a heated pad. After a midline incision of the skin above the skull, a 6.0-mm hole was drilled in the parietal bone, and the dura mater was gently removed. The multiprobe assembly was held by a micromanipulator attached to the operating table and was lowered to make gentle contact with the brain. The MPA was cemented to the skull with dental acrylic cement. The animal was connected to the measuring system 10–20 min after the operation. All parameters were monitored and recorded on a 12-channel Grass polygraph and simultaneously recorded on the computer (CODAS A/D and software, DATAQ, USA. 2.3). Experimental protocol: The gerbils were allowed to recover from surgery, and all parameters measured reached steady-state values. Body temperature was monitored and remained constant during the entire experimental procedure. Brain temperature was monitored, although absolute brain temperature could not be measured due to technical problems. After thiopental injection, brain temperature remained unchanged or demonstrated a small, non-significant increase. This trend was also observed upon inducing cerebral ischemia. Various doses of thiopental were administered IP until it was found that a dose of 7.5 mg thiopental induced burst suppression of the ECoG in the gerbil similar to that observed during surgical clipping of aneurysms. After determining the effective dose of thiopental, this dose (7.5 mg gerbil) was administered to seven gerbils. Cerebral ischemia was induced after burst suppression of the ECoG by irreversible occlusion of one or both carotid arteries until cerebral blood flow was reduced by approximately 90%. Control animals ($N = 7$) underwent a similar occlusion without prior thiopental administration, that is, a total of 14 animals.

#### 6.2.2.3 Statistical Analysis

Statistical analysis was performed using the student's *t*-test. Values are expressed as means ± S.E.M. Significance was determined at $p < 0.05$. For the calculation of NADH, a normalization technique was used since this parameter could not be calculated quantitatively. Each animal was allowed to breathe nitrogen anoxia at the beginning of the experiment for 30 s, and the value of NADH reached during this anoxic period was considered the 100% for this animal. All changes in NADH levels for each animal were calculated relative to the maximum level reached during anoxia.

### *6.2.3 Results*

Thiopental administration at a dose of 7.5 mg induced burst suppression of the ECoG in all animals. A typical reaction of the ECoG following thiopental administration may be observed in the polygraph trace presented in Fig. 6.6. It can be seen that the ECoG was normal before thiopental administration, whereas by 3.5 min after thiopental administration the ECoG exhibited burst suppression. In control animals, which were not given thiopental, the ECoG remained unchanged. However, ischemia resulted in a marked decrease in the ECoG in the control animals. Burst suppression of the ECoG was accompanied by a significant decrease in the CBF compared to the control animals (77 ± 3% vs. 100%, $p < 0.05$) (Fig. 6.7b). No changes were recorded in the other measured parameters (such as NADH, extracellular ion concentrations, and DC potentials) following the administration of thiopental. In all animals (controls and thiopental-injected animals), occlusion of one or

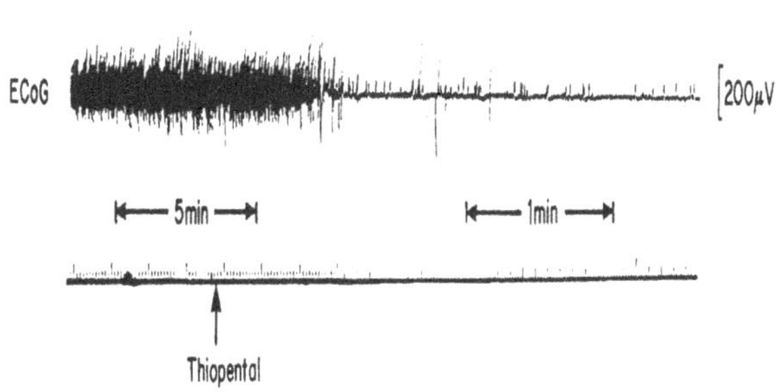


**Fig. 6.6** Polygraph tracing of ECoG before, during (arrow), and after IP administration of 7.5 mg thiopental. Burst suppression occurred 3.5 min following thiopental administration [42]

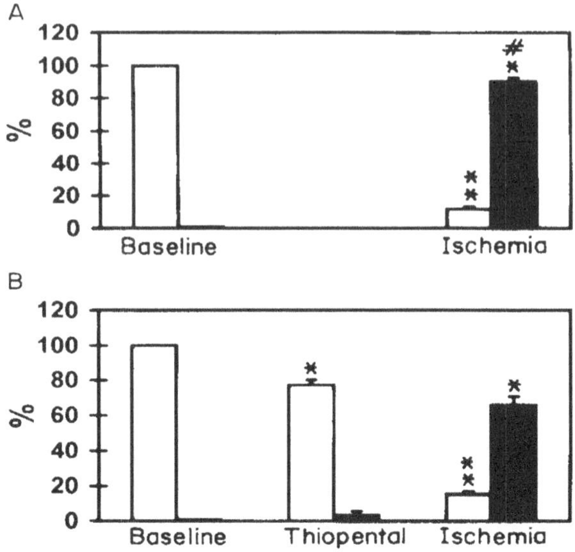


**Fig. 6.7** Histogram presenting the changes in CBF (empty bars) and NADH (filled bars) in control (**a**) and in the experimental animals (**b**). *$p < 0.05$ compared to the baseline; **$p < 0.01$ compared to the baseline; #$p < 0.05$ control group compared to the experimental group [42]

both carotid arteries resulted in an immediate, similar decrease in CBF. A polygraph trace exhibiting the reactions of a control and an experimental animal to carotid artery occlusion is presented in Fig. 6.8. It can be seen that in the control animal (A), occlusion resulted in an immediate decrease in CBF followed by an increase in NADH. An initial gradual increase in the extracellular potassium ion concentration without a change in the DC potential was recorded. Two minutes after the occlusion, a large increase in extracellular $K^+$ was recorded, together with a negative shift in the DC potential. This event is typical of ischemic depolarization. A large increase in the reflectance signal R represents massive vasoconstriction. In contradistinction, occlusion in animals in which ECoG burst suppression thiopental anesthesia preceded the occlusion B, NADH increased to a lesser extent, and extracellular potassium ion concentrations exhibited only a minor and gradual increase, which was followed by a small negative shift in the DC potential. The large increase in extracellular K was not recorded. Figure 6.8 demonstrates that the reaction to carotid artery occlusion was not identical in the control and experimental animals. Ischemia in the control animals resulted in a decrease in blood flow to a level that was not statistically different in the animals in which ECoG burst suppression was induced by thiopental administration (11.5 ± 1.5% vs. 15.7 ± 1.5%) (Fig. 6.7). However, the increase in NADH levels following this decrease in cerebral blood flow was significantly higher in the control group than in the experimental group (90 ± 2.4% vs. 65.75 ± 4.8%, $p < 0.05$) (Fig. 6.7). The leakage of potassium ions out of the cells following ischemia was significantly greater in the control animals than in those in which thiopental was administered prior to occlusion (19.7 ± 4 vs. 9.3 ± 1.3 mM, $p < 0.05$) (Fig. 6.9a). The time until a large increase in extracellular potassium ion concentrations was observed was significantly longer in the experimental animals (1.8 ± 0.75 vs. 3.97 ± 0.65 min, $p < 0.05$) (Fig. 6.9b). Changes in the DC potential

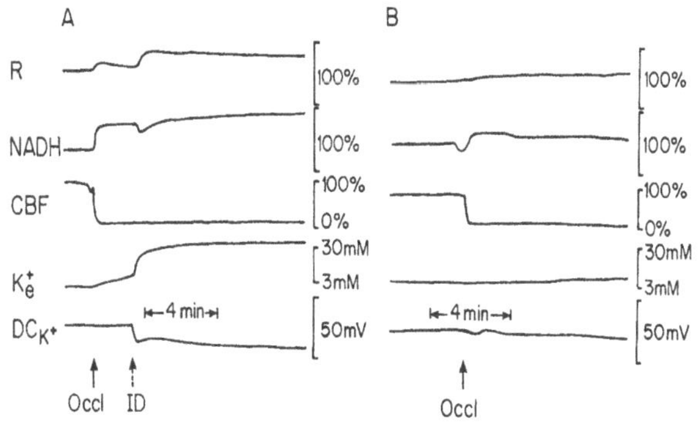


**Fig. 6.8** Polygraph tracing of the effects of unilateral carotid artery occlusion on the responses of the gerbil brain in a control animal (**a**) and after thiopental administration (**b**). R, reflectance; NADH, intramitochondrial NADH levels; CBF, laser Doppler cerebral blood flow; $K_e^+$, extracellular potassium ion concentration; DC $K^+$, direct current potential concentric to the potassium ion electrode; Occl, occlusion of carotid arteries; ID, ischemic depolarization [42]

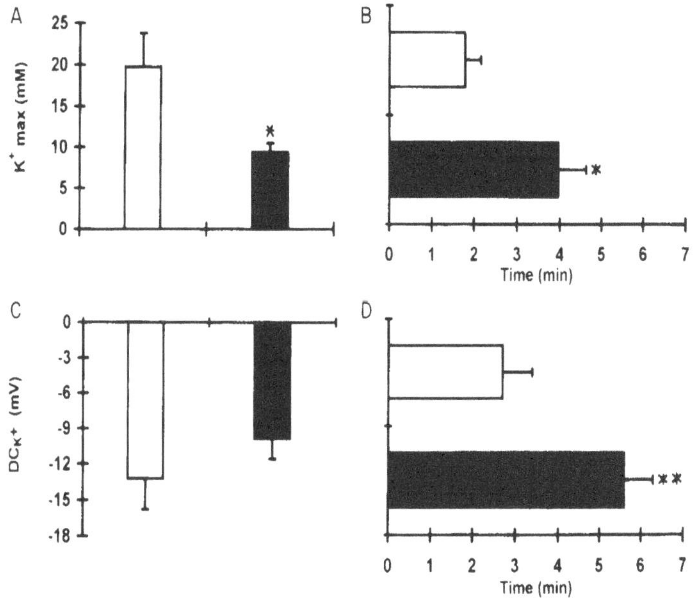


**Fig. 6.9** Histogram depicting the changes in extracellular $K^+$ concentrations and $DCK^+$ after carotid artery occlusion in the control group (empty bars) and in the experimental group (filled bars). (**a**) The maximal value of extracellular $K^+$ (mM) after carotid artery occlusion; (**b**) the time until the large increase in the extracellular $K^+$ concentration ($n = 5$) (in two of the gerbils it was not possible to use the $K^+$ measurements); (**c**) the maximal negative shift in the $DCK^+$ (mV) after carotid artery occlusion; and (**d**) the time until the large decrease in $DCK^+$. $^*p < 0.05$; $^{**}p < 0.01$ [42]

were not statistically significant. However, the time to reach the minimum $DC_{K+}$ was significantly longer in the $K^+$ experimental group. It should be noted that the time to the secondary increase in reflectance [39] in the control animals was $1.47 \pm 0.49$ min, and the maximal level of the reflectance reached 53% above baseline. Under the effect of thiopental, seven out of the ten experimental animals did not exhibit a secondary increase in reflectance at all, and in the three animals in which this increase was observed, it increased by 10–50% ($p < 0.05$).

### *6.2.4 Discussion*

Thiopental anesthesia is frequently used during neurosurgery due to its cerebroprotective effect [43]. However, the mechanism for this protection has not yet been elucidated [31]. This study supports previous studies indicating a cerebroprotective effect of thiopental anesthesia. Thiopental induced its protective effect by beneficially affecting cerebral energy metabolism. It was shown that NADH levels, which have been shown to be inversely proportional to the oxygen balance [44], did not increase to the same extent in ischemia-induced following thiopental as they did in control animals (Fig. 6.7b), although blood flow decreased to the same extent. This indicates that the oxygen balance was not impaired to the same extent following thiopental anesthesia. This is in agreement with Drummond and Shapiro [25], who stated that the component of cerebral metabolism on which anesthetics act is predominantly that associated with electrical activity. Furthermore, Musch et al. [32] reported a decrease in the metabolic activity of the brain following thiopental anesthesia, and Runciman et al. [45] observed that thiopental anesthesia resulted in a 24% decrease in oxygen consumption.

It has been reported that barbiturates also modulate glutamate receptors [46], indicating that these drugs may affect the ischemic membrane depolarization and thus reduce energy demand. However, a major part of the ATP formed in the brain is used for maintaining ion homeostasis, and during "unprotected" ischemia, the oxygen supply to the brain may be decreased to such an extent as to result in irreversible damage. In this study, the improvement in oxygen balance caused by thiopental anesthesia apparently prevented the depletion of high energy compounds during ischemia, as expressed by the slower leakage of potassium out of the cells and the lower level of this leakage. These results suggest that the inhibition of $Na^+$-$K^+$-ATPase under the thiopental treatment was not as extensive as in the control animals. We have previously shown [6] that the slope of K leakage was correlated to the extent of ischemia-induced by unilateral carotid artery occlusion. It is possible that this is due to changes in membrane properties, enabling increased $K^+$ leakage out of the cells. Similar results were obtained in this laboratory when comparing the effects of 2000 ppm CO on brain function. Pentobarbital anesthesia was found to have a protective effect on the brain as opposed to the unanesthetized brain exposed to 2000 ppm CO [47]. This indicates the importance of measuring extracellular $K^+$ for determining the effects of thiopental anesthesia on the brain.

Thiopental administration results in an improved oxygen balance or energy state of the ischemic brain. Therefore, it is considered a suitable agent for anesthesia for procedures involving a risk of decreased oxygen supply to the cells. Thiopental is also known to cause cardiovascular depression and hypotension [48, 49]. This may be beneficial under conditions such as intracranial aneurysm clipping but may be harmful in patients in whom cardiovascular depression is counter-indicated due to a decrease in cardiovascular reserve. In this study, blood pressure was not measured due to technical difficulties of measurement in the gerbil. Experiments are now planned in which blood pressure, as well as blood gases, will be measured in the rat under thiopental anesthesia. Awareness of the oxygen balance and the metabolic state of the brain during neurosurgery is important since the brain is a completely aerobic organ and depends upon an adequate blood and oxygen supply for its normal function. If it is not possible to monitor brain function during surgical interventions, irreversible damage to neural tissue may ensue and may counteract the beneficial effects of the surgery, leading to more serious damage. In this laboratory, we have developed a multiprobe assembly that can monitor brain function in experimental animals [38, 39] and in patients in the operating theater using the same monitoring principles [40]. The model used to examine the feasibility of using this multiprobe assembly for monitoring brain function after burst suppression of the EEG was carotid artery occlusion, which resulted in a decreased blood supply to the cerebral cortex and increased levels of intramitochondrial NADH (indicating a shortage of oxygen). The reason that carotid artery occlusion (cerebral ischemia) was used for determining whether this assembly can monitor brain function even after the ECoG was no longer an available tool is twofold: (1) This is a model that we have studied extensively [6, 39] and have enough data for comparing the ischemic response with and without therapeutic intervention, in this case, thiopental administration. (2) Cerebral ischemia is one of the more serious complications that can arise during neurosurgery, and it is important to determine whether this assembly can detect this ischemia even after the suppression of the spontaneous electrical activity. Our results demonstrate that the MPA detects changes in brain parameters even after ECoG suppression by thiopental.

In conclusion, this study supports the claim that thiopental affords protection to the brain under ischemic conditions and that this is due, at least partially, to improved energy metabolism, as indicated by the slower and smaller leakage of potassium ions out of the cells and a smaller increase in NADH levels.

## 6.3 Flunarizine and Gerbil's Brain Ischemia and CSD

*In the current study, the following protocol was used—Protocol 3.12 (Chap. 3, Table 3.1)*

The comparison between SD and ischemia was also performed in gerbils. The effects of a calcium channel blocker-flunarizine injection on the responses of the gerbil brain were tested.

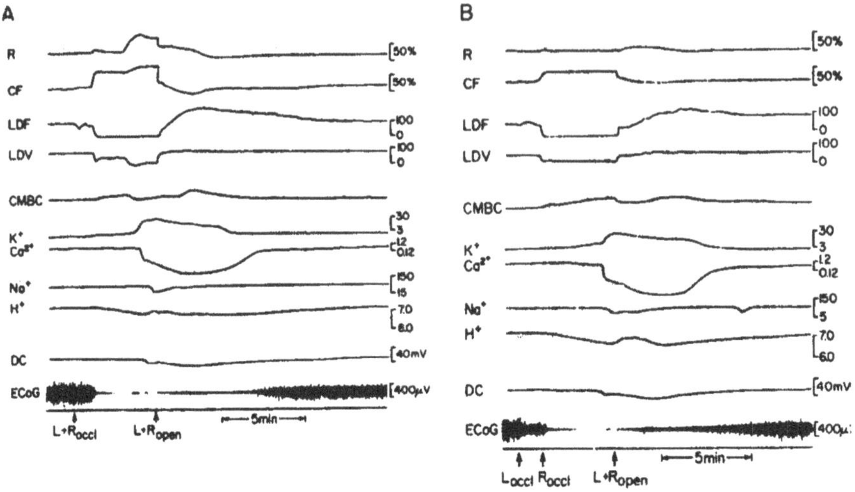


**Fig. 6.10** The effects of flunarizine (40 mg/kg i.p.) on the responses to cerebral ischemia in the Mongolian gerbil [50]. (**a**) Before and (**b**) after injection of Flunarizine

Figure 6.10a shows the effects of bilateral carotid artery occlusion (4 min), and in Fig. 6.10b, the protective effect of flunarizine is shown. Only five experiments using flunarizine were performed; therefore, the results are only preliminary ones. In this specific gerbil, the connection between the two anterior cerebral arteries [51] was very good; therefore, the occlusion of one artery was not enough to create ischemia. Under complete ischemia, the CBF reached very low values, as also indicated by the maximum level of the NADH redox state. An ischemic depolarization (ID) develops within 2–2.5 min characterized by the massive vasoconstriction (large increase in R signal and a decrease in blood volume (LDV)). The ionic responses to the ID were typical as described before [6, 52]. After the injection of flunarizine, the responses to the same length of ischemia were different, as shown in Fig. 6.10b. The main effect of flunarizine is in the delay of the appearance of the ID event. The changes in blood volume were blocked almost completely, as seen during the 4 min of ischemia. Also, the responses of the various ions were not identical to those recorded before the flunarizine injection. Due to the flunarizine injection, the hyperemic response after the ischemic episode (reperfusion hyperemia) was also smaller.

The effects of flunarizine on the responses to SD are shown in Fig. 6.11a, b. It seems that the leakage of $K^+$ and the large increase in CBF during the SD wave were attenuated by flunarizine. Using the multisite monitoring system [44] for monitoring NADH and DC potential, flunarizine showed clear effects.

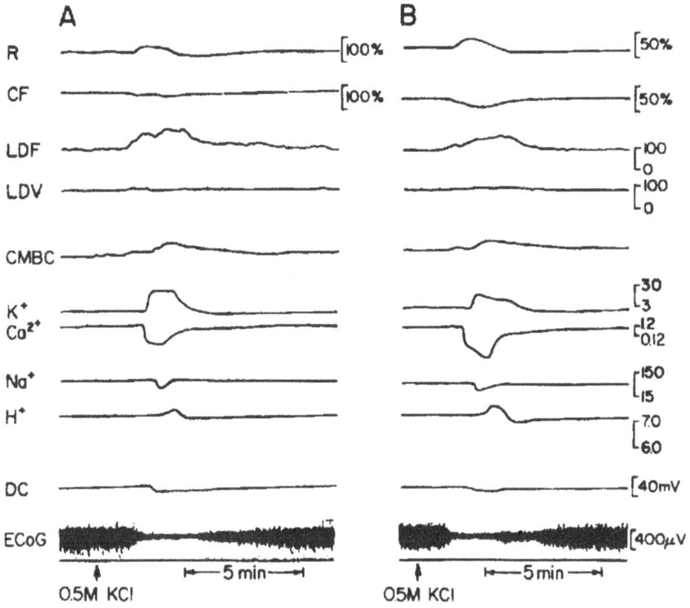


**Fig. 6.11** Metabolic-hemodynamic, ionic, and electrical responses to spreading depression before (**a**) and after (**b**) flunanizine (40 mg/kg i.p.) injection in the Mongolian gerbil [50]

# References


1. Mayevsky A (1978) Pyridine nucleotide oxidation reduction state of the cerebral cortex in the awake gerbil. J Neurosci Res 3:369–374
2. Levy DE, Brierley JB (1974) Communications between vertebro-basilar and carotid arterial circulations in the gerbil. Exp Neurol 45:503–508
3. Berry K, Wisniewski HM, Svarzbein L, Baez S (1975) On the relationship of brain vasculature to production of neurological deficit and morphological changes following acute unilateral common carotid artery ligation in gerbils. J Neurol Sci 25:75–92
4. Kobayashi M, Lust WD, Passonneau JV (1977) Concentrations of energy metabolites and cyclic nucleotides during and after bilateral ischemia in the gerbil cerebral cortex. J Neurochem 29:53–59
5. Mies G, Hossmann KA (1983) A method for regional evaluation of potassium depletion during graded focal ischemia in gerbils. J Cereb Blood Flow Metab 3:S313–S314
6. Mayevsky A, Friedli CM, Reivich M (1985) Metabolic, ionic and electrical responses of the gerbil brain to ischemia. Am J Phys 248:R99–R107
7. Choki J, Greenberg J, Sclarsky D, Reivich M (1984) Correlation between brain surface potassium and glucose utilization after bilateral cerebral ischemia in the gerbil. Stroke 15:851–857
8. Harris RJ, Symon L, Branston NM, Bayhan M (1981) Changes in extracellular calcium activity in cerebral ischaemia. J Cereb Blood Flow Metab 1:203–209
9. Branston NM, Strong AJ, Symon L (1977) Extracellular potassium activity, evoked potential and tissue blood flow. Relationships during progressive ischaemia in baboon cerebral cortex. J Neurol Sci 32:305–321
10. Hansen AJ, Zeuthen T (1981) Extracellular ion concentrations during spreading depression and ischemia in the rat brain cortex. Acta Physiol Scand 113:437–445

11. Mayevsky A, Zarchin N (1987) Metabolic ionic and electrical activities during and after incomplete or complete cerebral ischemia in the Mongolian gerbil. Adv Exp Med Biol 215:265–273
12. Farber JL (1981) The role of calcium in cell death. Life Sci 29:1289–1295
13. Nowicki JP, MacKenzie ET, Young AR (1982) Brain ischaemia, calcium and calcium antagonists. Pathol Biol (Paris) 30:282–288
14. Siesjo BK (1981) Cell damage in the brain: a speculative synthesis. J Cereb Blood Flow Metab 1:155–185
15. Gelmers HJ (1987) Nimodipine in ischemic stroke. Clin Neuropharmacol 10:412–422
16. Heffez DS, Passonneau JV (1985) Effect of nimodipine on cerebral metabolism during ischemia and recirculation in the mongolian gerbil. J Cereb Blood Flow Metab 5:523–528
17. Milde LN, Milde JH, Michenfelder JD (1986) Delayed treatment with nimodipine improves cerebral blood flow after complete cerebral ischemia in the dog. J Cereb Blood Flow Metab 6:332–337
18. Vibulsresth S, Dietrich WD, Busto R, Ginsberg MD (1987) Failure of nimodipine to prevent ischemic neuronal damage in rats. Stroke 18:210–216
19. Mayevsky A, Zarchin N (1981) The effects of unilateral carotid occlusion on the responses to decapitation in the gerbil brain. Brain Res 206:155–160
20. Cohen S, Mayevsky A (1989) Effects of nimodipine on the responses to cerebral ischemia in the Mongolian gerbil. Adv Exp Med Biol 248:429–438
21. Mayevsky A (1984) Brain NADH redox state monitored *in vivo* by fiber optic surface fluorometry. Brain Res Rev 7:49–68
22. Mayevsky A, Yoles E, Zarchin N (1986) Metabolic, ionic and electrical responses to oxygen deficiency in the newborn dog *in vivo*. Adv Exp Med Biol 200:261–269
23. Dierking H, Tegtmeier E, Holler M, Peters TL (1987) Effects of flunarizine, nimodipine, and diphenylhydantoine on K, Ca and DC in the flunarizine, nimodipine, and diphenylhydantoine on hypoxic cortex of the rat brain *in vivo*. J Cereb Blood Flow Metab 7:S 156
24. Mori K, Iwayama K, Kawano T, Kaminogo M (1978) DC potential and extracellular $K^+$ and $Ca^{2+}$ at critical levels of brain ischemia in cats. J Cereb Blood Flow Metab 7:s112
25. Drummond JC, Shapiro HM (1990) Cerebral physiology. In: Miller RD (ed) Anesthesia. Churchill Livingstone, New York, pp 621–658
26. Donegan J (1991) Effect of anaesthesia on cerebral physiology and metabolism. In: Newfield P, Cottrell JE, Wilson CB (eds) Neuroanaesthesia: handbook of clinical and physiologic essential. Little Brown, Boston, pp 17–29
27. Schregel W, Bredenkotter U, Sihle-Wissel M, Cunitz G (1995) Transcranial Doppler ultrasound: effects of intravenous anesthetics in neurosurgical patients. Ultraschall Med 16:60–64
28. Smith AL (1977) Barbiturate protection in cerebral hypoxia. Anesthesiology 47:285–293
29. Drummond JC, Cole DJ, Patel PM, Reynolds LW (1995) Focal cerebral ischemia during anesthesia with etomidate, isoflurane, or thiopental: a comparison of the extent of cerebral injury. Neurosurgery 37:742–748; discussion 748–749
30. Shapiro HM (1985) Barbiturates in brain ischaemia. Br J Anaesth 57:82–95
31. Guo J, White JA, Batjer HH (1995) The protective effects of thiopental on brain stem ischemia. Neurosurgery 37:490–495
32. Musch TI, Pelligrino A, Dempsey JA (1980) Effects of prolonged $N_2O$ and barbiturate anesthesia on brain metabolism and pH in the dog. Respir Physiol 39:121–131
33. Sano T, Patel PM, Drummond JC, Cole DJ (1993) A comparison of the cerebral protective effects of etomidate, thiopental, and isoflurane in a model of forebrain ischemia in the rat. Anesth Analg 76:990–997
34. Shimoji K, Takahata Y, Fujiwara N, Endoh H, Taga K, Ohama E (1987) Effects of pentobarbital and ketamine on brain injury-induced anti-ischemic activity. Brain Res 408:385–388
35. Clark DL, Rosner BS (1973) Neurophysiologic effects of general anesthetics. I. The electroencephalogram and sensory evoked responses in man. Anesthesiology 38:564–582
36. Fragen RJ, Avram MJ (1990) Barbiturates. In: Miller RD (ed) Anaesthesia. Churchill Livingstone, New York, pp 225–242

37. MacIver MB, Mandema JW, Stanski DR, Bland BH (1996) Thiopental uncouples hippocampal and cortical synchronized electroencephalographic activity. Anesthesiology 84:1411–1424
38. Mayevsky A, Yoles E, Zarchin N, Kaushansky D (1990) Brain vascular ionic and metabolic responses to ischemia in the Mongolian gerbil. J Basic Clin Physiol Pharmacol 1:207–220
39. Mayevsky A (1990) Level of ischemia and brain functions in the Mongolian gerbil *in vivo*. Brain Res 524:1–9
40. Mayevsky A, Doron A, Manor T, Meilin S, Zarchin N, Ouaknine GE (1996) Cortical spreading depression recorded from the human brain using a multiparametric monitoring system. Brain Res 740:268–274
41. Mayevsky A, Meilin S, Rogatsky GG, Zarchin N, Sonn J (1995) Multiparametric monitoring of the awake brain exposed to carbon monoxide. J Appl Physiol 78:1188–1196
42. Mayevsky A, Meilin S, Manor T, Ornstein E, Zarchin N, Sonn J (1998) Multiparametric monitoring of brain oxygen balance under experimental and clinical conditions. Neurol Res 20:S76–S80
43. Belopavlovic M, Buchthal A (1980) Barbiturate therapy in the management of cerebral ischaemia. Anaesthesia 35:271–278
44. Mayevsky A, Chance B (1982) Intracellular oxidation-reduction state measured *in situ* by a multicannel fiber-optic surface fluorometer. Science 217:537–540
45. Runciman WB, Mather LE, Selby DG (1990) Cardiovascular effects of propofol and of thiopentone anaesthesia in the sheep. Br J Anaesth 65:353–359
46. Charlesworth P, Jacobson I, Richards CD (1995) Pentobarbitone modulation of NMDA receptors in neurones isolated from the rat olfactory brain. Br J Pharmacol 116:3005–3013
47. Mendelman A, Guggenheimer-Furman E, Zarchin N, Meilin S, Mayevsky A (1996) Protection of rat brain against exposure to CO afforded by barbiturate anaesthesia. Isr J Med Sci 32(Supplement):S32
48. Batjer HH (1993) Cerebral protective effects of etomidate: experimental and clinical aspects. Cerebrovasc Brain Metab Rev 5:17–32
49. Wada DR, Harashima H, Ebling W, Osaki EW, Stanski DR (1996) Effects of thiopental on regional blood flows in the rat. Anesthesiology 84:596–604
50. Mayevsky A, Zarchin N, Sonn J (1993) Brain redox state and $O_2$ balance in experimental spreading depression and ischemia. In: Lehmenkuhler A, Grotemeyer K-H, Tegtmeier F (eds) Migraine – basic mechanisms and treatment. Urban & Schwarzenberg, Munchen-Wier, Baltimore, pp 379–393
51. Mayevsky A, Breuer Z (1990) In. Schurr A, Rigor BM (eds) The Mongolian gerbil as a model for cerebral ischemia. CRC Press, pp 27–46
52. Mayevsky A, Zeuthen T, Chance B (1974) Measurements of extracellular potassium, ECoG and pyridine nucleotide levels during cortical spreading depression in rats. Brain Res 76:347–349

# Chapter 7
# Responses to Complete Ischemia of the Partial Ischemic Gerbil Brain

**Abstract** The effects of partial ischemia (unilateral carotid artery occlusion) on the metabolic and electrical responses to decapitation were studied in the awake anesthetized gerbils' brains. The gerbil was connected to the two-channel fluorometer/reflectometer by flexible light guides implanted above the two hemispheres of the brain. The results show that (1) the metabolic rate as measured by this present technique is lower in the gerbil than in the rat and (2) the metabolic rate of the ischemic hemisphere is lower as compared to that of the normoxic contralateral hemisphere.

**Keywords** Gerbil brain · Partial ischemia · Complete ischemia · NADH redox state

## 7.1 Introduction

In order to study the effects of complete ischemia on brain energy metabolism, several animal models were developed. One of the models that is in use is decapitation, whereby the brain is transferred into a closed system [1–3] and the rate of utilization of various metabolites can be calculated. Partial ischemia was also used as an experimental model by many investigators [3–5] for studying changes in brain energy metabolism. The effects of partial ischemia on the metabolic and electrical responses of the brain to complete ischemia in the same animal model were described previously in the awake and anesthetized rat [3]. The main disadvantage of the rat model is that the comparison was done between two groups of animals (experimental and control).

 
A. Mayevsky, *The Mongolian Gerbil Brain*,
https://doi.org/10.1007/978-3-031-69549-0_7

## 7.2 The Two-Channel Monitoring Model

In the Mongolian gerbil, the circle of Willis is not complete [6]. Therefore, with the development of a two-channel DC fluorometer [7] for monitoring NADH redox state, it was possible to study the effects of partial ischemia in one hemisphere on the response to decapitation, using a second hemisphere as a control. In this communication, we present the effects of unilateral carotid artery occlusion (partial ischemia) on the hemodynamic, metabolic, and electrical responses to decapitation (complete ischemia) in the awake and anesthetized gerbil.

## 7.3 Methods

The subjects for the study were 33 male Mongolian gerbils (*Meriones unguiculatus*). The animal was anesthetized by i.p. injection of Equithesin (a mixture of chloral hydrate, pentobarbital, magnesium sulfate, propylene glycol, and alcohol) for the duration of the operation (0.3 mL/100 g). The head was placed in a special holder, and a mid-line incision was made so that the frontal and parietal bones were exposed. A hole of 5 mm diameter was drilled in the two parietal bones to fit specially designed light guide holders located epidurally. A stainless steel screw was placed in the skull on two sides of the light guide holder and used as an electrocorticogram (ECoG) electrode. The two cannulae and the four screws were cemented to the skull with dental acrylic cement. One of the common carotid arteries was isolated in the neck region, and a suture was carefully placed around it for the ligation to be done during the experiment. The gerbil was connected to the fluorometer/reflectometer 10 min after the end of the operation, and measurements of NADH and ECoG were obtained from the two brain hemispheres. The calibration of the signals was done as previously described [8]. The decapitation was done in the fully awake control gerbil or 10–15 min after the ligation of the carotid artery in the awake or anesthetized gerbil.

## 7.4 Results and Discussion

A typical response to decapitation recorded from an awake gerbil (right carotid artery occluded) is shown in Fig. 7.1 [9]. The three traces from the top were recorded from the right hemisphere (partially ischemic), and the three traces at the bottom were obtained from the left hemisphere (normoxic, control side). The changes in the reflectance signals immediately after decapitation are small in both hemispheres. The secondary reflectance increase (SRI) appeared in both hemispheres but not at the same time. The response of the NADH fluorescence (corrected fluorescence, CF) to decapitation is different in two hemispheres. The control side shows a larger

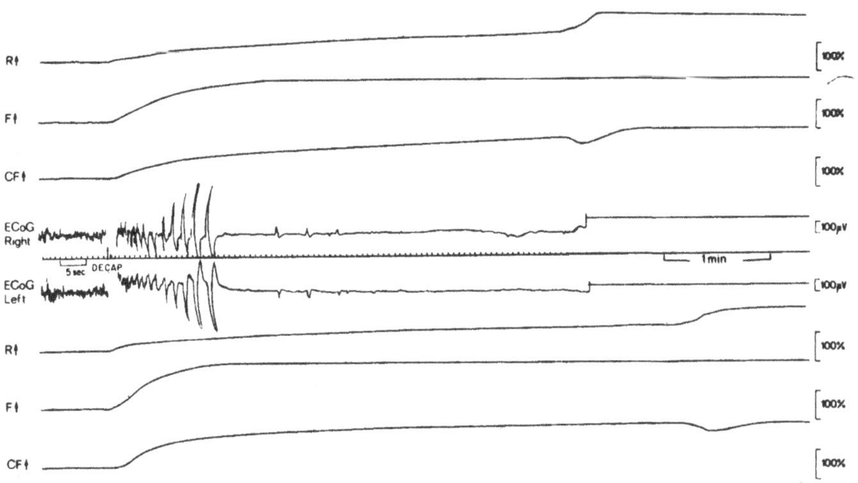


**Fig. 7.1** The effect of unilateral carotid artery ligation (right side) on the metabolic and electrical responses to decapitation in the awake gerbil [9]

CF increase with faster kinetics. The right (ischemic) hemisphere has a smaller rate of oxygen consumption, as evaluated by the changes in the CF trace. In order to analyze the results more accurately, the analog signals were used, and Table 7.1 summarizes the results. All parameters were defined in our previous publication [3]. Four sets of experiments were performed. Group 1 was an awake control group (two hemispheres). In Groups 2 and 3 (awake gerbils), one of the hemispheres was occluded 10–15 min prior to decapitation. The results obtained from the awake gerbils included in Groups 2 and 3 were divided according to the level of percentage increase of the corrected fluorescence (CF) signal during the ligation of the carotid artery. In Group 2, we included gerbils that showed a 20–50% increase in the CF, while in Group 3, only gerbils that showed a CF increase above 50% were included. Group 4 was the same as Group 2, except that the animals were anesthetized. All the animals were exposed to short anoxia at the beginning of the experiment, and the results show (last column in Table 7.1) the identical metabolic state of the two hemispheres in each group. All statistical analyses were done between the two hemispheres of the same brain using the paired Student's t-test. The results show that the two hemispheres of the brain in the control group (1) were identical in all parameters analyzed. In the three treated groups (2–4), significant differences were found in three parameters between the control and the ischemic hemisphere. The percentage of CF increase after decapitation was larger in the control hemisphere as compared to the occluded one due to the initial CF increase in response to the ligation. The same type of difference was found in the parameter showing the interrelation between energy availability and electrical activity ($CF_0/CF_{max}$). The most significant difference between the normoxic and the ischemic brain was found in the rate of oxygen consumption (time to $CF_{max}/2$). Partial ischemia led to a decrease in oxygen consumption in the cortex of the awake and anesthetized brain.

**Table 7.1** The effect of partial ischemia on the hemodynamic, metabolic, and electrical responses to the decapitation of the awake and anesthetized gerbil's brain (mean ± S.E.M.) [9] [ID = 329]

| Group No. | Treatment (No. of gerbils) | Hemisphere | $T_0$ (seconds) | $T_1$ (seconds) | $T_2$ (seconds) | % $CF_{max}$ increase after decapitation | $CF_0$/$CF_{max}$ | $T_3$ (seconds) | $T_4$ (seconds) | % CF increase during anoxia |
|---|---|---|---|---|---|---|---|---|---|---|
| 1 | Awake (7) | Control | 23.9 ± 1.9 | 1.1 ± 0.9 | 60 ± 5 | 113 ± 7 | 88 ± 2 | 7.3 ± 0.5 | 120 ± 17 | 69 ± 4 |
| | | Control | 23 ± 1.9 | 1.1 ± 0.1 | 62 ± 6 | 107 ± 7 | 86 ± 2 | 7.4 ± 0.5 | 148 ± 20 | 74 ± 3 |
| 2 | Awake (14) | Control | 27.1 ± 1.4 | 1.1 ± 0.1 | 60 ± 5 | 105 ± 6*** | 87 ± 2 | 7.2 ± 0.4** | 127 ± 16 | 66 ± 3 |
| | Unilateral occlusion | Treated | 26.5 ± 1.5 | 1.3 ± 0.1 | 64 ± 4 | 71 ± 7 | 77 ± 3 | 9.3 ± 0.7 | 96 ± 16 | 66 ± 4 |
| 3 | Awake (5) | Control | 23 ± 0.9 | 1.1 ± 0.3* | 70 ± 5 | 98 ± 14*** | 76 ± 11* | 7.8 ± 1.1*** | 180 ± 25 | 73 ± 2 |
| | Unilateral occlusion | Treated | 26.6 ± 2.9 | 6.0 ± 2.2 | 91 ± 24 | 31 ± 9 | 46 ± 11 | 19.5 ± 2.8 | 90 ± 11 | 71 ± 6 |
| 4 | Anesthetized (7) | Control | 39.9 ± 2.6 | 1.8 ± 0.2 | 80 ± 6 | 77 ± 8** | 85 ± 3 | 12.9 ± 1.2* | 155 ± 19 | 75 ± 7 |
| | Unilateral occlusion | Treated | 37.4 ± 2.1 | 3.1 ± 0.7 | 93 ± 7 | 45 ± 6 | 69 ± 5 | 19.4 ± 2.3 | 119 ± 24 | 75 ± 5 |

*CF* corrected fluorescence, $CF_0$ CF at time 0, $T_0$ time to ECoG = 0, $T_1$ time to CF increase, $T_2$ time to $CF_{max}$, $T_3$ time to $CF_{max}/2$, $T_4$ time to secondary reflectance increase

*$p < 0.05$; **$p < 0.01$; ***$p < 0.001$

Another comparison was made for the four parameters: $T_0$, $T_1$, $T_3$, and % $CF_{max}$ increase between the various animal groups. Analyses were done between the following:

1. The control hemispheres of the awake and anesthetized gerbils (Groups 2–4 in Table 7.1)
2. The treated hemispheres of the awake and anesthetized gerbils (Groups 2–4 in Table 7.1)
3. The treated hemispheres of the two awake, partially ischemic groups (Groups 2 and 3 in Table 7.1)

The statistical analysis (Student's t-test) showed a significant difference between all the compared groups. The level of significance varied between $p < 0.01$ and $p < 0.001$. Only the comparison between Groups 2 and 3 did not show a significant difference for $T_0$. The results presented in this study make a contribution to the understanding of the metabolic and electrical responses to complete ischemia. The decapitation model was used by Lowry in mice [2], Nordstrom and Siesjo [10] in rats. In our present study, we used the normoxic or ischemic awake or anesthetized gerbil brain in order to evaluate the kinetic changes in energy metabolism taking place after decapitation. The results show that there are a few differences between the rat [3] and the gerbil exposed to decapitation. In the awake rat, we found that the electrical activity (ECoG) disappeared within 10–15 s, while in the awake gerbil, it disappeared after 22–26 s (Table 7.1, Group 1). The rate of oxygen utilization, as evaluated as the time to reach $CF_{max/2}$, was lower in the gerbil as compared to the rat. According to the two numbers calculated for the rat and the gerbil (Ref. [3] and Table 7.1), it seems that the rate of oxygen utilization in the gerbil brain is 50% of that found in the rat. The same conclusion was drawn by Siesjo [11] comparing the rate of ~P utilization in the rat and the gerbil brain. The advantages of using the gerbil brain in decapitation studies are more pronounced when testing the effects of incomplete ischemia (unilateral artery ligation) on the metabolic activity in the awake and anesthetized animal. The gerbil's incomplete circle of Willis resulted in an uneven impact of unilateral carotid occlusion on the two hemispheres, as indicated by the increased level of NADH after the occlusion. These results were described previously [3, 12] and are revealed in this study by comparing the level of $CF_{max}$ increase during the decapitation of the unilaterally ligated gerbils (Table 7.1, Groups 2 and 3). In Group 2, the control hemisphere showed a $105 \pm 6\%$ CF increase as compared to the $71 \pm 7\%$ measured in the partial ischemic hemisphere. This difference between the responses of the two hemispheres was larger in Group 3, where the occlusion induced a more severe unilateral ischemia ($98 \pm 14\%$ vs. $31 \pm 9\%$). The same effect was found in the anesthetized gerbil ($77 \pm 8\%$ vs. $45 \pm 6\%$), while anesthesia itself (Group 4) caused a smaller increase of CF as compared to the awake gerbil brain (Group 1). This was due to the initial increase of NADH in the transition from the awake to the anesthetized state. The indications that deep anesthesia decreases the rate of oxygen utilization, as found in other studies [2, 3, 13], are very clear in our gerbil model. The time to reach $CF_{max/2}$ was significantly longer in the occluded hemisphere as compared to the contralateral normoxic hemisphere

(Groups 2 and 3 in Table 7.1). This indicates that in the ischemic awake or anesthetized brain the rate of oxygen utilization was slower than in the normoxic brain. Comparison between the two hemispheres of the same animal shows the uniqueness of our model and its usefulness for various studies.

## References

1. Samson FE Jr, Balfour WM, Dahl NA (1960) Rate of cerebral ATP utilization in rats. Am J Phys 198:213–216
2. Lowry OH, Passonneau JV, Hasselberger FX, Schulz DW (1964) Effect of ischemia on known substrates and cofactors of the glycolytic pathway in brain. J Biol Chem 239:18–30
3. Mayevsky A (1978) Ischemia in the brain: the effects of carotid artery ligation and decapitation on the energy state of the awake and anesthetized rat. Brain Res 140:217–230
4. Eklof B, Siesjo BK (1972) The effect of bilateral carotid artery ligation upon the blood flow and the energy state of the rat brain. Acta Physiol Scand 86:155–165
5. Payan HM (1967) Ill-effect of carotid artery ligation: experimental study on influence of age. Ann Surg 165:544–550
6. Levine S, Payan H (1966) Effects of ischemia and other procedures on the brain and retina of the gerbil (*Meriones unguiculatus*). Exp Neurol 16:255–262
7. Mayevsky A, Bar-Sagie D (1978) The interrelation between CBF, energy metabolism and ECoG in a new awake brain model. Adv Exp Med Biol 92:761–768
8. Mayevsky A (1976) Brain energy metabolism of the conscious rat exposed to various physiological and pathological situations. Brain Res 113:327–338
9. Mayevsky A, Zarchin N (1981) The effects of unilateral carotid occlusion on the responses to decapitation in the gerbil brain. Brain Res 206:155–160
10. Nordstrom CH, Siesjo BK (1978) Influence of phenobarbital on changes in the metabolites of the energy reserve of the cerebral cortex following complete ischemia. Acta Physiol Scand 104:271–280
11. Siesjo BK (1978) Ischemia (studies in gerbils). In: Siesjo BK (ed) Brain energy metabolism. Wiley, New York, pp 497–498
12. Mayevsky A (1978) Pyridine nucleotide oxidation reduction state of the cerebral cortex in the awake gerbil. J Neurosci Res 3:369–374
13. Siesjo BK (1978) Anaesthesia, analgesia and sedation. In: Siesjo BK (ed) Brain energy metabolism. Wiley, Chichester, New York, Brisbane, Toronto, pp 233–255

# Chapter 8
# Brain Vasculature in the Mongolian Gerbil

**Abstract** A unique blood supply to the brain, the "Circle of Willis" (COW), exists in all mammals except for the Mongolian gerbil (*Meriones unguiculatus*). The posterior connection between the basilar artery and the carotid artery system has been found to be missing in most gerbils. Furthermore, in some of the animals, the anterior communication was not complete, thus leading to partial ischemia following unilateral carotid artery occlusion. Due to those peculiar characteristics, the Mongolian gerbil today has become a widespread animal model for cerebral ischemia studies. *M. unguiculatus* has been used in most of the studies, while the level of ischemia has been evaluated by the development of neurological symptoms created by the occlusion of the carotid arteries. In this study, we investigated the vasculature structure of the commonly used gerbil, *M. unguiculatus* (MU-TF), and compared it to the vasculature of the Israeli gerbil, *Meriones tristrami*, as well as to that of the Albino rat. We determined the correlation between the anatomical pattern of the blood vessels and the biochemical responses during partial or complete ischemia and anoxia by monitoring the oxidation-reduction state of the intramitochondrial NADH using an in vivo surface fluorescence technique. The corrected fluorescence signal was found to be inversely correlated with oxygen availability. This is the first time that the brain vasculature of two lines of *M. tristrami* (MT-HU, MT-BD) has been investigated. Gerbils belonging to *Meriones unguiculatus* obtained from Tumblebrook Farms and from the NY Institute for Basic Research were compared to two lines of *Meriones tristrami* obtained from two different breeders in Israel. In all groups of gerbils and in the albino rats, a significant correlation was found between the anatomical patterns of the Circle of Willis and the metabolic responses to ischemia as evaluated by NADH redox state measurements, namely, that with a complete anterior anatomical Circle of Willis, the metabolic response to unilateral occlusion was minimal.

**Keywords** Circle of Willis · Mitochondrial NADH · Brain ischemia · Gerbil strains · Metabolic mapping



A. Mayevsky, *The Mongolian Gerbil Brain*,
https://doi.org/10.1007/978-3-031-69549-0_8

## 8.1 Historical Overview

The aim of this chapter is to deal with the specific issues of brain ischemia that are unique to the gerbil model. The anatomy of blood vessels to the brain will be discussed in detail, and a few of the biochemical and physiological responses to ischemia will be correlated to it.

Although the gerbil was involved in a few hundred studies in many laboratories, a detailed anatomy of the blood vessels to the brain was reported only in a few published papers [1–7]. Based on those papers as well as our own work, the following vasculature of the gerbil brain was established.

Figure 8.1 shows the general anatomy of the blood vessels of a typical mammal, including the human, arising from the aorta and reaching the various parts of the CNS. Blood reaches the brain through two main pairs of arteries, namely, the

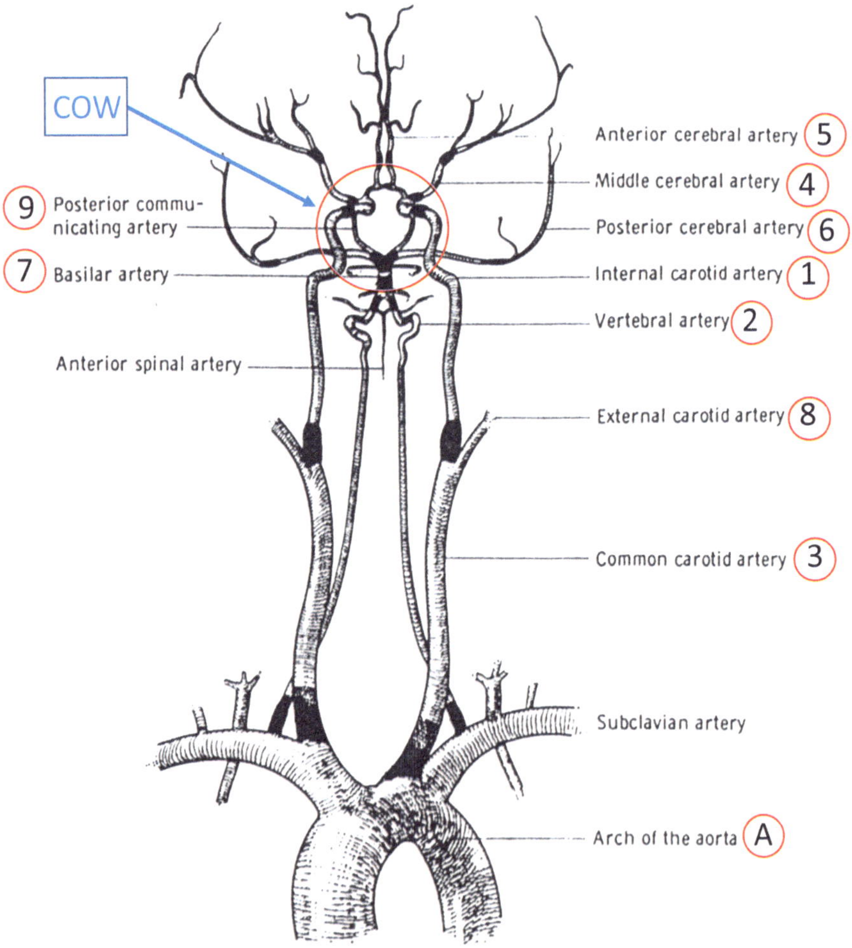


**Fig. 8.1** Schematic presentation of the blood vessels supply to the human brain area. COW Circle of Willis

internal carotid (ICA-1) and the vertebral (VA-2). The internal carotid artery rises from the common carotid artery (CCA-3) and reaches the base of the brain lateral to the tuber cinereum. Three main branches of the ICA supply various parts of the cerebral hemisphere, namely, the middle (4), anterior (5), and posterior (6) cerebral arteries. The two vertebral arteries (2) are merged to form the basilar artery (BA-7), which in all mammals, together with the branches of the two internal carotid arteries (including the 2 posterior communicating arteries (9)), forms the "Circle of Willis" (COW). The stapedial artery (not shown) and the external carotid artery (8) are supplying other head structures.

Figure 8.2 shows the schematic ventral view of the rat brain [8] and arterial vasculature, which are very similar to the one presented in Fig. 8.1. The Circle of Willis is presented. Figure 8.3a shows the picture of the ventral view of the perfused rat brain [9]. The components of the Circle of Willis are very clear. Figure 8.3b shows the area of the medulla oblongata (MO) of the same brain presented in Part A. The two vertebral arteries and the basilar artery are labeled [9].

In their original paper on the gerbil as a model for ischemia, Levine and Payan [1] did not analyze the anatomy of the blood vessels but rather behaviorally concluded that *"Bilateral ligation was uniformly lethal. Unilateral ligation produced ischemia lesions sometimes lethal."* Only in [6], Levine and Sohn checked the brain vasculature anatomy and found *"inadequate anastomosis between the basilar and carotid arteries."* They postulated that the *"lack of a significant posterior communicating artery prevents redistribution of blood flow from the vertebral-basilar circulation to the carotid circulation."* Their finding basically fits the main scheme shown in Fig. 8.2 [10] Part A (center), but suggested also at least one of the exceptions shown in Fig. 8.2 [10] Part $P_2$, namely that a small PCOA connected the PCA to the SCeA (the main branches of the BA). In a short abstract, Kahn et al. [11] also reported that a connection between the carotid and the basilar systems was not found.

In the first detailed anatomical study, Kahn [7] was able to demonstrate the anatomy of the anterior as well as the posterior section of the COW. He reported that 53% of the adult gerbils showed evidence of infraction and died within 5 days after unilateral artery occlusion. His other conclusion was that "the gerbil has no significance connections between the basilar-vertebral system and the carotid system" as seen in Fig. 8.4. Harrison et al. [12] came to the same conclusion regarding the lack of a major posterior communicating artery (PCOA). In a detailed study, as seen in Figs. 8.5 and 8.6, Levy and Brierley [2] concluded that the variability in ischemia due to bilateral carotid occlusion was "possibly related to differences in the number and size of these communicating vessels."

In Ref. [4], Berry et al. published a schematic presentation of the brain vasculature in the gerbil, seen in Fig. 8.7, which served as a basic concept developed later by other investigators. They claimed that "All animals displayed the expected lack of an anastomosis between the anterior and posterior circulations," as seen in Fig. 8.8.

The other published papers on the gerbils assumed that bilateral occlusion led to complete ischemia while unilateral occlusion created various levels of partial

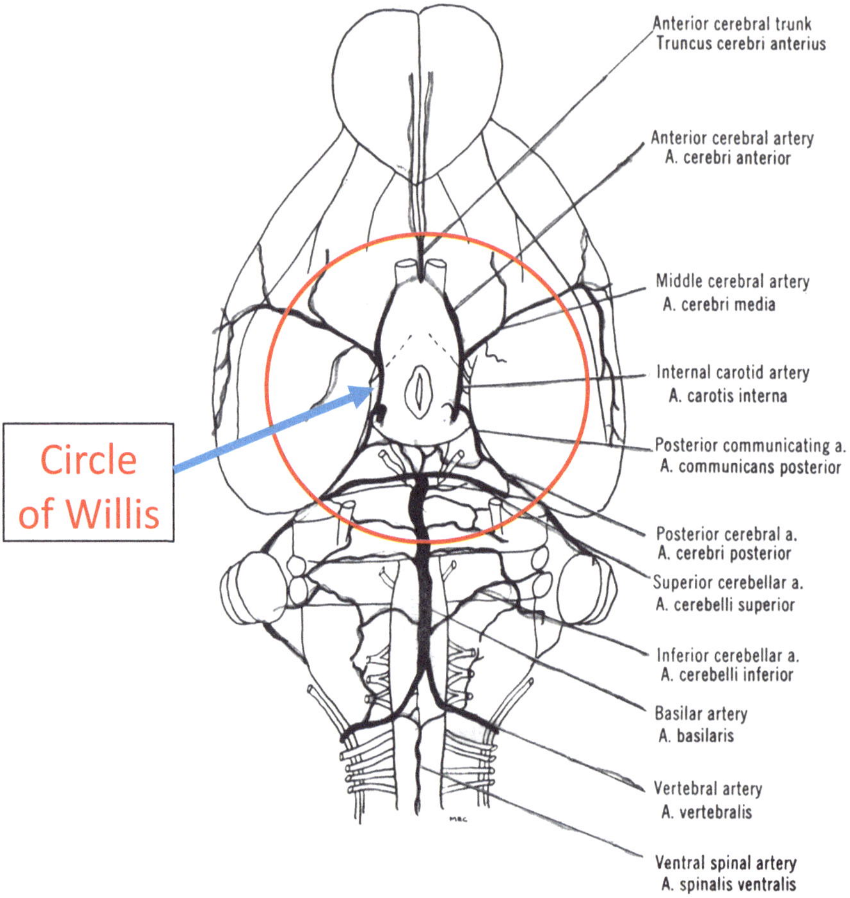


**Fig. 8.2** Ventral view of the rat's brain and its arterial system including the Circle of Willis [8]

ischemia depending upon the degree of communication between the anterior cerebral arteries.

In a short communication, Schonfeld and Glick [13] described a correlation between cerebral vasculature, seizures, and strokes developed in the Mongolian gerbils. This susceptibility of the gerbils to seizures is another characteristic of this animal, which was studied by various groups [13–16] (Fig. 8.9).

In a detailed study published in Ref. [3], Donadio et al. could not confirm the results of Schonfeld and Glick and concluded that "neither the vascular pattern of the ACA nor the frequency occurrence of stroke following unilateral ligation was related to seizure propensity." In their two differentiated colonies of seizure prone (SP) and nonseizure prone (NSP), a detailed anatomical features of the anterior circulation were given and served as a reference to our recent study [9].

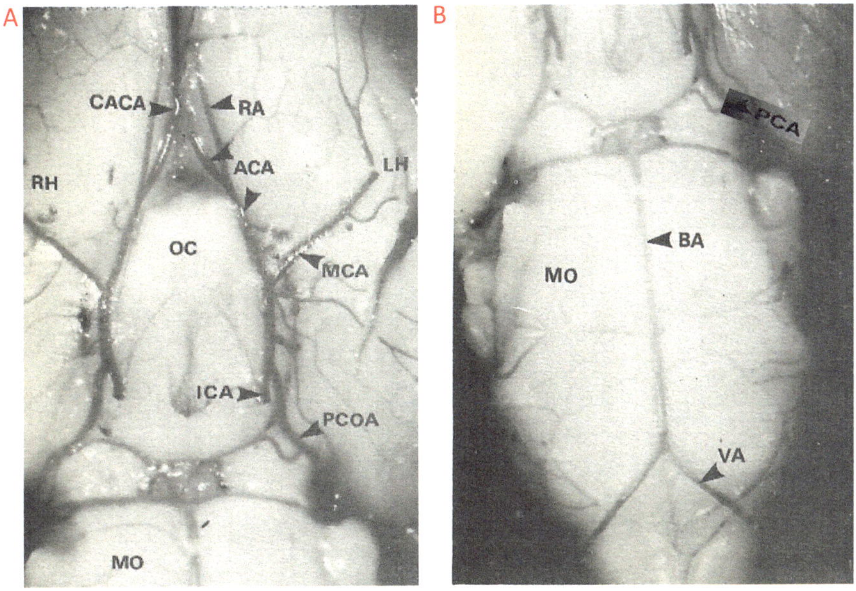


**Fig. 8.3** (**a**) Circle of Willis in the rat brain. RH and LH, the right and left hemispheres. CACA central anterior communicating artery, RA rostral artery, ACA anterior cerebral artery, MCA middle cerebral artery, ICA internal carotid artery, PCOA posterior communicating artery. (**b**) The posterior part of the rat brain. PCA posterior cerebral artery, BA basilar artery, MO medulla oblongata, VA vertebral artery [9]

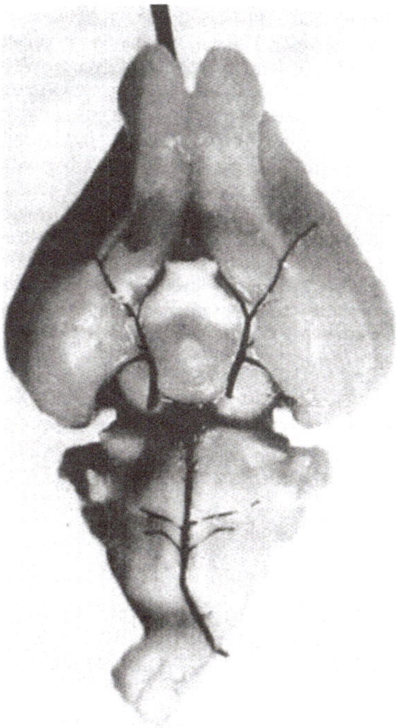

**Fig. 8.4** Base of gerbil brain after latex injection of arterial system. The posterior communicating arteries are absent [7]. (Reprinted with permission from Wolters Kluwer Health, Inc.)

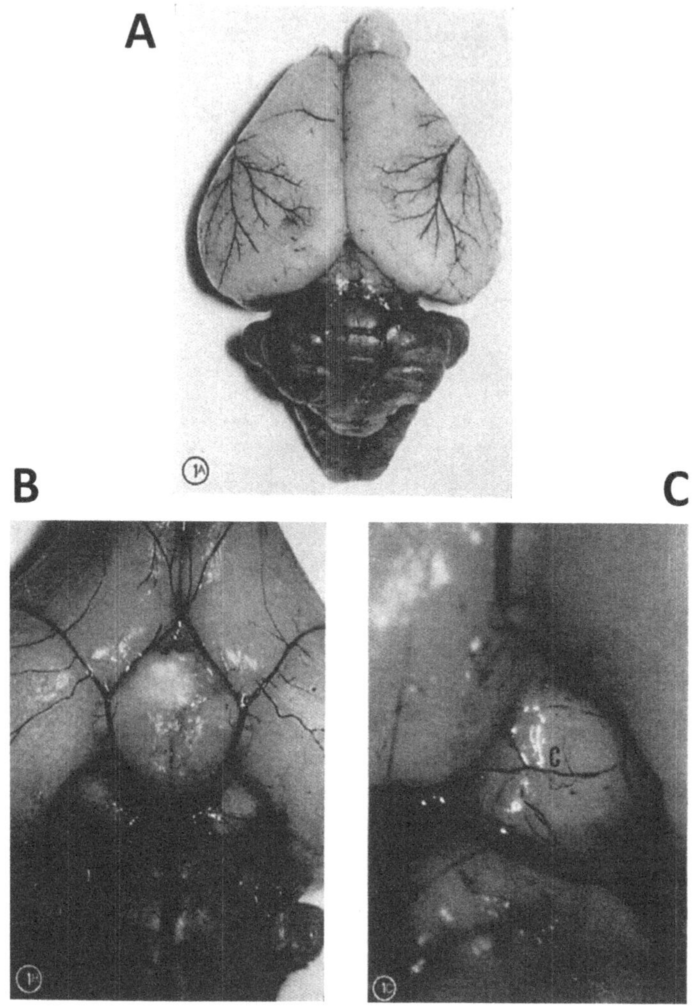


**Fig. 8.5** Vascular anatomy of the gerbil outlined by intra-aortic dye injection. (**a**) Tributaries of the middle cerebral arteries fill despite bilateral common carotid artery occlusion (×3.3). (**b**) Base of the brain. The internal carotid arteries divide into posterior, middle, and anterior cerebral arteries. The two anterior cerebral arteries join rostral to the optic chiasm. The basilar artery bifurcates into two superior cerebellar arteries. There is no posterior communicating artery (×3.3). (**c**) A 40 μm diameter communicating vessel (C) links the bifurcation of the basilar artery (B) with the left posterior cerebral artery (P) (×12) [2]. (Reprinted with permission from Elsevier)

The various anatomical configurations shown in Fig. 8.5 are based on the various publications mentioned. The anterior configurations shown in Parts A, $A_1$, and $A_2$ are parallel to the I, II, and III patterns of Donadio et al. [3], and in our experiments, they had a ratio of 50%, 25%, and 25%, respectively. The other two patterns they

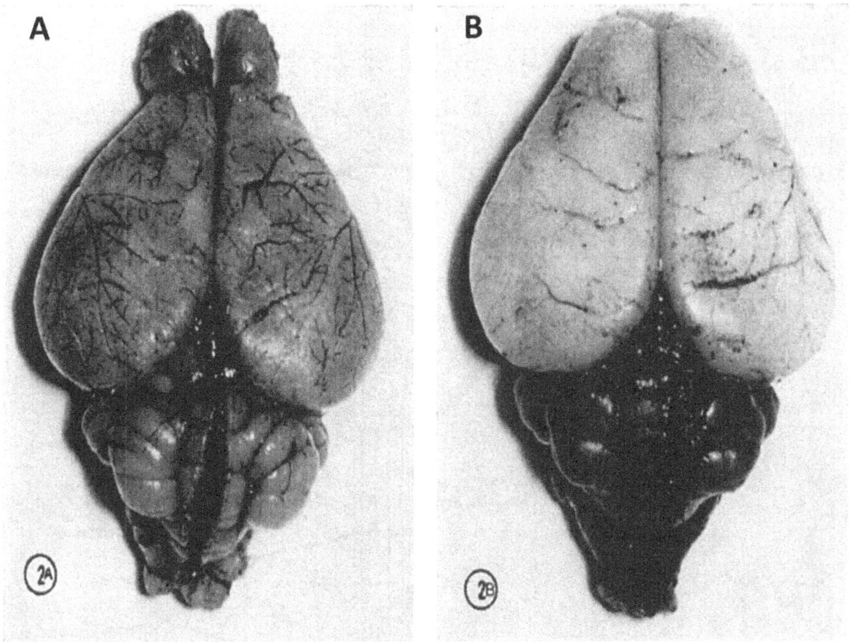


**Fig. 8.6** Intravenous carbon black staining the brain parenchyma in gerbils with bilateral carotid artery occlusion. (**a**). Nearly uniform staining of cerebral and cerebellar surfaces means that complete cerebral ischemia was not present, whereas in (**b**), marked bilateral cerebral pallor suggests the presence of cerebral ischemia (×3.7) [2]. (Reprinted with permission from Elsevier)

described appeared at very low frequencies (IV and V). The details of these patterns will be given in the next section. Donadio et al. [3] gave the name "rostral arteries" to the two arteries branched from the ACA below the olfactory bulbs. The same pattern of anterior configuration as well as the posterior one was published recently by Zheng et al. [17]. The three possible patterns of the posterior connection between the carotid and the vertebral arteries (P, $P_1$, and $P_2$) are based on our own studies and will be discussed later on.

To summarize the anatomical patterns of the blood vessels to the brain, I want to speculate that small variations in the anatomy occurred during the years, probably due to genetic factors. If years ago, we found very rarely a posterior communication type of connection (as in Figs. 8.10 [13] and 8.17 Parts $P_1$ and $P_2$ [19]), it seems that in our work [9], those connections were in up to about 20% of the gerbils tested (Fig. 8.11).

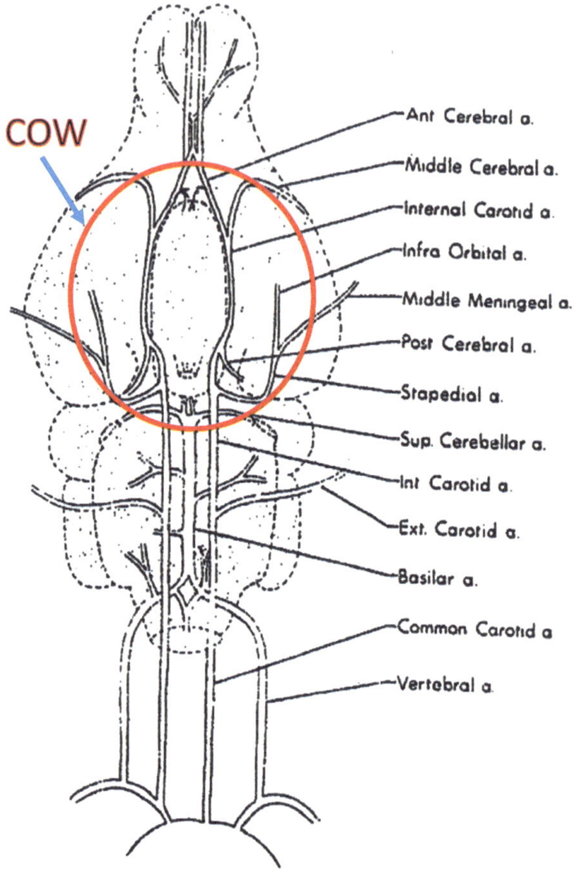


**Fig. 8.7** General schematic presentation of the blood vessels originating from the aorta and supplying the gerbil head and brain [4]. (Reprinted with permission from Elsevier)

## 8.2 Brain Vasculature and Mitochondrial NADH Responses to Ischemia

### *8.2.1 Anatomical Patterns and Biochemical Correlates*

The unclear anatomical patterns of the anterior as well as posterior circulations described by various investigators and mentioned earlier led us to use a new approach to solve this question. As we showed in the past, it was possible to evaluate the level of anatomical connection in the anterior and posterior circulation by monitoring the redox state of the mitochondria (intracellular oxygen levels) during ischemia and anoxia using surface fluorometry [20–23]. We extended those ideas and techniques to clarify the correlation between the anatomical pattern of the "Circle of Willis" and the biochemical responses of the same brain to unilateral and bilateral carotid artery occlusion as well as exposure to anoxia (Figs. 8.12 and 8.13).

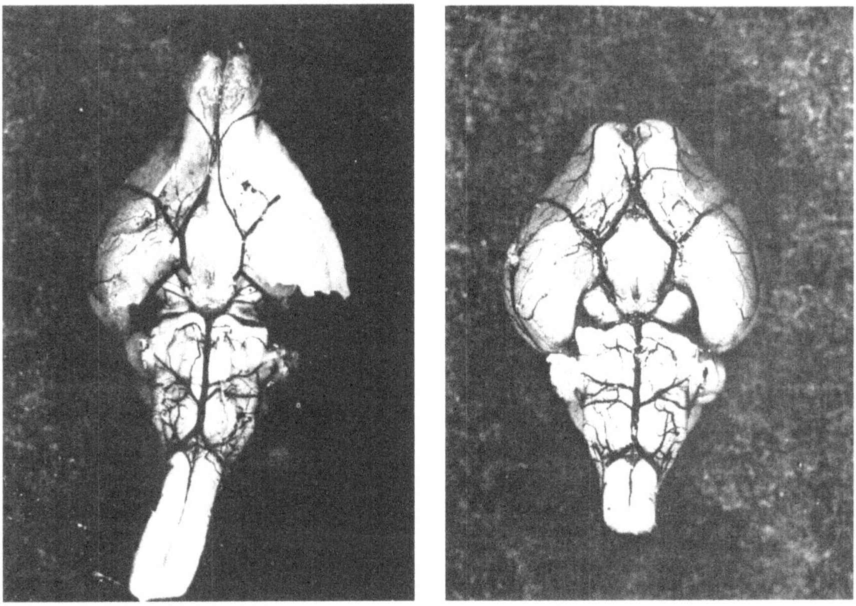

**Fig. 8.8** (**a**) Brain of "susceptible" gerbil showing separate anterior cerebral arteries with no communication between them. Note also lack of communication between posterior and anterior circulations, and artifactually traumatized left cerebral hemisphere [4]. (Reprinted with permission from Elsevier). (**b**) Brain of "resistant" gerbil showing anterior cerebral arteries fusing shortly after their origin to form single or azygos vessel. Note absence of connection between anterior and posterior circulations. The olfactory bulbs have been inadvertently removed [4]. (Reprinted with permission from Elsevier)

The aims of this study were as follows:

(i) To study the anatomy of the brain vasculature in the following groups (Fig. 8.14):

  (a) The most commonly used *Meriones unguiculatus* bred in Tumblebrook Farms (MUTF), MA, USA
  (b) The seizure-prone (IBR/SP) strain bred at the NYS Institute for Basic Research in Developmental Disabilities as a separate colony for more than 15 generations of sib-sib matings [3]
  (c) *Meriones tristrami* obtained from two sources in Israel, namely, Beit Dagan (MT-BD) and Hebrew University (MT-HU)
  (d) The Albino rat

(ii) To study the anatomical biochemical (metabolic) correlation in response to unilateral or bilateral carotid artery occlusion in the various strains.
(iii) To study the interhemispheric heterogeneity responses to unilateral carotid occlusion in the MU-TF. The first section [18] will present the basic qualitative data, while all the statistical analyses and the main discussion will be topics for the second presentation [24] (Figs. 8.15 and 8.16).

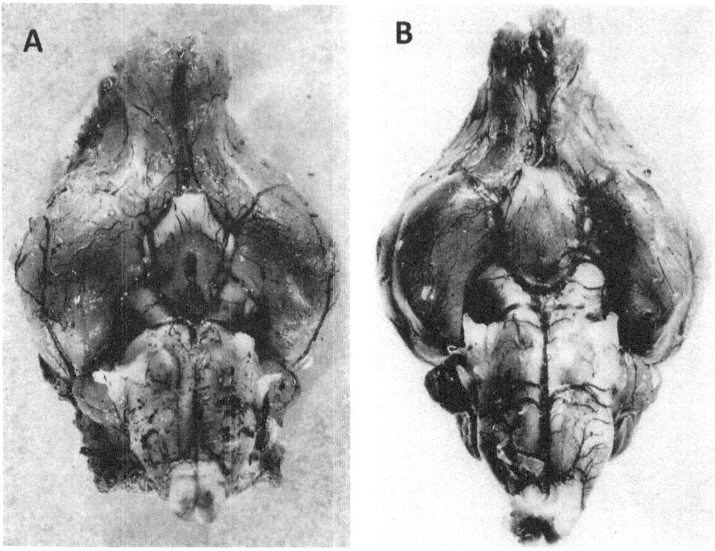


**Fig. 8.9** (**a**) Brain of seizure-susceptible gerbil, which is characterized by two separate anterior cerebral arteries. (**b**) Brain of seizure-resistant gerbil which is characterized by fusion of anterior cerebral arteries below optic chiasm which then course anteriorly as one large midline vessel [13]. (Reprinted with permission from Elsevier)

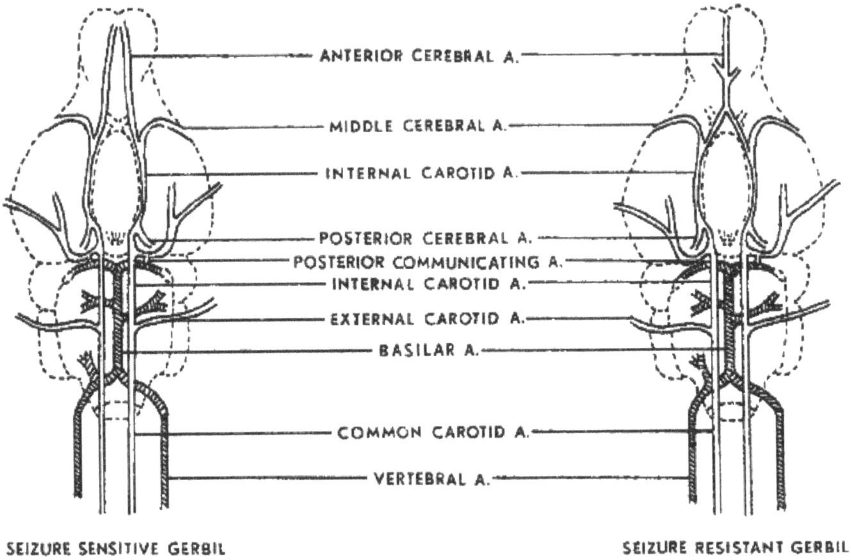


**Fig. 8.10** Schematic representation of seizure-sensitive and seizure-resistant gerbil cerebrovasculature, respectively (see text) [13]. (Reprinted with permission from Elsevier)

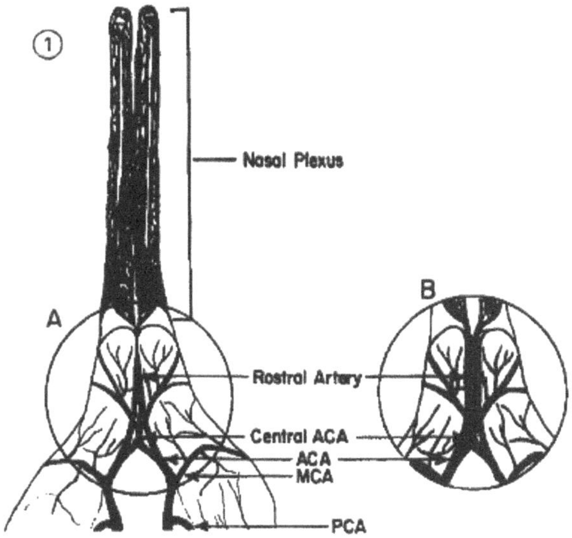


**Fig. 8.11** (**a**) Schematic representation of the dorsal view of the gerbil brain revealing the vascular arrangement of the carotid circulation. (**b**) The median fissure has been separated to reveal how the rostral arteries remain separate as they course down through the fissure. ACA anterior cerebral artery, MCA middle cerebral artery, PCA posterior cerebral artery [3]. (Reprinted with permission from Elsevier)

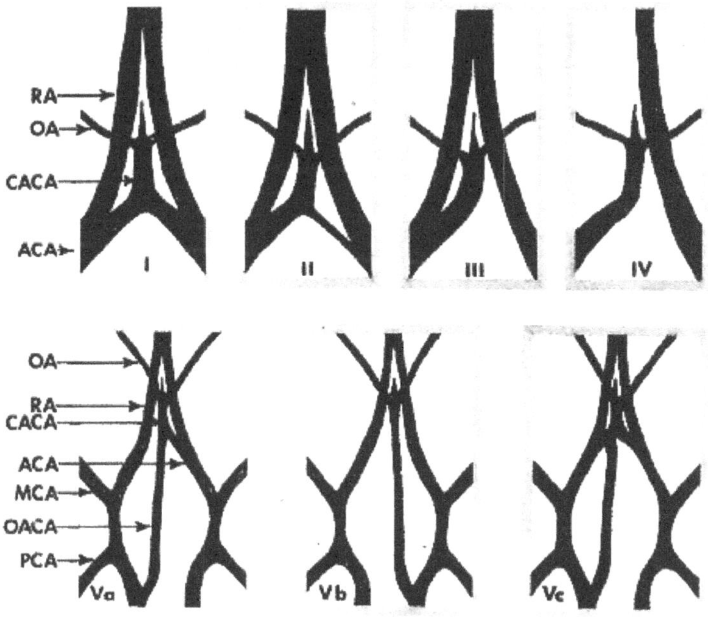


**Fig. 8.12** Schematic representation of the different patterns seen in the carotid circulation. OA olfactory artery, RA rostral artery, OACA supraoptic anteior cerebral artery, ACA anterior cerebral artery. See text for explanation [3]. (Reprinted with permission from Elsevier)

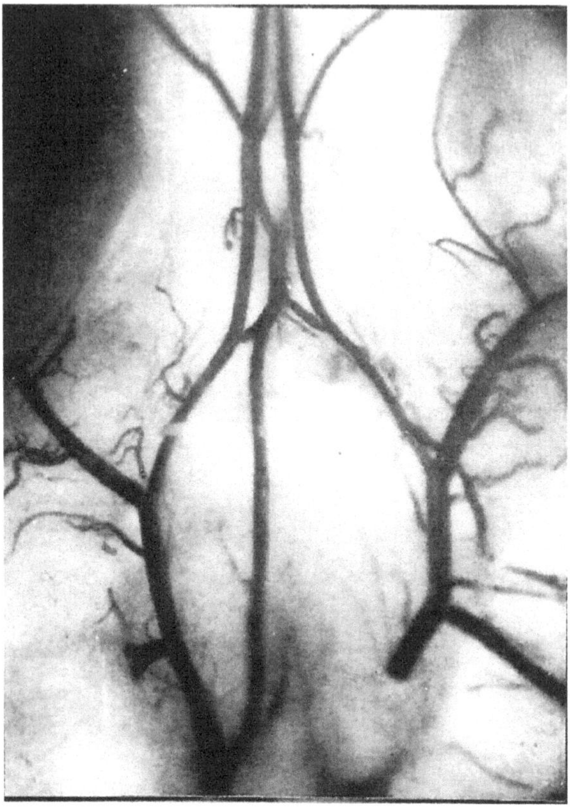

**Fig. 8.13** Photograph of the dorsal portion of the gerbil brain showing the presence of an artery arising from the internal carotid prior to the formation of the posterior cerebral artery. This artery (supraoptic anterior cerebral artery) crosses over the optic area and then joins the anterior cerebral arteries [3]. (Reprinted with permission from Elsevier)

Figure 8.17 shows, schematically, the various brain vasculature patterns found in different gerbils. The anterior configurations and nomenclature shown in areas $A_1$, $A_2$, and $A_3$ are parallel to the I, II, and III patterns described by Donadio et al. [3]. The three possible patterns of the posterior connection between the carotid and the vertebral arteries (P, $P_1$, and $P_2$) have never been reported and are based on our own studies, as will be discussed. In the first set of experiments, we used 53 Mongolian gerbils (*M. unguiculatus*) from Tumblebrook Farms (MU-TF) as well as 8 seizure-prone (IBR-SP) gerbils from the IBR (New York State Institute for Basic Research in Developmental Disabilities). Each gerbil was anesthetized and operated as previously described in detail [20, 23]. NADH redox state was monitored from each hemisphere and was correlated to the electrical activity (ECoG). The gerbil was exposed to a unilateral occlusion (1–2 min), which was followed by the occlusion of the other artery (also for 1–2 min). After reaching a high NADH level steady state, the gerbil was exposed to pure $N_2$ in order to determine the maximum NADH level. This level was used as a comparison level, and all calculations were performed relative to this value. The animal was allowed to recover completely from the combined ischemia and anoxia for at least 30 min before repeating the ischemia.

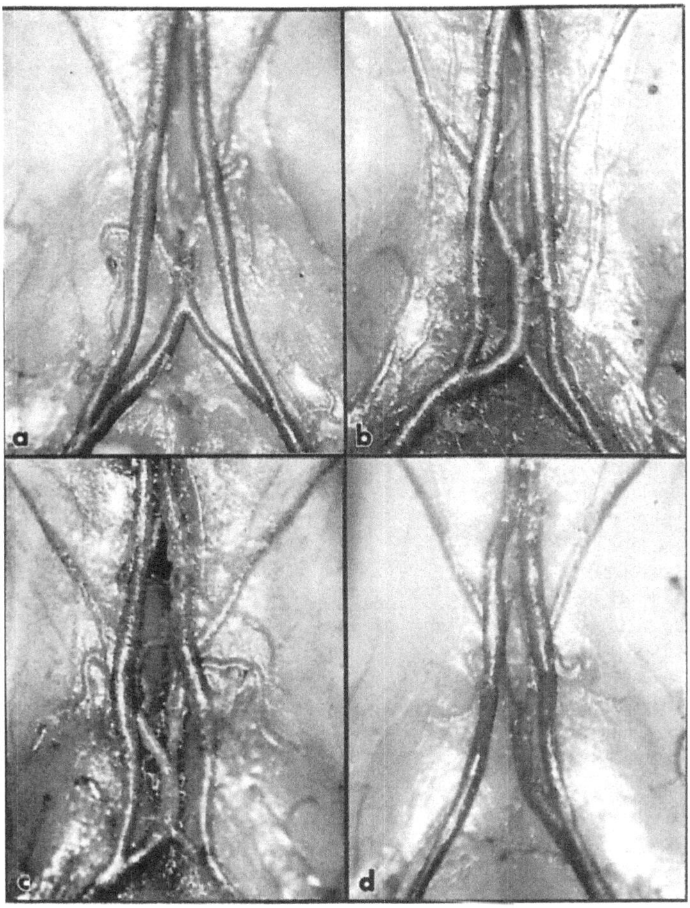


**Fig. 8.14** Four photographs of the varying vascular patterns at the junction of the left and right anterior cerebral arteries: (**a**) nearly equal contributions—Type I; (**b**, **c**) one side gives small contribution—Type II; (**d**) one side gives no contribution—Type III [3]. (Reprinted with permission from Elsevier)

At the end of the biochemical monitoring, the brain was perfused [3] and analyzed for the anatomical structure of the Circle of Willis (anterior and posterior).

A complete Circle of Willis including the anterior connection and the posterior connection via the PCOA may be seen in Fig. 8.18. In the brain shown in Part A, the two optic nerves have been removed. This is the first report containing such anatomical information.

In order to compare the results obtained from the *M. unguiculatus*, similar experiments were performed using the Israeli gerbil, *M. tristrami*, as well as the common Albino rat. The effects of unilateral and bilateral carotid occlusion on the redox state

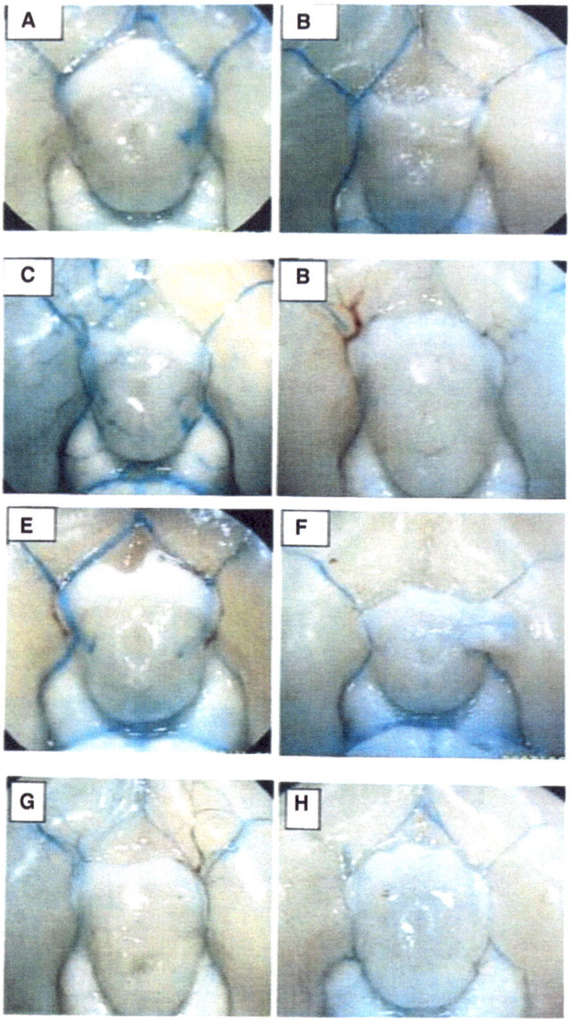


**Fig. 8.15** The eight CoW patterns of anterior (ACoA) and posterior communicating arteries (PCoA). (**a**) Type A-I, complete PCoA and ACoA. (**b**) Type A-II, complete PCoA and missing left ACoA. (**c**) Type A-III, complete PCoA and an ACoA in which the left branch of the ACA is smaller. (**d**) Type A-IV, complete PCoA with an ACoA in which the right branch of the ACA is smaller. (**e**) Type B-I, PCoA is missing and ACoA is complete. (**f**) Type B-II, both the PCoA and ACoA are missing. (**g**) Type B-III, missing PCoA and ACoA in which the left branch of the ACA is smaller. (**h**) Type B-IV, missing PCoA and ACoA in which the right branch of the ACA is smaller [5]. (Reprinted with permission from Dr. Xiaoyan Du—the first author of the publication)

of the *M. tristrami* obtained from Belt Dagan (MT-BD) are presented in Fig. 8.19b. Figure 8.19c demonstrates the responses of the Albino rat to the same procedure. The responses of these two animals were very similar to those obtained with *M. unguiculatus* Type $A_1$ anterior and Type $P_1$ posterior. These responses suggest that the Circle of Willis is well developed in these two animals. The anatomical features of the vasculature to the brain in these two animals correspond significantly with the metabolic responses.

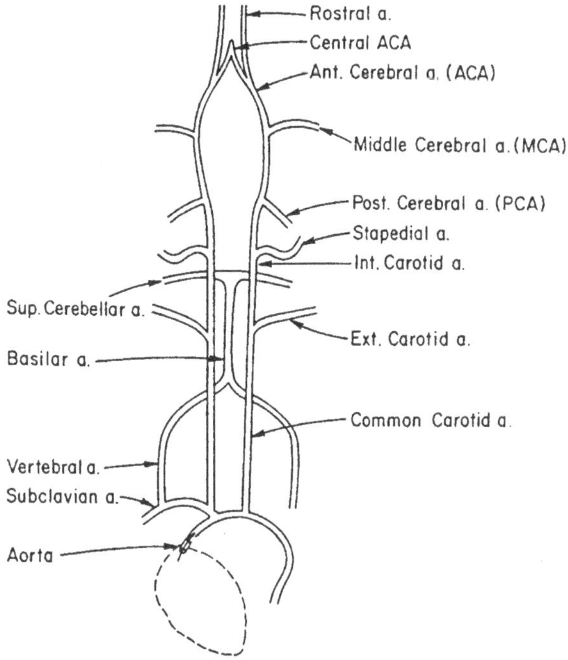


**Fig. 8.16** General schematic presentation of the blood vessels originating from the aorta and supplying the gerbil head and brain [10]

## *8.2.2 Anterior Communications*

### 8.2.2.1 Type I Connection

As defined by Donadio et al. [3], in this type of connection, as seen in Fig. 8.20, the two ACAs are fully connected and have created a central ACA (CACA).

Figure 8.21 shows the metabolic responses to ischemia alone and addition of anoxia to the ischemic brain in the gerbil having Type I anterior communication (Fig. 8.17, Part A) in the Circle of Willis. As can be seen in Part C, the two hemispheres had a similar typical response to anoxia, namely an increase of about 50% in CF was recorded. This increase of the CF traces represents the intramitochondrial redox state shifted toward State 5 of Chance and Williams [25] due to the decrease in oxygen supply to the brain. The reflectance traces ($R_R$ and $R_L$) show a steady decrease during the anoxic cycle due to an increase in blood volume in the tissue as an autoregulatory response. As shown in Part A of Fig. 8.21, occlusion of the right common carotid artery led to an initial increase in NADH, but the compensation through the anterior communication supply part of the blood and therefore the CF showed partial recovery. When the left artery was also occluded, both hemispheres reached their maximum level of NADH, and the addition of $N_2$ did not affect the redox state further. The meaning of this response is also that the posterior circulation did not contribute blood to the cerebral cortex under bilateral carotid artery

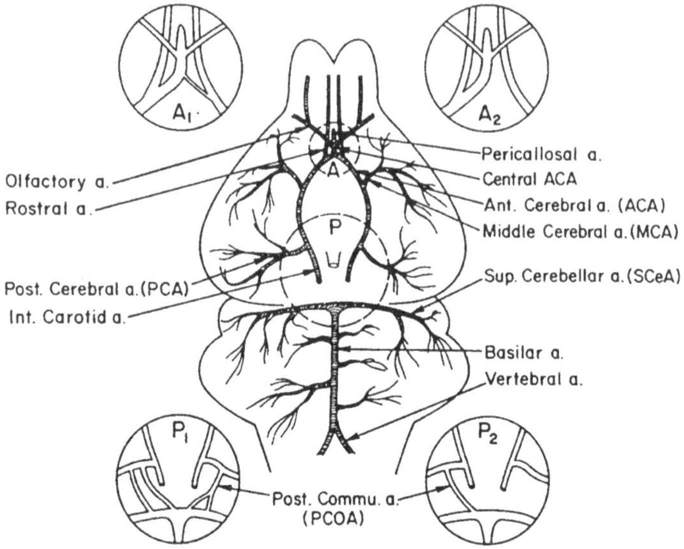


**Fig. 8.17** Schematic presentation of the blood vessels reaching the brain of the Mongolian gerbil and forming the Circle of Willis. Three anterior (A) and three posterior (P) vasculature patterns are shown. Blood reaches the brain via two main pairs of arteries, the internal carotid (ICA) and the vertebral. The internal carotid artery rises from the common carotid artery and reaches the base of the brain lateral to the tuber cinereum. Three main branches of the ICA supply various parts of the cerebral hemisphere, that is, the middle, anterior, and posterior cerebral arteries. The two vertebral arteries merge to form the basilar artery (BA) which, together with the branches of the two internal carotid arteries, forms the "Circle of Willis" (COW) in all mammals. The stapedial artery and the external carotid artery (not shown in the figure) supply other head structures. Each of the ICA has three branches, that is, the middle cerebral artery (MCA), and the posterior and anterior cerebral arteries (PCA and ACA, respectively). In a large proportion of the gerbils (Part A) the two ACAs merge to form the central ACA (CACA), also called the anterior communicating artery in other vertebrates. The two vertebral arteries form the basilar artery which is itself divided into the superior cerebellar artery (SCeA). In most gerbils, no connections were found between the SCeA and PCA (Part P). As will be discussed later on, the two patterns shown in parts P1 and P2 were found in a growing number of gerbils. In these gerbils the posterior communicating artery connects the PCA and the SCeA [18]

occlusion. In Part B of Fig. 8.21, the left carotid was occluded first, followed by the right occlusion and anoxia as done in Part A. The responses were very similar to those recorded when the right carotid artery was occluded initially.

The typical vasculature found at the end of the in vivo study is shown in Fig. 8.20. The CACA is seen very clearly as a continuation of the two ACA. The biochemical responses were correlated to the anatomical structure. Due to unilateral occlusion, the NADH in the ipsilateral hemisphere was about 20% elevated (more reduced) above the normoxic level.

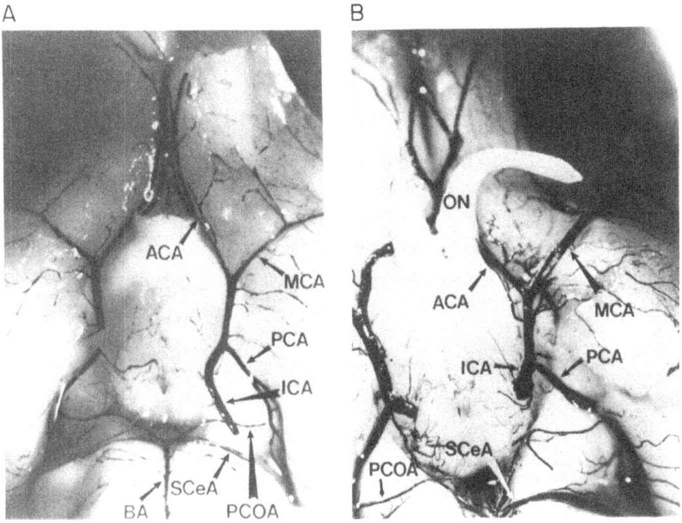


**Fig. 8.18** A representative complete circle of Willis seen in two different gerbils in part (**a**) and (**b**). ACA anterior cerebral artery, MCA middle cerebral artery, PCA Posterior cerebral artery, ICA internal carotid artery, ON optic nerve, SCeA superior cerebellar artery, BA basilar artery, PCOA posterior communicating artery [18]

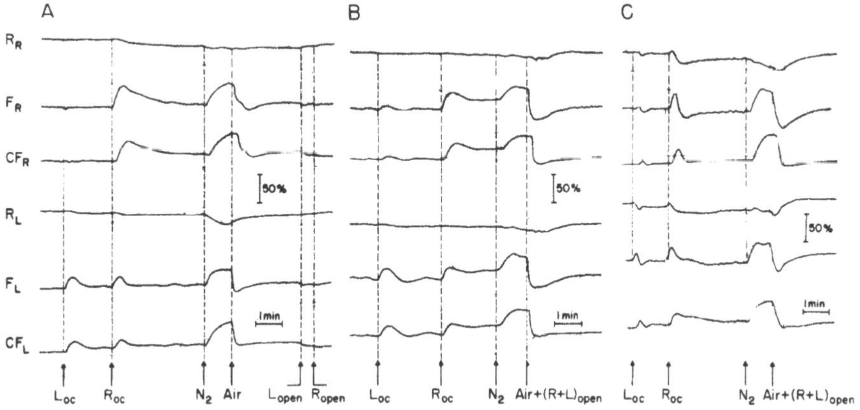


**Fig. 8.19** A Metabolic and hemodynamic responses to unilateral and bilateral carotid artery occlusion (Loc, Roc) and anoxia ($N_2$) measured from a gerbil (MU-TF) with an anatomy as shown in Fig. 8.18, namely, Type $A_I$ and $P_I$ of Fig. 8.17. Part (**b**) shows the typical responses of the Israeli gerbil (MT-HU or MT-BD). Part (**c**) represents the typical responses of the Albino rat to the same ischemic and anoxic procedures [18]

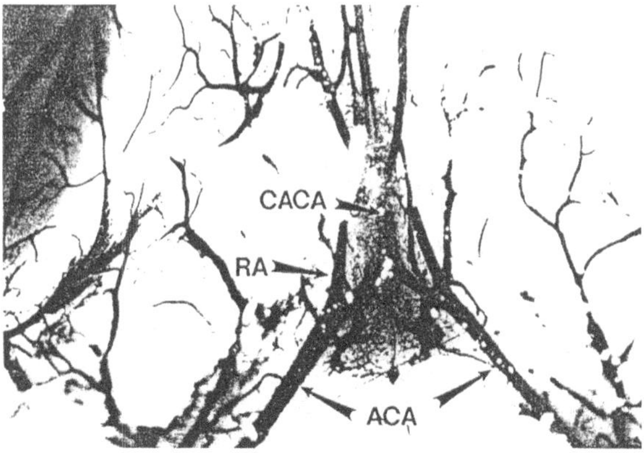


**Fig. 8.20** The anterior part of the Circle of Willis in a Type I gerbil. ACA anterior cerebral artery, RA rostral artery, CACA central anterior cerebral artery [10, 18]

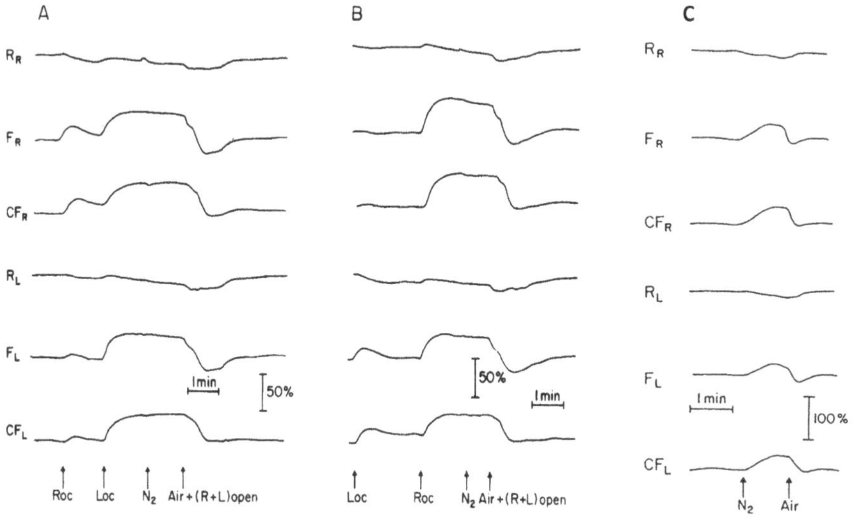


**Fig. 8.21** Effects of ischemia and anoxia on the metabolic and hemodynamic states of a Type I gerbil brain. $R_R$ and $R_L$, reflectance at 366 nm of the right and left hemispheres, respectively. $F_R$, $F_L$, $CF_R$ and $CF_L$, 450 nm fluorescence (F) and corrected fluorescence (CF) measured from the right and left hemispheres. Roc and Loc, right or left carotid artery occlusion. The amplitude scale is common to all traces. Part (**c**) shows the responses to anoxia measured in the same gerbil. In Part (**a**), ischemia was begun by occlusion of the right carotid artery while in Part (**b**) the left carotid artery was occluded first [18]

#### 8.2.2.2 Type II Communication

In this group of gerbils (shown schematically in Fig. 8.17, Part $A_1$), the two ACA that merged to form the CACA are not symmetrical in size (Fig. 8.22), and therefore the compensation capacity under unilateral ischemia was very limited. As seen, the right ACA is much smaller than the left one at the merging point into the CACA.

Figure 8.23 shows the response to unilateral, bilateral occlusion, and the addition of $N_2$ to the occluded brain. Under left occlusion, the left hemisphere's NADH became more reduced to 60–70% of the maximal NADH level achieved under bilateral occlusion. Here again, the addition of $N_2$ did not affect the redox state due to the lack of posterior communication. The same results were obtained when the right carotid artery was occluded initially, followed by the left artery occlusion (result not shown).

#### 8.2.2.3 Type III Connection

Figure 8.24 shows the anatomical arrangement of Type III connection gerbil. As seen, the left branch of the ACA that forms the CACA is missing, and only the right ACA has a branch in the level of the CACA that is missing in this animal. This rare anatomical arrangement, namely, the missing of a regular CACA (Fig. 8.17, Part $A_2$), may lead to complete ischemia in the ipsilateral hemisphere under unilateral carotid artery occlusion. As seen in Figure 8.25a, the right hemisphere became maximally reduced under right artery occlusion without any effect on the left hemisphere. The same response evidently occurred under the left artery occlusion. Also, this gerbil was missing the posterior communication, as seen under the addition of $N_2$ to the gerbil.

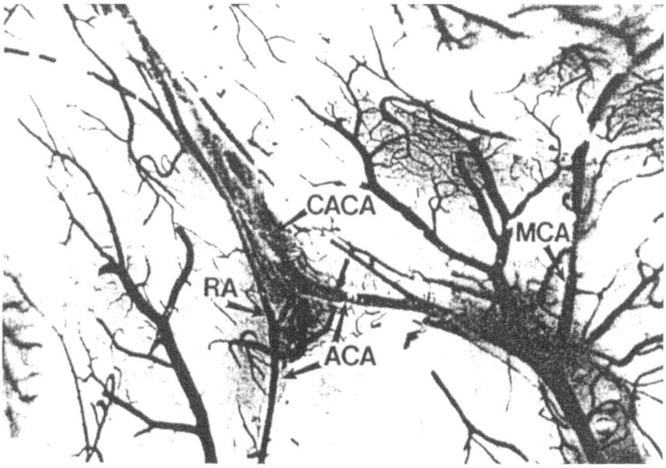


**Fig. 8.22** The anterior part of the Circle of Willis in a Type II gerbil. MCA middle cerebral artery, ACA anterior cerebral artery, RA rostral artery, CACA central anterior cerebral artery [10]

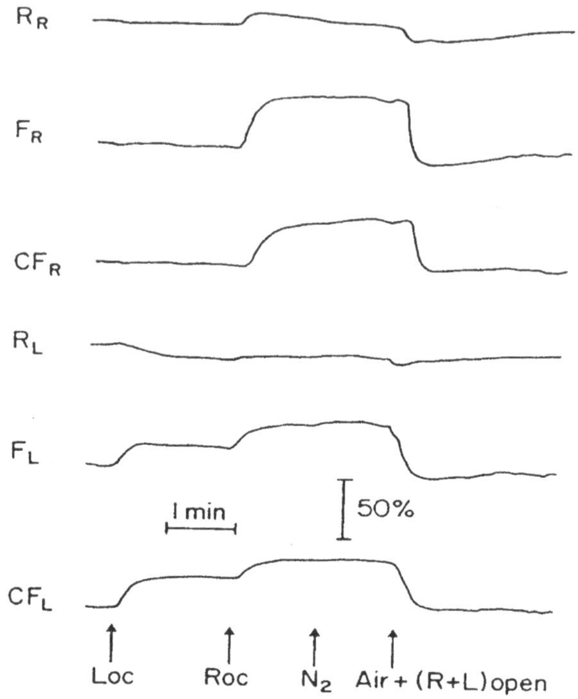


**Fig. 8.23** Brain responses to unilateral and bilateral carotid artery occlusion (Loc Roc) and anoxia ($N_2$) recorded from the brain of a Type II gerbil. All abbreviations are as in Fig. 8.21 [10]

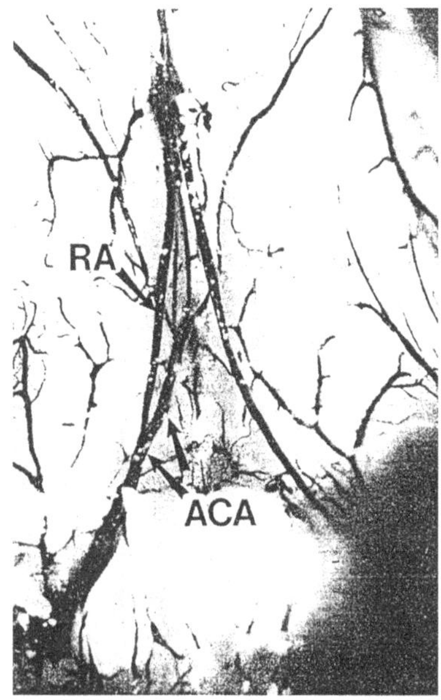


**Fig. 8.24** The anterior part of a Type III Circle of Willis in a Mongolian gerbil. The left ACA changed into the Rostral artery without giving off a branch into the CACA [10]

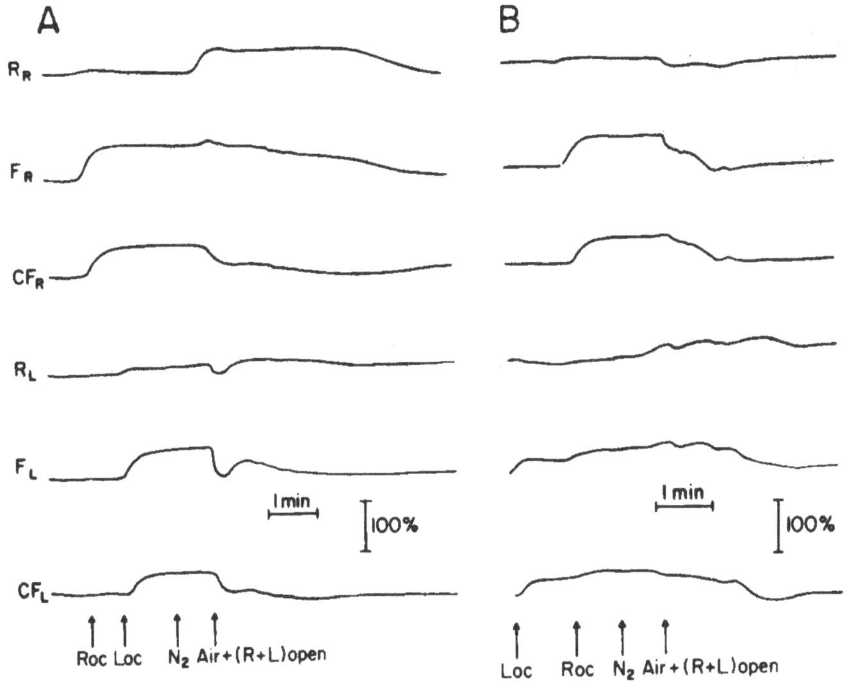


**Fig. 8.25** The effects of right and left carotid artery occlusion (Roc. Loc) and anoxia ($N_2$) on the metabolic and hemodynamic signals monitored from the two hemispheres of a Type III gerbil. Abbreviations as in Fig. 8.21. The order of the right and left occlusions were reversed in parts (**a**, **b**) [18]

### *8.2.3 Posterior Communication*

As shown in Fig. 8.17, Part P, most of the gerbils do not have connections between the PCA and SCeA, but in a small number of animals, the configuration $P_1$ or $P_2$ may exist. The anatomical evidence for this type of gerbil is shown in Fig. 8.18. As seen, the right PCOA is intact, and the left one is probably disconnected from the SCeA but is still connected to the PCA. The rest of the blood vessels typical of Type I (anterior area) are also seen very clearly in this preparation. The connection between the two ACA is covered by the branches of the optic nerve (ON), which normally were removed before the brain was photographed. As seen in Fig. 8.26, the NADH fluorescence responses to carotid occlusions were correlated to the anatomical arrangements. When the right carotid artery was occluded, the NADH in the ipsilateral hemisphere showed an increase followed by a partial recovery during the occlusion, suggesting compensation via vessels of the contralateral hemisphere. When the left artery was also occluded, the initial increase recorded from the left hemisphere as well as the additional increase in the right one ($CF_R$) showed a partial recovery. Only when $N_2$ was applied, both hemispheres reached the maximal NADH level, suggesting the existence of the posterior communicating artery, compensating for blood when the two carotid arteries were occluded.

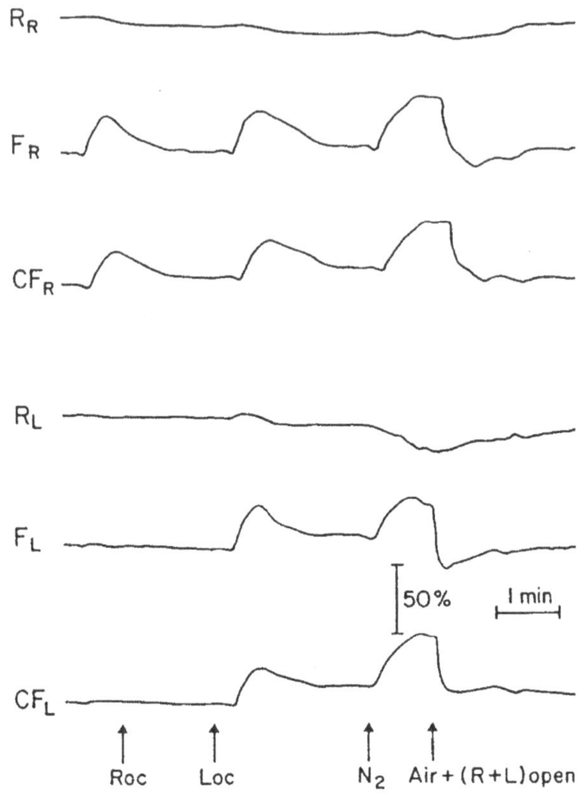


**Fig. 8.26** Metabolic and hemodynamic responses to unilateral and bilateral carotid artery occlusion (Roc,Loc) and anoxia ($N_2$) measured from a gerbil having the anatomy as shown in Fig. 8.18b, namely, Type A and P of Fig. 8.17 [10]

## 8.2.4 Intrahemisphere Heterogeneity

In 1983, we [23] described a unique pattern of responses to unilateral carotid occlusion suggesting an interhemispheric transfer of blood via an unknown anatomical pathway. Figure 8.27 shows the expected response to unilateral occlusion as measured in vivo as well as in the frozen brain. As seen under unilateral occlusion (Locc), the NADH level ($CF_R$) was increased in vivo (right part of the figure). When the same brain was frozen and analyzed for the fluorescence of flavoproteins (FP) and NADH levels (PN) using the low temperature brain scanner [26], a clear difference between the two hemispheres was detected. The left occluded hemisphere showed a fully reduced state of FP (dark area) and NADH (bright area). Both fluorochromes are in the maximally reduced metabolic state, but due to the differences in the fluorescing component, they seem to be showing the opposite response. The MR signal represents the calculated ratio between FP signal and FP + PN signal. The left hemisphere showed also the secondary reflectance increase response (SRI) after 1.5–2 min of ischemia [21].

Another peculiarity of the gerbil brain is the response of the olfactory bulbs to unilateral occlusion. As seen in Fig. 8.27, the two bulbs were not ischemic and

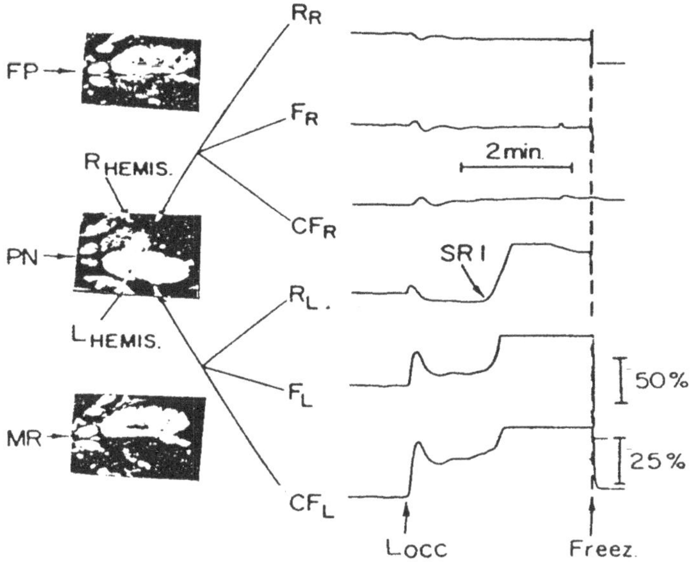


**Fig. 8.27** Unilateral ischemia induced by left carotid artery occlusion in seizure-prone (SP) gerbil. In this gerbil, the two olfactory bulbs obtain the blood from the same artery and were not affected by the unilateral occlusion. During the occlusion, a large increase in reflectance was recorded in vivo prior to the freezing of the brain. SRI secondary reflectance increase [10]

probably received the blood supply via the same anatomical route as this of the right hemisphere. The two bulbs obtained blood supply via the same artery connected to the right ACA.

Figure 8.28 shows the intrahemisphere heterogeneity response to right carotid occlusion. Four light guides [20] were located above the two hemispheres, recording simultaneously the NADH redox *state* in vivo, and later on the same brain was frozen and analyzed for the 2-D distribution of the FP and PN. As seen, measurement at site 3, which was on the left hemisphere showed the same response as points 1 and 2 of the right hemisphere. As seen in the 2-D pictures, this area near the midline was supplied by the other hemisphere and also in the various depths of the brain as shown by us [27]. By measuring the reflectance at the two excitation wavelengths (366 and 460 nm), we excluded the effects of changes in blood volume on the redox state of the frozen brain.

The heterogeneity in the hemisphere response to unilateral carotid occlusion shown in Fig. 8.28 led us to the conclusion that the two hemispheres of the gerbil brain in the ischemia study are interconnected and therefore cannot be used (in most gerbils) in the same experiment as the experimental side versus the control one. This heterogeneity can be seen in various brain pictures and descriptions published by various investigators [3, 28]. It is not clear as yet whether the pericallosal artery is the reason for the intrahemispheric heterogeneity.

Another example of the intrahemispheric heterogeneity is shown in Fig. 8.29. As seen, a typical *M. unguiculatus* (MU-TF) response to various perturbations while

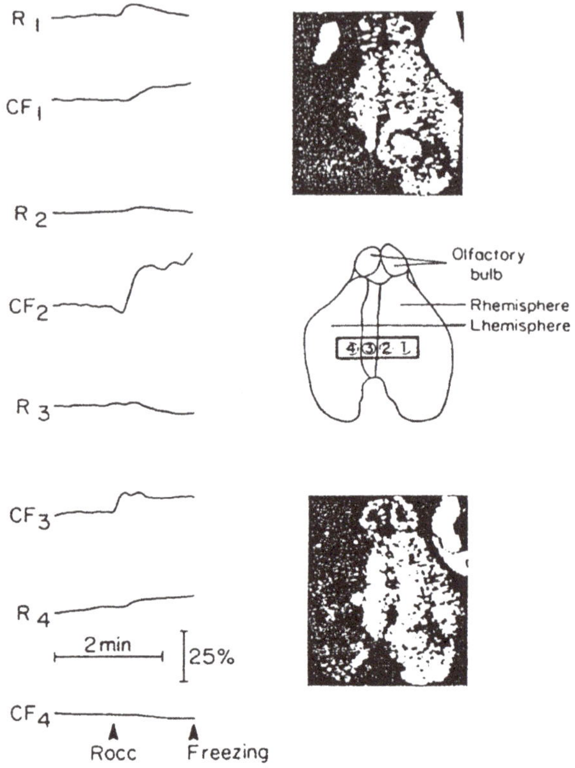


**Fig. 8.28** The effects of right carotid occlusion on NADH fluorescence measured in vivo and in the frozen brain of a normal (*Meriones unguiculatus*) gerbil. Site 3 that was located above the midline area of the left hemisphere did respond to the contralateral right carotid occlusion. The two scans shown are of the PN and represent different depths in the brain [10]

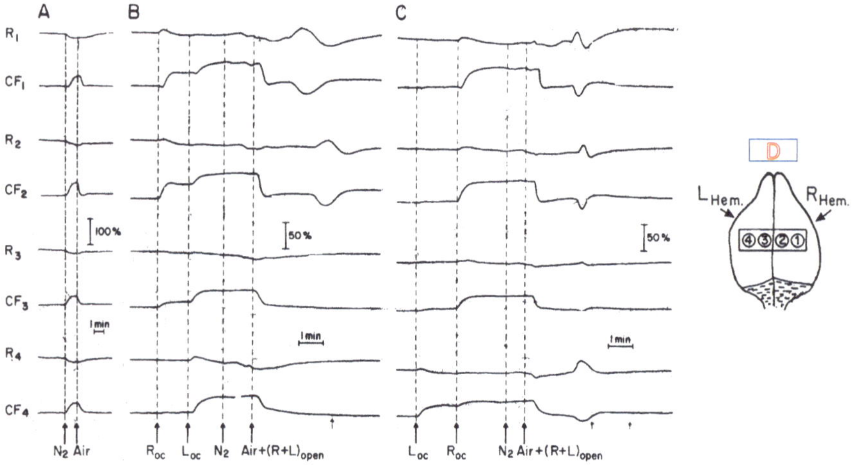


**Fig. 8.29** The effects of anoxia (**a**), unilateral (**b**) and bilateral carotid occlusion (**c**) on the NADH redox state as measured from 4 different sites in the Mongolian Gerbil Type All, $R_1$–$R_4$, 366 nm reflectance measured from points 1 to 4, respectively, shown in (**d**). CFI-CF4, NADH corrected fluorescence measured from the same sites as the reflectance [18]

simultaneously monitoring the NADH from four sites (as shown in Part D of Fig. 8.29). The anatomical pattern of this gerbil determined that it was typical of an $A_{II}$ vasculature, namely that the connection between the two ACAs was incomplete (Fig. 8.17, $A_2$). Part A of the figure demonstrates that all four monitoring sites responded similarly to anoxia as evaluated by the corrected NADH fluorescence ($CF_I$–$CF_4$) as well as by the reflectance ($R_1$–$R_4$). When the right carotid artery was occluded (Roc in Part B), sites 1 and 2 measuring from the right hemisphere exhibited a clear increase in NADH (CF1 + CF2). Site 3, measuring from the midline area of the left hemisphere, also exhibited an increase in NADH, while site 4 was unchanged. The same pattern of responses was found when the left carotid artery was occluded initially (span C Fig. 8.29). As expected, all 4 sites showed an increase in NADH during the bilateral occlusion (Loc in Part B or Roc in Part C). Exposure of the bilateral occluded gerbil to $N_2$ did not result in a further increase in NADH. Some sites showed a response to spreading depression, which developed during the ischemic-anoxic period during the recovery period. These results suggest that site 3 was supplied via the same artery as sites 1 and 2, although it is located on the contralateral hemisphere. When the same type of monitoring was performed using the *M. Tristrami* (Beit Dagan line), or the rat, no difference was found in the response to unilateral occlusion between the four monitored sites.

## 8.3 Strain Differences and Statistical Evaluation

### *8.3.1 Introduction*

Our previous papers [9, 10] and the preceding chapter show that a good correlation was found between the anatomical arrangement of the brain vasculature and the mitochondrial redox state responses to ischemia in the various strains of gerbils tested. In order to evaluate the significance of the differences between the carious strains of gerbils, we converted the analog signals into a digital form and ran the data through various statistical tests. We attempted to answer the following questions.

(i) What is the distribution of the anterior as well as the posterior patterns in the *M. unguiculatus* obtained from Tumblebrook Farms in comparison to that of the *M. tristrami* and the Albino rat?
(ii) What is the correlation between the NADH responses to unilateral and bilateral carotid artery occlusion in the various strains and the anatomical patterns?
(iii) Is there a correlation between the anterior pattern and the posterior pattern as evaluated by the anatomical and metabolic responses?
(iv) Is there a genetic factor in the determination of the anatomical patterns?
(v) What is the nature of the intrahemispheric heterogeneity responses to unilateral occlusion?

### *8.3.2 Materials and Methods*

All the details regarding animals, operation procedures, and experimental setup were described in our preceding paper [18].

### *8.3.3 Results*

#### 8.3.3.1 Anatomical Patterns

In order to analyze the distribution of the various anterior and posterior patterns in the four types of gerbils, we examined the anatomical patterns in a total of 97 animals. Table 8.1 shows the summary of the patterns of vascular arrangements in the different groups of gerbils. Eleven Albino rats were used as a comparison group. The major anterior patterns found in the *M. unguiculatus* bred in Tumblebrook Farms (MU-TF) were I, II, and III [3], and only one pattern of Type V was found in the *M. unguiculatus* bred in the Institute for Basic Research in NY (IBR/SP). The chi-square test ($X^2$) showed that there were no significant differences between the anterior and the posterior distribution in the two sources of the MU. Therefore, one may consider these results as coming from the same population. Patterns I and III, representing complete connection and disconnection, appeared in approximately 50% and 30%, respectively, in gerbils of the MU. The pattern II was found in approximately 25% of the gerbils having two ACAs of unequal diameters, for example, diameters of 86/21 microns or 110/10, 160/20, and 100/40 in other gerbils. The posterior connection appeared in a small fraction of the MU gerbils, approximately 10% bilateral and 20% unilateral. The two groups of *M. tristrami* tested (HU and BD) exhibited only the Type I anterior connection, namely a good compensation mechanism between the two hemispheres, as was also found in the Albino rat. Furthermore, the posterior bilateral connection appeared in 100% of the MT gerbils

**Table 8.1** Frequency distribution (percent) of the various anterior and posterior connections in the five groups of animals tested anatomically [24]

| | | Anterior connection | | | | Posterior connection | |
|---|---|---|---|---|---|---|---|
| | | I | II | III | V | Bilateral | Unilateral |
| *M. unguiculatus* TF | 53 | 47 | 23 | 30 | 0 | 9 | 20 |
| *M. unguiculatus* (IBR/SP) | 8 | 66 | 11 | 11 | 11 | 12 | 25 |
| *M. tristrami* (HU) | 9 | 100 | 0 | 0 | 0 | 199 | 0 |
| *M. tristrami* (BD) | 27 | 100 | 0 | 0 | 0 | 100 | 0 |
| Albino rat (Wistar strain) | 11 | 100 | 0 | 0 | 0 | 100 | 0 |

as well as in the Albino rats. A significant difference was found in the various gerbils between the diameter of the posterior communicating arteries (PCOA). This difference was in the range of 20–78 micro-m in the MU, while in the MT-HU, it was 46–110/micro-m, and in the MT-BD, 35–110 micro-m.

#### 8.3.3.2 Anatomical-Biochemical Correlations

Our previous paper [18] has demonstrated a good correlation between the anatomical patterns (anterior and posterior) and the metabolic responses to ischemia. All gerbils were tested in vivo for the metabolic responses and were later perfused and analyzed for brain vasculature.

#### 8.3.3.3 Anterior Circulation

The metabolic responses were measured simultaneously from the two hemispheres during unilateral carotid occlusion (right and left alternatively). The results presented in Table 8.2 include measurements from the MU, MT, and Albino rats. The percent change in NADH during the occlusion was calculated relative to the percentage change obtained during complete anoxia. It may be seen that the results obtained under right or left occlusion were very similar when compared to the responses measured in the ipsilateral or contralateral hemispheres. Furthermore, using the Student's t-test, we found that the differences between the MU-TF and the IBR/SP were not significant.

Table 8.3 demonstrates that the changes found in the MT (HU or BD) and Albino rats under unilateral occlusion were smaller than those recorded in the Type I of the MU (TF or IBR/SP). Using the two-way analysis of variance test (ANOVA), a significant difference was found between the occluded and unoccluded hemispheres in the rat ($F = 29.44$, df = 1, $p < 0.0001$). No significant difference was found between the effects of unilateral occlusion on the response of the ipsilateral hemisphere (left or right). The responses to unilateral occlusion of the MT-HU or MT-BD as well as

**Table 8.2** The effect of unilateral carotid artery occlusion on the NADH increase in the ipsilateral and contralateral hemisphere in various gerbils and in the Albino rat [24]

| Brain hemisphere | | *Meriones unguiculatus (TF)* | | | *Meriones unguiculatus (IBR)* | | | Meriones testrami | | Albino rat Wistar |
|---|---|---|---|---|---|---|---|---|---|---|
| | | I(13) | II(4) | III(9) | I(4) | II(1) | III(2) | HU(9) | BD(24) | (12) |
| R. occl. | Right | 40 ± 5 | 63 ± 5 | 86 ± 5 | 33 | 76 | 93 | 21 ± 5 | 18 ± 2 | 7 ± 1 |
| | Left | 2 ± 0.8 | 18 ± 1.7 | 0.6 ± 9.15 | 13 | 11 | 0 | 0.8 ± 0.15 | 2.3 ± 1 | 1 ± 0.1 |
| L. occl. | Right | 1.3 ± 0.8 | 0 | 0 | 4 | 0 | 0 | 1.2 ± 1 | 1.9 ± 0.6 | 1 ± 1 |
| | Left | 35 ± 4 | 59 ± 11 | 83 ± 3 | 42 | 87 | 100 | 13 ± 3 | 24 ± 3 | 6.7 ± 1 |

Number in parentheses represents the number of animals tested. Results are expressed in mean % change (± S.E.M.) during the occlusion calculated relative in change during anoxia

**Table 8.3** The variations in responses to unilateral occlusion between pairs of the various gerbil groups [24]

| | MT-BD | MT-HU | Rat |
|---|---|---|---|
| MU-TF | NSP | $p < 0.005$ | $p < 0.005$ |
| MT-BD | X | NSP | $p < 0.005$ |
| MT-UH | X | X | NSP |

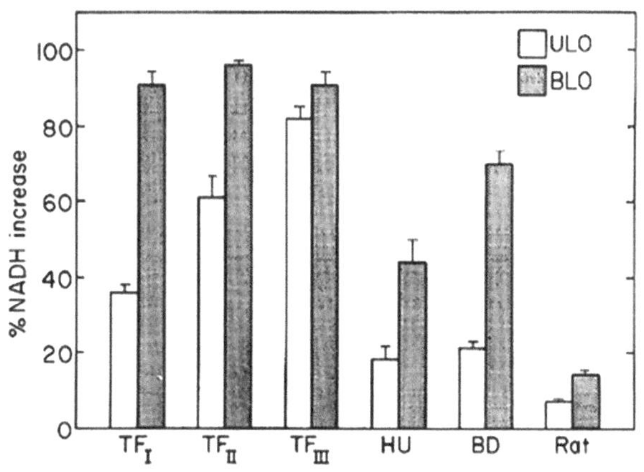


**Fig. 8.30** Mitochondrial redox state responses to unilateral (ULO) and bilateral carotid occlusions (BLO) in the 6 groups of animals tested. $TF_I$, $TF_{II}$, $TF_{III}$—*M. unguiculats* Tumblebrook Farms, HU *M. tristrami* Hebrew University, BD *M. tristrami* Beit Dagan, Rat Albino rat. The % change in NADH increase was normalized for the changes in response to $N_2$ exposure (100%) in each animal [24]

MU-TF and the Albino rat were analyzed by the ANOVA test and found to be significant ($F = 11.56$, df = 3, $p < 0.0001$). In order to compare the sensitivity of the four groups tested with unilateral occlusion, we used the Tukey test. The results show that the maximal sensitivity to ULO was in the MU-TF group, followed by the MT-BD and MT-HU, while the rat exhibited the minimal change. The MT-BD was closed in the MU-TF while the MT-HU was closed in the rat. The data shown in Fig. 8.30 summarizes the effects of ULO (unilateral occlusion) and BLO (bilateral occlusion) on the changes in NADH measured in the various groups of animals. It may be seen that in the three groups of TF gerbils, BLO produced above 90% ischemia while ULO led to a gradual increase in NADH and was correlated to the anatomical pattern (I, II, or III). The rat showed a very small response to ULO or BLO, while the two groups of *M. Tristrami* were in an intermediate state as described in the previous paragraph. The histogram shown in Fig. 8.31 indicates that the responses of the two *M. Unguiculatus* strains used were not statistically different (two-way ANOVA) when exposed to unilateral occlusion. However, the three anatomical patterns were significantly different ($F = 44.63$, df = 2, $p < 0.0001$) in their response to ULO. We also tested the variance between each pair of anterior

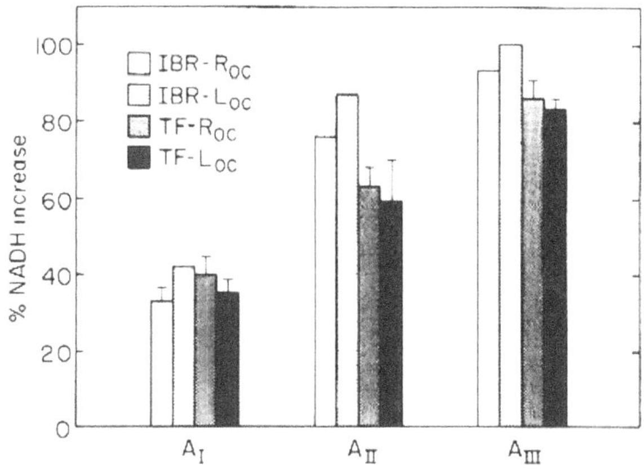


**Fig. 8.31** Changes in NADH responses to unilateral carotid occlusion in the *M. unguiculatus* (TF and IBR) as correlated to the anterior anatomical patterns. $R_{oc}$, $L_{oc}$, right or left carotid artery occlusion. In the IBR groups only $R_{oc}$–$A_I$ had enough results to calculate the standard error of the mean (see Table 8.1) [24]

anatomical patterns using the Tukey test. Significant differences were found between pattern I and II ($p < 0.05$) and between I and III ($p < 0.05$). The difference between II and III was insignificant. The Duncan test showed all three pairs to be significantly different.

#### 8.3.3.4 Posterior Circulation

The best test for the existence of posterior communication in the Circle of Willis is to expose the bilateral carotid occluded gerbil to $N_2$. The increment in NADH during such a procedure is correlated to the anatomical connection in the posterior part of the brain. In this part of the study, we measured the diameter of the posterior communicating artery (PCOA) in the fixated brain (Table 8.4). In the Albino rat possessing a complete Circle of Willis, the increase in NADH due to BLO was minimal and only exposure to $N_2$ led to maximal accumulation of NADH. The diameter of the PCOA was in the range of 140–210 μ. In *M. unguiculatus*, the level of NADH increased under BLO and was above 90% (both in the TF and IBR strains). The diameter of the PCOA was the smallest, and it was missing in most gerbils. The *M. tristrami* of both strains showed an intermediate status in the PCOA diameter as well as in the level of NADH increase.

A significant correlation ($p < 0.01$) was found between the diameter of the PCOA and the NADH changes measured in the MU-TF ($R = -0.910$) and MT-BD ($R = -0.928$). In the MT-HU, the correlation was insignificant. Calculations of the PCOA area (based on the diameter) showed the connection between the posterior blood supply and the percentage change in NADH under BLO, as shown in Fig. 8.32. The difference between the two regression lines was shown to be minimal.

**Table 8.4** The relationship between the size of the posterior communicating artery and the changes in NADH (mean ± S.E.M.) under bilateral carotid occlusion calculated as percent of the level found under exposure to $N_2$ [24]

| | MU-TF | MU-IBR | MT-BD | MT-HU | Rat |
|---|---|---|---|---|---|
| NADH % of max. value | 91.7 ± 2 | 98 ± 1.7 | 70 ± 3.6 | 44 ± 5.9 | 14 ± 1 |
| PCoA diameter (μm) | 30–78 | – | 35–110 | 46–110 | 140–210 |

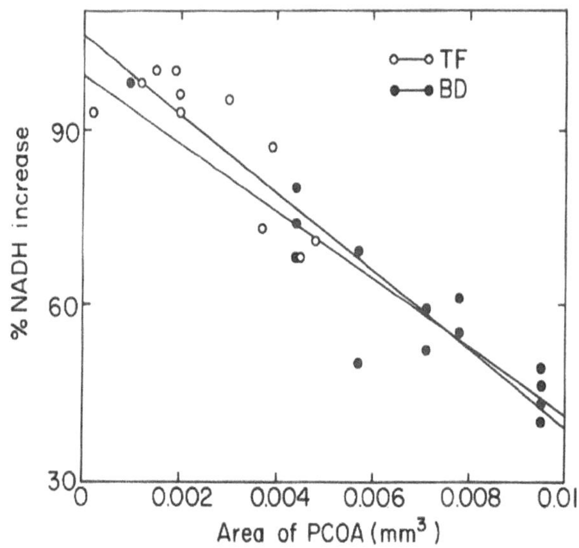


**Fig. 8.32** The correlation between the area of the posterior communicating artery (PCoA) and the percent change in NADH under bilateral carotid artery occlusion in two strains. TF *M. unguiculatus* Tumblebrook Farms, BD *M. tristrami* Beit Dagan [24]

#### 8.3.3.5 Relationship Between Anterior and Posterior Circulations

To test this point, we compared the correlation between the changes in NADH under ULO and BLO in gerbils possessing a complete (or nearly complete) anterior connection. Since the two hemispheres responded similarly to ULO, we calculated the average for the correlation calculations. Figure 8.33 shows the two regression lines obtained in the two MT strains. In the MT-HU and MT-BD, the values were $R = 0.895$, $P < 0.01$ and $R = 0.603$, $P < 0.05$, respectively. The correlation between the anatomical patterns of the posterior communication and the various anterior vasculature is shown in Fig. 8.34. Using a chi-square test ($X^2$), it was found that the differences in the frequency distribution of the posterior connection were not significant in the $A_I$–$A_{III}$ patterns.

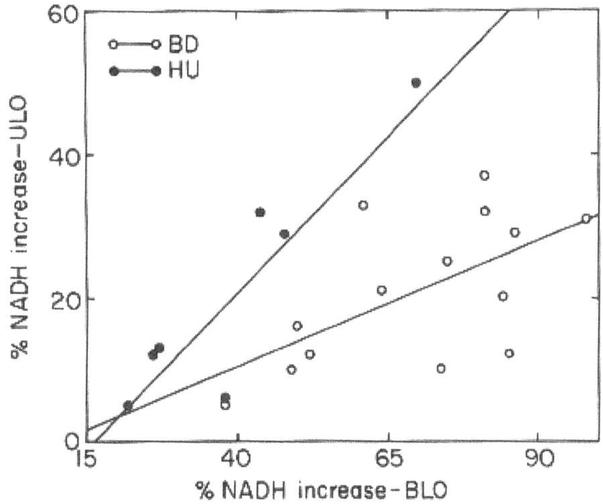


**Fig. 8.33** The interrelation between the metabolic response to unilateral occlusion and that of bilateral occlusion measured in two lines of the *M. tristrami* [24]

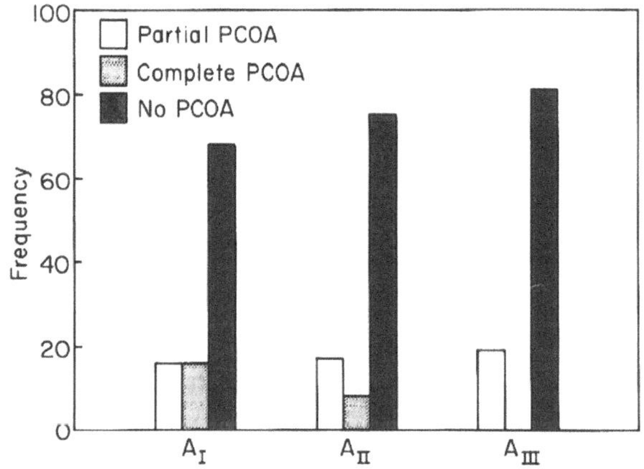


**Fig. 8.34** Frequency distribution of the three posterior communications pattern in the three anterior vasculative patterns in the *M. unguiculatus* [24]

## *8.3.4 Discussion*

The assumption that bilateral carotid artery occlusion creates complete ischemia in all gerbils is no longer valid. We believe that during the years, the level of communication between the basilar and carotid circulations in the gerbil population changed. In our laboratory years ago, we found very few gerbils with posterior communication (3–5% of total). In our recent study, we found unilateral or bilateral posterior communication in 20% or 10% of the gerbils tested, respectively. This anatomical evidence was supported by the changes in the NADH redox state measured under bilateral carotid artery occlusion as compared to the exposure of the occluded gerbil to $N_2$. Levy and Brierley also concluded, without giving actual numbers, that the level of communication in this anterior and posterior circulation varied between gerbils. Barzaghi et al. reported that only 75% of the gerbils had a fiat EEG under bilateral carotid occlusion. The remaining 25% were discarded due to the incompleteness of ischemia. We believe that although a posterior communication may exist in 10–20% of the gerbils, the actual level of ischemia is close to 100% in all gerbils due to the secondary vasoconstriction response to ischemia (ID), which was recorded in all gerbils exposed to 5 min of bilateral carotid occlusion [29–31]. Under unilateral occlusion, the level of ischemia covers the entire range from a minimal level, if at all, in Type I configuration (Fig. 8.31, Part A) to a maximal level obtained in Type III vasculature in the anterior part. It is very difficult to predict the level of ischemia that will be reached during the occlusion unless a specific parameter is monitored during the experiment. We monitored the NADH redox state during the occlusion and found a significant correlation between this parameter and the anatomical features determined in the same gerbil. Another technique for prediction of the level of ischemia after unilateral carotid artery occlusion [32, 33] by measuring the diameter of the distal common carotid artery (CCA) before and during the ligation was recently described. These investigators showed a good correlation between the decrease in the CCA diameter, the neurological symptoms developed, and the type of brain vasculature determined anatomically. The facility of stroke induction in the gerbil by unilateral or bilateral carotid artery occlusion accounts for its application in pharmacological studies, as has been discussed in detail [34]. Jarot and Domer used the mortality rate as an index for evaluating drug effects and claimed that small groups of animals could be used due to the small variation in individual responses. During the last years, we have measured relative CBF using a laser Doppler flowmeter combined with NADH monitoring and found a significant correlation between these two parameters in gerbils exposed to ischemia ills. In this study, we compared two kinds of Mongolian gerbils, the commonly used *M. unguiculatus* and *M. tristrami*, which have not been used in ischemia experiments until now. One of the aims of this study was to compare the responses to ischemia and anoxia in the various gerbils and those measured in the rat. In order to verify the mitochondrial redox state responses, we correlated them to the anatomical pattern of the vasculature in the same monitored brain. The typical responses to ischemia and anoxia, shown in the first paper, were significantly correlated to the anatomical patterns. This was true in all groups tested. Therefore, we can conclude that monitoring NADH under ischemia in the gerbil is a significant indicator of the anatomical pattern of brain vasculature. Evaluation of the differences between *M. unguiculatus* and *M. tristrami* must be divided into discussion of the anterior and posterior circulations. Figure 8.30 demonstrates the anterior communication in TF. The two MTs (HU and BD) provide sufficient compensation during unilateral occlusion, similarly to that found in the rat. Nevertheless, the MT was significantly different from all three groups of the MU ($TF_I$, $TF_{II}$, and $TF_{III}$). The results shown in Fig. 8.31 support the conclusion that unilateral ligation was not different in the two sources of the MU. This was true when comparing the two sources of the MT, namely, the HU and BD. The posterior circulation in the three groups of TF was identical, and very little compensation was achieved under bilateral occlusion. This was different in the MT, which exhibited good compensation under bilateral occlusion, as was also found in the rat. Here again, the anatomical-biochemical correlation was significant (Fig. 8.32). The advantage of the gerbil as a small animal model for ischemia was demonstrated a few years ago [35, 36] using the $^{31}$P NMR approach. The non-destructive technique of phosphorus-31 nuclear magnetic reso nance was combined with non-invasive surface coils to follow changes in phosphocreatine and adenine nucleotide concentration as well as the intracellular pH of the two hemispheres of the gerbil brain after carotid ligation.

The main disadvantage of using the gerbil as an animal model is its high sensitivity to experimental procedures. We were able to show that the various levels of anesthetics used led to changes in responses to activation of the brain [22, 37]. When spreading depression was induced by an epidural application of a KCl solution, the typical oxidation response of the NADH was replaced by a reduction cycle, suggesting that blood flow compensation was limited under an increased anesthetics level. In the rat brain, we were unable to change the response to SD by an overdose of anesthetics, and only partial ischemia induced such changes as reported in the gerbil. Therefore, we believe that the level of anesthesia is a critical factor affecting the functions of the gerbil brain, and this has to be taken into consideration.

Monitoring the physiology of the brain is one solution for the prevention of pathological states in the brain during the experiment. The phenomenon of "No-reflow" was described by Ito et al. [38] in the gerbil brain ischemia experiments. They were able to show that the decrease in blood pressure in the early stage of reperfusion was the main reason for this phenomenon. In our laboratory, we found that only deeply anesthetized gerbils or other "not delicate" manipulations of the carotid arteries during the operation or the occlusion resulted in a "no-reflow" phenomenon. The noreflow phenomenon was also reported by Levy et al [39]. Recently, Tomida et al. [40] showed the large effects of electrode placement on cortical blood flow using the hydrogen clearance techniques, suggesting that the gerbil is a more sensitive model to brain manipulation. Therefore, it seems to us that appropriate handling of the gerbil during the preparation as well as online monitoring of the physiological state of the brain will lead to better experimental results in the gerbil mode.

## References


1. Levine S, Payan H (1966) Effects of ischemia and other procedures on the brain and retina of the gerbil (*Meriones unguiculatus*). Exp Neurol 16:255–262
2. Levy DE, Brierley JB (1974) Communications between vertebro-basilar and carotid arterial circulations in the gerbil. Exp Neurol 45:503–508
3. Donadio MF, Kozlowski PB, Kaplan H, Wisniewski HM, Majkowski J (1982) Brain vasculature and induced ischemia in seizure-prone and non-seizure-prone gerbils. Brain Res 234:263–273
4. Berry K, Wisniewski HM, Svarzbein L, Baez S (1975) On the relationship of brain vasculature to production of neurological deficit and morphological changes following acute unilateral common carotid artery ligation in gerbils. J Neurol Sci 25:75–92
5. Du XY, Zhu XD, Dong G, Lu J, Wang Y, Zeng L, Zhao TY, Ye HH, Li RS, Bai JY, Chen ZW (2011) Characteristics of circle of Willis variations in the mongolian gerbil and a newly established ischemia-prone gerbil group. ILAR J 52:E1–E7

6. Levine S, Sohn D (1969) Cerebral ischemia in infant and adult gerbils. Relation to incomplete circle of Willis. Arch Pathol 87:315–317
7. Kahn K (1972) The natural course of experimental cerebral infarction in the gerbil. Neurology 22:510–515
8. Zeman W, Innes JRM (1963) Craigie's neuroanatomy of the rat. Academic Press, New York
9. Breuer Z (1988) Determination of brain vasculature in the Mongolian gerbil and the biochemical correlated responses under ischemia. Department of Life Sciences, Bar-Ilan University, Ramat Gan
10. Mayevsky A, Breuer Z (1990) In: Schurr A, Rigor BM (eds) The Mongolian gerbil as a model for cerebral ischemia. CRC Press, pp 27–46
11. Kahn KJ, Lawrence EP, Pranzarone GF (1971) The use of dextran-40 in modifying the course of experimental intraciion in the gerbil. Neurology 21:400
12. Harrison MJ, Brownbill D, Lewis PD, Russell RW (1973) Cerebral edema following carotid artery ligation in the gerbil. Arch Neurol 28:389–391
13. Schonfeld AR, Glick SD (1979) Cerebrovascular abnormalities associated with seizure susceptibility in the Mongolian gerbil. Brain Res 173:147–151
14. Thiessen DD, Lindsey G, Friend HC (1968) Spontaneous seizures in the Mongolian gerbil. Psychon Sci 11:227
15. Cohn R (1979) Convulsive activity in gerbils subjected to cerebral ischemia. Exp Neurol 65:391–397
16. Loskota WJ, Lomax P (1975) The Mongolian gerbil (*Meriones unguiculatus*) as a model for the study of the epilepsies: EEG records of seizures. Electroencephalogr Clin Neurophysiol 38:597–604
17. Zheng Y, Yingchang H, Mengzhong Y (1986) The Willis' circle in Mongolian gerbils. Acta Anat Sin 17:360
18. Mayevsky A, Breuer Z (1992) Brain vasculature and mitochondrial responses to ischemia in gerbils. I. Basic anatomical patterns and biochemical correlates. Brain Res 58:242–250
19. Mayevsky A (1978) Pyridine nucleotide oxidation reduction state of the cerebral cortex in the awake gerbil. J Neurosci Res 3:369–374
20. Mayevsky A, Chance B (1983) Multisite measurements of NADH redox state from cerebral cortex of the awake animal. Adv Exp Med Biol 159:143–155
21. Mayevsky A, Zarchin N (1981) The effects of unilateral carotid occlusion on the responses to decapitation in the gerbil brain. Brain Res 206:155–160
22. Mayevsky A, Lebourdais S, Chance B (1980) The interrelation between brain $PO_2$ and NADH oxidation- reduction state in the gerbil. J Neurosci Res 5:173–182
23. Mayevsky A, Zarchin N, Kaplan H, Haveri J, Haselgrove J, Chance B (1983) Brain metabolic responses to ischemia in the Mongolian gerbil: *in vivo* and freeze trapped redox state scanning. Brain Res 276:95–107
24. Breuer Z, Mayevsky A (1992) Brain vasculature and mitochondrial responses to ischemia in gerbils: II. Strain differences and statistical evaluation. Brain Res 598:251–256
25. Chance B, Williams GR (1955) Respiratory enzymes in oxidative phosphorylation (I- kinetics of oxygen utilization). J Biol Chem 217:383–393
26. Haselgrove J, Barlow CH, Chance B (1980) The 3D distribution of metabolic states in the gerbil brain during the course of spreading depression. In: Passoneau JV, Hawkins RA, Lust WD, Welsh FA (eds) Cerebral metabolism and neural function. Williams & Wilkins, Baltimore/London, pp 72–76
27. Mayevsky A, Kaplan H, Haveri J, Haselgrove J, Chance B (1986) Three-dimensional metabolic mapping of the freeze-trapped brain: effects of ischemia on the Mongolian gerbil. Brain Res 367:63–72
28. Levy DE, Duffy TE (1977) Cerebral energy metabolism during transient ischemia and recovery in the gerbil. J Neurochem 28:63–70
29. Mayevsky A, Zarchin N, Tannenbaum B (1984) Brain responses to experimental oxygen deficiency in the Mongolian gerbil. Adv Exp Med Biol 180:191–202

30. Mayevsky A, Zarchin N (1987) Metabolic ionic and electrical activities during and after incomplete or complete cerebral ischemia in the Mongolian gerbil. Adv Exp Med Biol 215:265–273
31. Mayevsky A, Friedli CM, Reivich M (1985) Metabolic, ionic and electrical responses of the gerbil brain to ischemia. Am J Phys 248:R99–R107
32. Kitagawa K, Matsumoto M, Handa N, Fukunaga R, Ueda H, Isaka Y, Kimura K, Kamada T (1989) Prediction of stroke-prone gerbils and their cerebral circulation. Brain Res 479:263–269
33. Matsumoto M, Hatakeyama T, Akai F, Brengman JM, Yanagihara T (1988) Prediction of stroke before and after unilateral occlusion of the common carotid artery in gerbils. Stroke 19:490–497
34. Jarrott DM, Domer FR (1980) A gerbil model of cerebral ischemia suitable for drug evaluation. Stroke 11:203–209
35. Thulborn KR, du Boulay GH, Duchen LW, Radda G (1982) A $^{31}$P nuclear magnetic resonance in vivo study of cerebral ischaemia in the gerbil. J Cereb Blood Flow Metab 2:299–306
36. Chance B, Barlow C, Nakase Y, Takeda H, Mayevsky A, Fischetti R, Graham N, Sorge J (1978) Heterogeneity of oxygen delivery in normoxic and hypoxic states: a fluorometer study. Am J Phys 235:H809–H820
37. Mayevsky A, Zarchin N, Friedli CM (1982) Factors affecting the oxygen balance in the awake cerebral cortex exposed to spreading depression. Brain Res 236:93–105
38. Ito U, Ohno K, Yamaguchi T, Tomita H, Inaba Y, Kashima M (1980) Transient appearance of "no-reflow" phenomenon in Mongolian gerbils. Stroke 11:517–521
39. Levy DE, Brierley JB, Plum F (1975) Ischaemic brain damage in the gerbil in the absence of 'no-reflow'. J Neurol Neurosurg Psychiatry 38:1197–1205
40. Tomida S, Wagner HG, Klatzo I, Nowak TS (1989) Effect of acute electrode placement on regional CBF in the gerbil: a comparision of blood flow measured by Hydrogen Clearance, {$^{3}$H} Nicotine and [$^{14}$C] lodoantipyrine techniques. J CBF Metabol 9:79

# Chapter 9
# Two- and Three-Dimensional Metabolic Mapping of the Ischemic Gerbil Brain

**Abstract** The effects of unilateral carotid artery (left or right) occlusion on the three-dimensional metabolic mapping were studied in the Mongolian gerbil (*Meriones unguiculatus*). The brain was freeze-trapped using the liquid nitrogen funnel technique and analyzed for the two-dimensional distribution of the reduced pyridine nucleotides and oxidized flavoprotein using a time-sharing micro-light guide surface fluorometer/reflectometer. The results could be summarized as follows. (1) Bilateral carotid artery occlusion induced uniform effects in terms of redox state in the cerebral cortex and the olfactory bulbs. (2) After unilateral carotid occlusion, the redox state of the ischemic ipsilateral hemisphere was higher and was not affected by blood volume changes as evaluated from the reflectance signal scanned in several brains. (3) The narrow band of tissue near the midline of the ischemic hemisphere connected to the contralateral hemisphere blood supply appeared in a large number of gerbils. (4) The anomaly in the redox state of the intracellular space also appeared in subcortical structures, as seen in all the two-dimensional mappings measured. (5) It seems that blood circulation to the two olfactory bulbs is connected to the same blood vessel. (6) It is suggested that unilateral carotid artery-occluded gerbils may show large variability inside the same hemisphere and therefore results must be evaluated very carefully.

**Keywords** Redox state mapping · Frozen brain · Brain circulation · Seizures · Brain ischemia

## 9.1 Introduction

During the past few years, the use of the Mongolian gerbil as an animal model for brain ischemia studies has expanded. In a large group of studies, unilateral ischemia was performed by occlusion of one of the carotid arteries, and the contralateral hemisphere served as a control to the occluded one. In our previous study [1–3], we described the history of this animal model and showed that, in parallel to its advantages as a model for ischemia, an unexplained response to unilateral carotid artery



A. Mayevsky, *The Mongolian Gerbil Brain*,
https://doi.org/10.1007/978-3-031-69549-0_9

occlusion was found [4]. In 1975, Berry et al. [5] showed the effects of unilateral carotid occlusion and concluded that the connection between the two anterior cerebral arteries varied in various gerbils. Donadio et al. [6] described the vasculature in the brain of their inbred IBR seizure-prone (SP) [7] and nonseizure-prone (NSP) colonies. The anatomical results published to date regarding the possible interhemispheric transfer of blood during unilateral carotid artery occlusion cannot explain the results obtained when the redox state was monitored in vivo or in the freeze-trapped brain [4]. In a large number of gerbils (regardless of origin or strain), we found a peculiar interhemispheric blood supply connection. Under unilateral carotid artery occlusion, a narrow band of tissue (about 1 mm in width) near the midline obtained its blood via the same vessels supplying the contralateral hemisphere. Due to the importance of this issue, we decided to investigate it in detail. In this study [8], we investigated the three-dimensional distribution of the metabolic state in gerbils exposed to unilateral carotid artery occlusion as evaluated by the low-temperature scanning of the pyridine nucleotides (PN) (NADH) and flavoproteins (FP) after in vivo freeze-trapping. Using this approach, one can conclude very definitively regarding the pattern of blood supply to various parts of the brain.

## 9.2 Methods

Two sources for the Mongolian gerbils, *Meriones unguiculatus*, were used. The random-bred strain was obtained from Tumblebrook Farms, West Brookfield, MA, United States, and the seizure-prone (IBR/SP) and nonseizure-prone (IBR/NSP) gerbils were obtained from the New York State Institute for Basic Research in Developmental Disabilities. Gerbils in the IBR colonies were tested for seizure activity, as described previously [5, 6]. The gerbil (50–70 g) was anesthetized by intraperitoneal injection of Equithesin (0.3 mL/100 g), mounted to a head-holder, and the parietal bone was removed bilaterally without removing the dura mater. The animal remained in the head-holder until the end of the freezing procedure.

### *9.2.1 Funnel Freezing of the Brain*

Because of the small size of the gerbil brain, a barrel of a 30-mL syringe was used as a funnel for the liquid $N_2$. The rim of the funnel was covered with Vaseline for better sealing during the freezing. When cessation of respiration was observed, the gerbil was removed from the holder and immersed in a container of liquid $N_2$. The head was severed by a bandsaw, and a block of the head was transferred to the low-temperature scanner for redox state analysis.

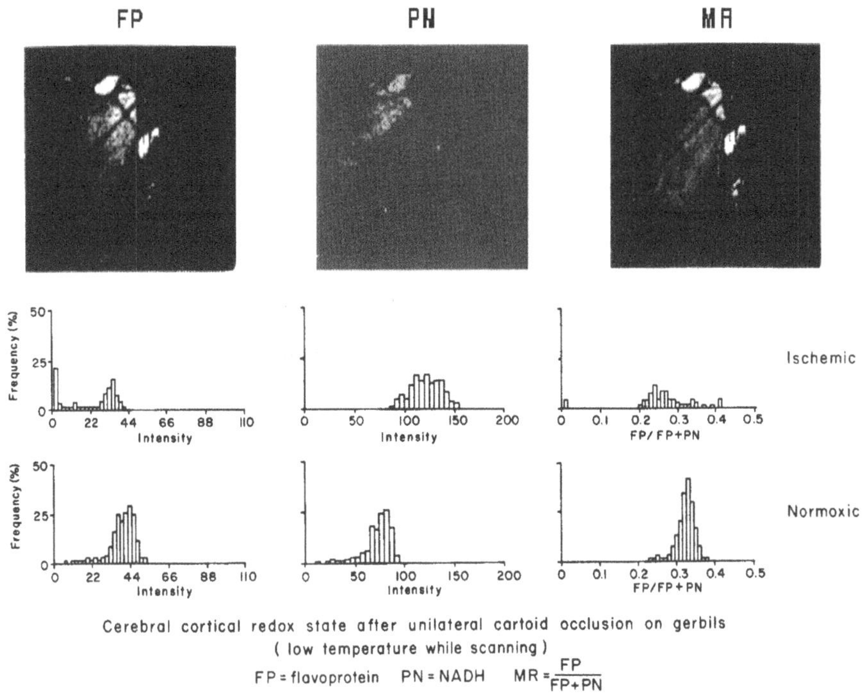


**Fig. 9.1** Frequency distribution of the metabolic redox state in the normoxic and ischemic cerebral hemispheres of the gerbil. FP, flavoprotein; PN, pyridine nucleotide [8]

### 9.2.2 *Frozen Brain Redox State Evaluation*

The metabolic state of the frozen brain was studied by recording the fluorescence signals of the mitochondrial NADH and FP. The excitation wavelengths were 366 nm and 460 nm, and the emitted lights were measured at 420–480 nm and 520–570 nm for reduced NADH and oxidized flavoproteins, respectively. The study of the fluorescent intensities of the enzymes will give a metabolic picture of the distributions of redox states, as has recently been described by Chance et al [9–11] The ratio between the two signals measured gives the corrected redox state relatively unaffected by changes in mitochondrial concentration or hemoglobin level in the tissue analyzed. The distribution of the fluorescence signals in three dimensions throughout the brain was recorded with an online scanner, as described in detail [9]. For quantitative analysis, a histogram of the frequency distribution of signal intensity (across any area of the sample) is plotted as shown in Fig. 9.1.

## 9.3 Results

The results presented were selected as representative of more than 50 gerbils included in the study. The figures shown include the effects of left or right artery occlusion; different strains (SP, NSP, or the random bred) and brains show the inter-hemispheric connection. In the first few experiments, we also measured the reflected signals (366 nm and 460 nm) from the same frozen sections; the results are shown in Figs. 9.2 and 9.3 (SP gerbil). The two reflectance spectra (PR and FR) shown in each figure are very similar, although they represent two different wavelengths (366 nm for PN and 460 for FP). Also, the effect of the ischemia was found only in the fluorescence signals, namely, the PN and FP, as well as in the calculated ratio (MR). The same pattern of changes was found in the two sections, representing different depths in the same brain. The ischemic hemisphere—the left one—shows the high fluorescence intensity of the NADH (PN) and a less clear picture in the FP signal, but the MR pattern shows a more reduced area in most of the left hemisphere (dark area). The midline area of the left hemisphere shows the same redox state as in the control hemisphere, suggesting a common source of blood supply. In other experiments, where the reflectance signal was measured, we never saw any effect of treatment (ischemia) on the reflectance map of the brain.

Therefore, in the remaining experiments, the reflectance results will not be presented. Another type of control was to measure the redox state after bilateral carotid artery occlusion, the results of which are shown in Fig. 9.4. The two hemispheres of each brain show the same pattern of redox state in the FP and PN signals, as well as

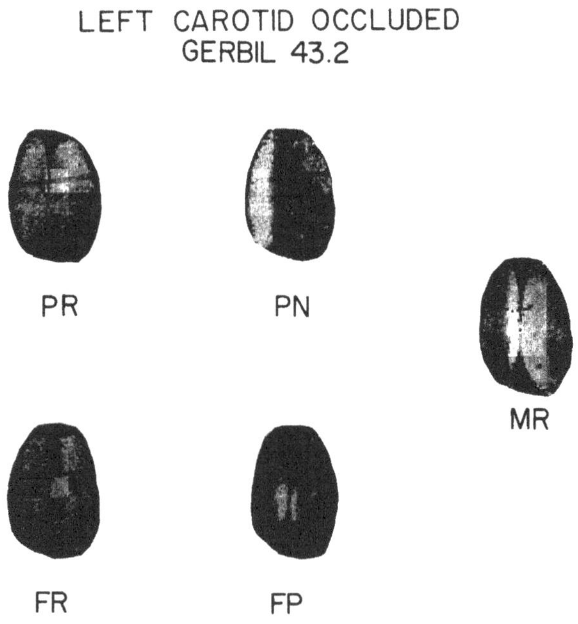


**Fig. 9.2** The effect of left carotid artery occlusion on the fluorescence and reflectance signals from the gerbil brain. PR and FR, pyridine nucleotide and flavoprotein reflectance measurements in the same brain [8]

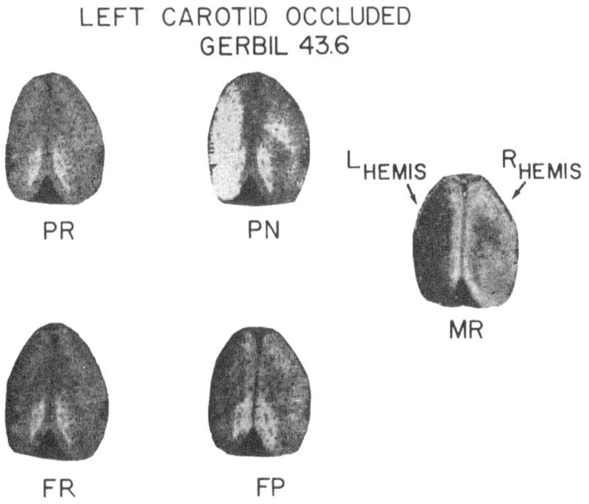


**Fig. 9.3** The metabolic responses to left carotid artery occlusion in the same gerbil shown in Fig. 9.2. The scans were obtained from deeper structures in the same brain [8]

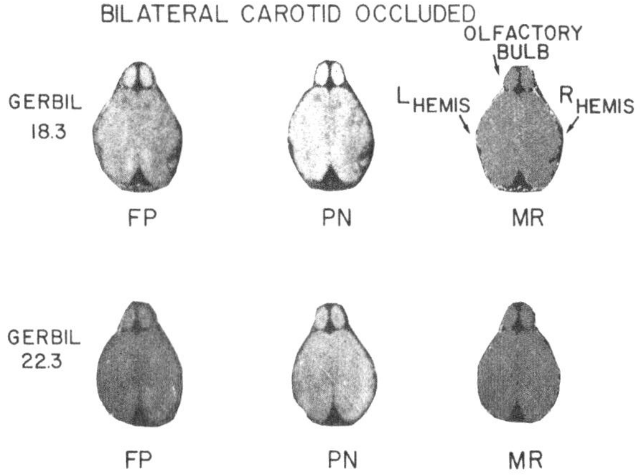


**Fig. 9.4** Effect of complete ischemia (bilateral carotid occlusion) on the redox state monitored in two gerbil brains [8]

the MR. The redox state of the olfactory bulbs is not different from that of the cerebral hemispheres. We expected theoretically that under unilateral carotid artery occlusion, the redox state of the two hemispheres would have the border zone at the midline region. However, only in a few gerbils did we obtain the expected results (Fig. 9.5). The entire left hemisphere in this subject is more reduced, as seen in the PN and MR, and less clear in the FP signal. Unexpectedly, the two olfactory bulbs are both oxidized to the same level as in the normoxic (right) hemisphere. It seems that the two olfactory bulbs are supplied by the same blood vessels, or the olfactory bulbs have bilateral supplies and are probably connected somehow to that of the right hemisphere in this specific gerbil. In most gerbils in which one carotid artery was occluded, the results did not confirm our expectations. Rather, the midline area of the two hemispheres showed the same redox state, which in some cases was reduced and in others normoxic. These results indicate that the same blood vessels

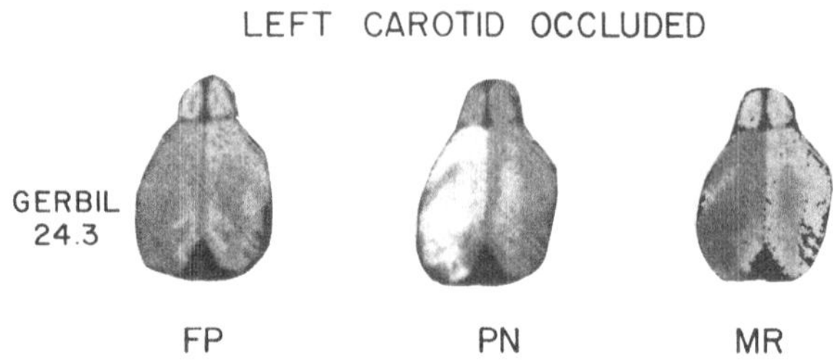


**Fig. 9.5** Metabolic response to partial ischemia induced by left carotid occlusion [8]

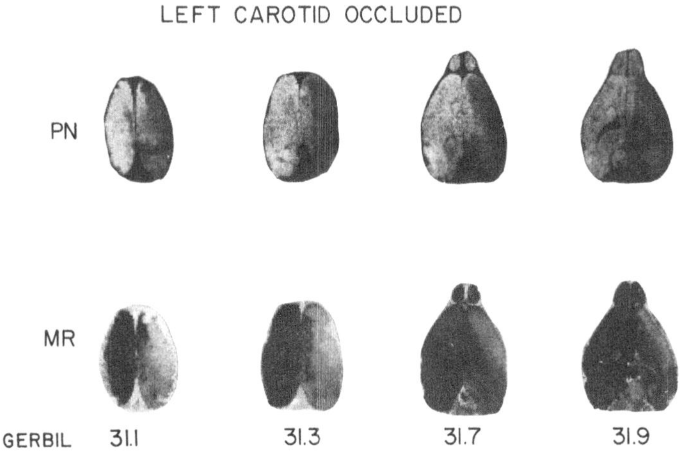


**Fig. 9.6** Effect of partial ischemia on the redox state measured at different depths of the gerbil brain. In each slice, the PN and the MR picture are shown [8]

supply those areas: the supply comes from the right hemisphere in some gerbils and from the left in others. In order to eliminate the possibility of an artifact, we tested the effects of depth on the appearance of the cross-hemispheres phenomenon. The effects of left carotid artery ligation on the PN fluorescence and MR are shown in Fig. 9.6. Four different depths are shown, representing 0.5, 1.0, 1.4, and 1.8 mm below the surface of the brain. In all depths measured, the cross-hemispheric phenomenon was found, eliminating the possibility of artifacts at a particular depth. In the brain pictured (Fig. 9.6), the olfactory bulb blood supply was connected to that of the left hemisphere. It seems that the existence of the interhemispheric connection is not related to the side of the ligation (left or right carotid) or to the gerbil's seizure propensity. Figures 9.7 and 9.8 represent carotid artery occlusion in nonseizure-prone gerbils. In these brains, the borderline zone was in the right (Fig. 9.7) or left (Fig. 9.8) normoxic hemispheres, and the olfactory bulb's blood supply was connected to that of the right and left hemispheres, respectively. The interhemispheric connection of blood supply in the midline area is not unique to the IBR seizure-prone and nonseizure-prone gerbils. Figure 9.9 shows the effect of unilateral ligation in two brains of gerbils obtained from Tumblebrook Farms (MA,

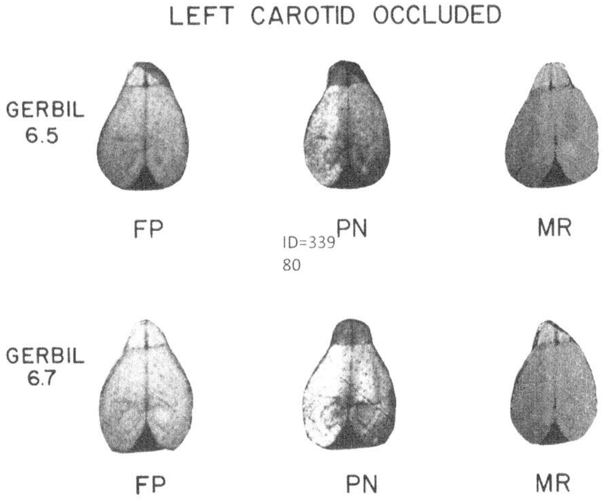


**Fig. 9.7** Redox states responses to left carotid occlusion in two depths of the same gerbil brain [8]

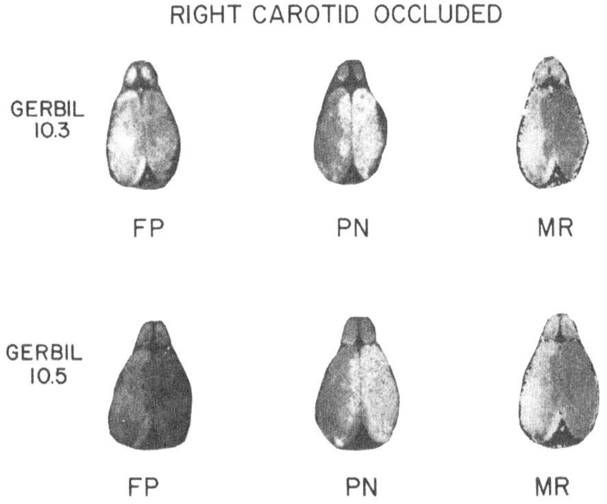


**Fig. 9.8** Effects of right carotid occlusion on the redox states measured in two brain depths [8]

United States). In gerbil 31, the left hemisphere was occluded, while in gerbil 49, the right carotid was ligated. As can be seen, the midline area obtained its blood supply from the other hemisphere. In brain no. 31, the olfactory bulbs were connected to the left hemisphere blood supply.

Quantitative analysis was done in order to test the interhemispheric connection in the three groups of gerbils used. Table 9.1 shows the occurrence of the "midline connection" in the seizure-prone and nonseizure-prone, as well as the random-bred animals. We divided each strain according to the carotid artery that was ligated. Most of the gerbils show the phenomenon of the "midline connection" (27 out of 41). Using the "Functions of Categorical Responses Funcut Procedure" [12], we found that the difference in the occurrence of the midline connection was not significant among the three groups of gerbils or when left or right carotid occlusion

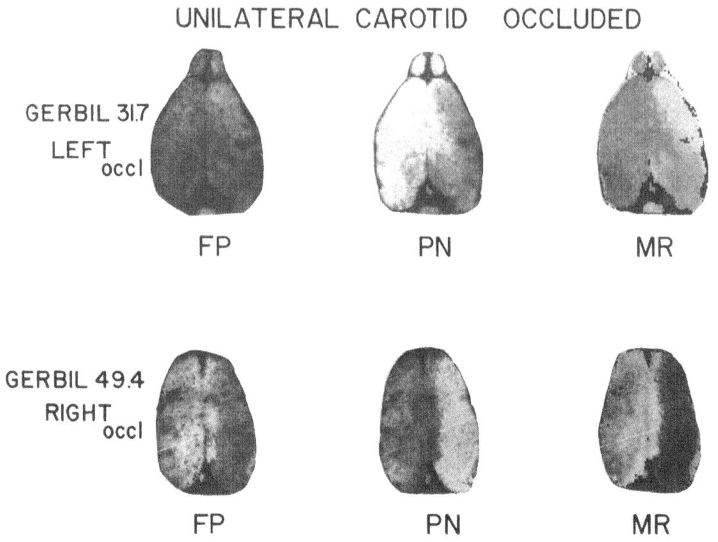


**Fig. 9.9** Effects of unilateral carotid occlusion on the redox state in the brains of two gerbils [8]

**Table 9.1** The effect of unilateral carotid artery ligation on the occurrence of interhemispheric blood transfer in the three groups of gerbils tested in this study [8]

| | Rocc | Locc | Total |
|---|---|---|---|
| SP | 6/7 | 3/7 | 9/14 |
| NSP | 6/7 | 4/6 | 10/13 |
| Random bred | 4/7 | 4/7 | 8/14 |

was compared. In a few animals, we obtained very unexpected results after unilateral carotid artery occlusion, as can be seen in Fig. 9.10. The right hemisphere was occluded, but part of the left hemisphere showed the same ischemic response, not at the midline area. This response was found in all depths of the brain that were scanned. The olfactory bulbs always had a uniform oxidative state, indicating that, like the midline region of the brain, the two bulbs are supplied by a common single source, but this source is not linked to that of the midline since the midline and the olfactory bulbs can have different oxidation levels (Fig. 9.8).

## 9.4 Discussion

This study [8] is an extension of the earlier one [4], in which we correlated the in vivo NADH responses to ischemia with those found in the freeze-trapped brain. During the past few years [13–20], more scientists have used the Mongolian gerbil

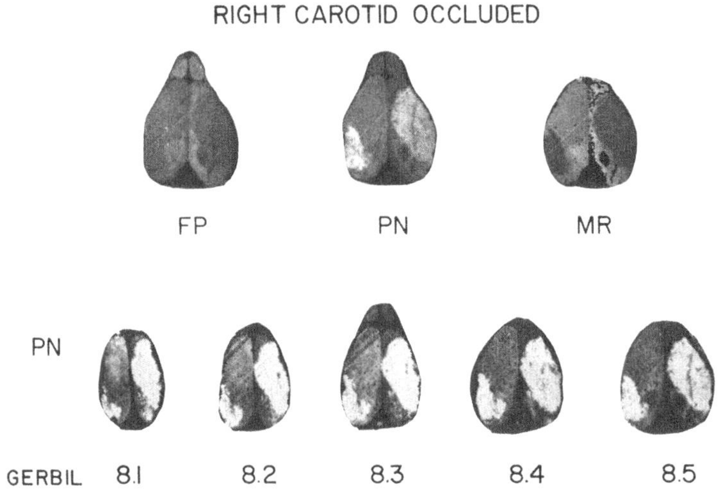


**Fig. 9.10** Effects of partial ischemia on the two-dimensional distribution of the redox state at various depths of the same gerbil brain [8]

brain as a stroke model, and in most studies published recently [21, 22], the interhemispheric blood connection was not discussed. In the paper published in 1977 by Levy and Duffy [22], this interhemispheric connection was seen but not mentioned as an exceptional finding. Similar pictures have been published recently but without adequate anatomical explanations. In view of the fact that the fluorodeoxyglucose labeling has now been seriously questioned on the basis of studies of the intracellular activity of glucose-6-phosphate phosphatase so that it is not clear in those cases studied that actual glucose uptake is being recorded, it becomes much more important to have an independent method of the rate of oxidative metabolism. We have selected the redox state of NADH as an indicator of the rate of oxidative metabolism under hypoxic conditions, a feature of NADH fluorescence documented by studies of isolated mitochondria [11]. We have used the freeze-trapped brain in ischemia studies in order to illustrate the heterogeneity of the redox state developed. As discussed previously [4], monitoring two fluorescence indicators (NADH and FP) has enabled us to calculate the ratio of fluorescence rather than the absolute values, which represent the real mitochondrial matrix space. Also, at low temperatures, enhancement is obtained of the fluorescence signals from pyridine nucleotide (PN) and flavoprotein (FP) by factors of 10 or more over the room temperature values [23]. The disadvantage of the freeze-trapped brain is that only one point in time per brain is possible. Therefore, we believe that both in vivo and freeze-trapped experiments are required in order to resolve the metabolic problem. The results obtained in this study show very clearly that unilateral carotid artery occlusion in the Mongolian gerbil results in an inhomogeneous response. The midline area between the two hemispheres presented unexpected behavior, namely, that the so-called "control" (unligated) hemisphere contained an ischemic area, and vice versa.

This phenomenon was found in the gerbils obtained from Tumblebrook Farms (Fig. 9.9), as well as from the IBR colony (Fig. 9.7), suggesting that it originates from a very basic blood vessel anatomy pattern. We have also excluded a possible superficial artifact since we have shown that this connection existed throughout the entire depth of the brain (Fig. 9.6). As we have previously demonstrated [4, 6], this phenomenon is not correlated to the seizure propensity of the animal, as suggested by others [24], indicating that further microanatomical studies must be done in order to explain it.

## References

1. Mayevsky A (1978) Shedding light on the awake brain. In: Dutton PL, Leigh J, Scarpa A (eds) Frontiers in bioenrgetics: from electrons to tissues. Academic Press, New York, pp 1467–1476
2. Mayevsky A, Zarchin N (1981) The effects of unilateral carotid occlusion on the responses to decapitation in the gerbil brain. Brain Res 206:155–160
3. Mayevsky A, Flamm ES, Pennie W, Chance B (1991) A fiber optic based multiprobes system for intraoperative monitoring of brain functions. SPIE 1431:303–313
4. Mayevsky A, Zarchin N, Kaplan H, Haveri J, Haselgrove J, Chance B (1983) Brain metabolic responses to ischemia in the Mongolian gerbil: *in vivo* and freeze trapped redox state scanning. Brain Res 276:95–107
5. Berry K, Wisniewski HM, Svarzbein L, Baez S (1975) On the relationship of brain vasculature to production of neurological deficit and morphological changes following acute unilateral common carotid artery ligation in gerbils. J Neurol Sci 25:75–92
6. Donadio MF, Kozlowski PB, Kaplan H, Wisniewski HM, Majkowski J (1982) Brain vasculature and induced ischemia in seizure-prone and non-seizure-prone gerbils. Brain Res 234:263–273
7. Kaplan H (1981) Effects of fostering on seizure activity in the Mongolian gerbil. Dev Psychobiol 14:565–570
8. Mayevsky A, Kaplan H, Haveri J, Haselgrove J, Chance B (1986) Three-dimensional metabolic mapping of the freeze-trapped brain: effects of ischemia on the Mongolian gerbil. Brain Res 367:63–72
9. Haselgrove J, Barlow CH, Chance B (1980) The 3D distribution of metabolic states in the gerbil brain during the course of spreading depression. In: Passoneau JV, Hawkins RA, Lust WD, Welsh FA (eds) Cerebral metabolism and neural function. Williams & Wilkins, Baltimore/London, pp 72–76
10. Haselgrove J, Barlow C, Eleff E, Chance B, Lebordais S (1981) In: Kovach AGB, Dora E, Kessler M, Silver IA (eds) Correlation of electrical signals and mitochondrial redox state during spreading depression. Pergamon Press, Budapest, pp 25–26
11. Chance B, Schoener B, Oshino R, Itshak F, Nakase Y (1979) Oxidation-reduction ratio studies of mitochondria in freeze- trapped samples. NADH and flavoprotein fluorescence signals. J Biol Chem 254:4764–4771
12. Grizzle JE, Starmer CF, Koch GG (1969) Analysis of categorical data by linear models. Biometrics 25:489–504
13. Levine S, Payan H (1966) Effects of ischemia and other procedures on the brain and retina of the gerbil (*Meriones unguiculatus*). Exp Neurol 16:255–262
14. Robinson DG (1980) The gerbil stroke model. The Gerbil Digest 7:1–4
15. Levine S, Sohn D (1969) Cerebral ischemia in infant and adult gerbils. Relation to incomplete circle of Willis. Arch Pathol 87:315–317

16. Kahn K (1972) The natural course of experimental cerebral infarction in the gerbil. Neurology 22:510–515
17. Kobayashi M, Lust WD, Passonneau JV (1977) Concentrations of energy metabolites and cyclic nucleotides during and after bilateral ischemia in the gerbil cerebral cortex. J Neurochem 29:53–59
18. Crockard A, Iannotti F, Hunstock AT, Smith RD, Harris RJ, Symon L (1980) Cerebral blood flow and edema following carotid occlusion in the gerbil. Stroke 11:494–498
19. Cohn R (1979) Convulsive activity in gerbils subjected to cerebral ischemia. Exp Neurol 65:391–397
20. Reivich M, Ginsberg M, Slater R, Jones S, Kovach A, Greenberg J, Goldberg H (1978) Alterations in regional cerebral hemodynamics and metabolism produced by focal cerebral ischemia. Eur Neurol 17(Suppl 1):9–16
21. Ishihara N, Gotoh F, Tachibana H, Ishikawa Y, Gomi S (1983) The effect of unilateral cerebral infarction on cerebral blood flow and glucose metabolism of the contralateral hemisphere in gerbils. J Cereb Blood Flow Metab 3:S321–S322
22. Levy DE, Duffy TE (1977) Cerebral energy metabolism during transient ischemia and recovery in the gerbil. J Neurochem 28:63–70
23. Chance B, Schoener B, Fergusson JJ (1962) *In vivo* induced oxidation by adrenocorticotrophic hormone of reduced pyridine nucleotide in the adrenal cortex of hypophysectomized rats. Nature 195:776–778
24. Schonfeld AR, Glick SD (1979) Cerebrovascular abnormalities associated with seizure susceptibility in the Mongolian gerbil. Brain Res 173:147–151

# Chapter 10
# Correlation Between Freeze-Trapped Brain and In Vivo Mitochondrial NADH Measurements

**Abstract** The interrelation between the metabolic responses to ischemia and seizure propensity was studied in two groups of seizure-prone (SP) and nonseizure-prone (NSP) gerbils. The metabolic state was evaluated in vivo using the light guide surface fluorometry as well as in the frozen brain scanned at liquid $N_2$ temperature for the flavoproteins/pyridine nucleotides ratio after monitoring the brain in vivo. The results could be summarized as follows. (1) Unilateral carotid artery occlusion led to partial ischemia in the ipsilateral hemisphere while the contralateral hemisphere remained normoxic. (2) Animal variability in the degree of the ischemia insult due to unilateral occlusion was not side-dependent or in correlation with seizure propensity. (3) A significant correlation was found between the NADH increase during ischemia and the redox state measurements done in the same brain after funnel freezing with liquid nitrogen. (4) In a large number of the gerbils (not depending on the origin or strain), a peculiar interhemispheric blood supply connection was found. A narrow band (1 mm in width) of tissue near the midline obtained its blood via the same vessels supplying the contralateral hemisphere. (5) We did not find any correlation between blood vessel anatomy in the brain and seizure propensity.

**Keywords** Gerbils · Seizures · NADH · Redox state · Ischemia · Frozen brain · Stroke · Redox state mapping

## 10.1 Introduction

The Mongolian gerbil, originally a desert rodent, has become a useful animal model for stroke studies in animals [1, 2]. The first report using the gerbil as a stroke model was published in 1966 by Levine and Payan [3]. Subsequently, in 1969, Levine and Sohn [4] postulated that the gerbil is lacking a significant posterior communicating artery and that the connection between the vertebral-basilar and the carotid circulation is minimal. Kahn in 1972 [5] and Levy and Brierley [6] reported these findings. A more detailed study was performed by Berry et al. [7], who showed and



A. Mayevsky, *The Mongolian Gerbil Brain*,
https://doi.org/10.1007/978-3-031-69549-0_10

concluded that the main factor affecting the variable susceptibility to unilateral occlusion is the architecture of the anterior cerebral arteries. The gerbil was also used as a special model for studying epilepsy triggered by a sudden change in the environment [8–11]. In 1979, Cohn [12] described convulsive activity in gerbils subjected to cerebral ischemia, and in 1979, Schonfeld and Glick [13] showed some evidence for the relationship between convulsive activity and ischemia in the gerbil. Using a perfusion technique, which allows for clear visualization of the vascular pathways, Donadio et al. [14] were unable to confirm this finding in their inbred IBR seizure-prone (SP) and nonseizure-prone (NSP) colonies. In their careful studies, they concluded that "Neither the vascular pattern of the anterior cerebral artery (ACA), nor the frequency of occurrence of stroke following unilateral ligation was related to seizure propensity." Another approach to the use of the gerbil as a stroke model was developed by us a few years ago. In 1978, we described the oxidation-reduction state of NADH in the gerbil brain under unilateral or bilateral carotid artery occlusion [15]. According to the metabolic responses measured, we concluded that the connection between the basilar and the carotid circulation was minimal, while the connection in the anterior circulation was varied among gerbils [16, 17].

The aim of this study [18] was to study in detail the correlation between the blood supply to the gerbil brain, as evaluated by metabolic responses (during ischemia), and the susceptibility to epilepsy.

## 10.2 Methods

Two sources for the Mongolian gerbils, *Meriones unguiculatus*, were used. The normal strain was obtained from Tumblebrook Farms, West Brookfield, MA, United States, and the seizure-prone (IBR/SP) and nonseizure-prone (IBR/NSP) gerbils were obtained from the New York State Institute for Basic Research in Developmental Disabilities. Gerbils in the IBR colonies were tested for seizure activity as described previously [11, 14].

### *10.2.1 In Vivo Surface Fluorometry/Reflectometry*

In order to evaluate the redox state of the brain, two-channel [19] and four-channel [16] DC fluorometers/reflectometers were used in various experiments. The light for the excitation of NADH (a 100 W water-cooled Hg arc) passed through a bundle of UV-transmitting fiber optics toward the brain. The emitted light was split in a 90:10 ratio for fluorescence and reflectance measurements as described previously [19]. In most of the experiments, the common part of the fiber optic had a 2 mm diameter. The diameter of the common part of the fiber optic in the last experiment to be shown was 1 mm out diameter (0.8 mm fiber diameter). The signals from the brain were amplified and calibrated as described previously [16, 19].

### 10.2.2 Frozen Brain Redox State Evaluation

The metabolic state of the frozen brain was studied by recording the fluorescence signals of the mitochondrial NADH and flavoproteins (FP). The excitation wavelengths were 366 and 460 nm, and the emitted light was measured at 440 and 540 nm for reduced NADH and oxidized flavoproteins (FP), respectively. The study of the fluorescent intensities of the enzymes will give a metabolic picture of the distributions of redox states as has recently been described by Chance et al. [20]. The ratio between the two signals measured gave the corrected redox state not affected by changes in mitochondrial concentration or hemoglobin level in the tissue analyzed. The distribution of the fluorescence signals in three dimensions throughout the brain was recorded with an online scanner as described in detail.

### 10.2.3 Animal Preparation

The gerbil (50–70 g) was anesthetized by intraperitoneal injection of Equithesin (0.3 mL/100 g) and mounted to a head holder for in vivo studies. Two or four 4-mm diameter holes were drilled in the appropriate location in the skull, and the dura mater remained intact. Special light guide holders were placed above the brain and cemented to the skull with dental acrylic. In gerbils used for the frozen brain studies, the parietal bone was removed bilaterally without removing the dura mater. The animal remained in the head holder until the end of the freezing procedure. The light guide was held by a special micromanipulator above the cortex for the in vivo monitoring and was removed before freezing the brain.

### 10.2.4 Funnel Freezing of the Brain

Because of the small size of the gerbil brain, a barrel of a 30-mL syringe was used as a funnel for the liquid $N_2$. The rim of the funnel was covered with Vaseline for better sealing during freezing. When cessation of respiration was observed, the gerbil was removed from the holder and immersed in a container of liquid $N_2$. The head was severed by a bandsaw and a block of the head was transferred to the low-temperature scanner for redox state analysis.

## 10.3 Results

Three sets of experiments were done in this study:

1. In vivo measurements in two and four sites in the slightly anesthetized gerbils
2. Comparison between in vivo and frozen brain measurements from two hemispheres
3. In vivo frozen brain analysis in the midline area of the gerbil brain

### *10.3.1 Two-Site Monitoring In Vivo*

In order to demonstrate the usage of fluorometry/reflectometry in our studies, the results shown in Fig. 10.1 were typical. We used a two-channel DC fluorometer, and the two bundles of the light guide were located above the parietal cortex of a normal gerbil. The electrocorticogram (ECoG) was obtained by using bipolar electrodes screwed into the skull before the cementation. The typical response to anoxia is shown in Fig. 10.1a, namely, NADH was increased to the maximal level (CF), and the electrical activity disappeared under complete depletion of oxygen. The recovery of the signals to the normoxic level was very fast simultaneously with the recovery of the normal electrical activity. The upper four traces were measured from the right hemisphere, while the other four were measured from the left one. The changes

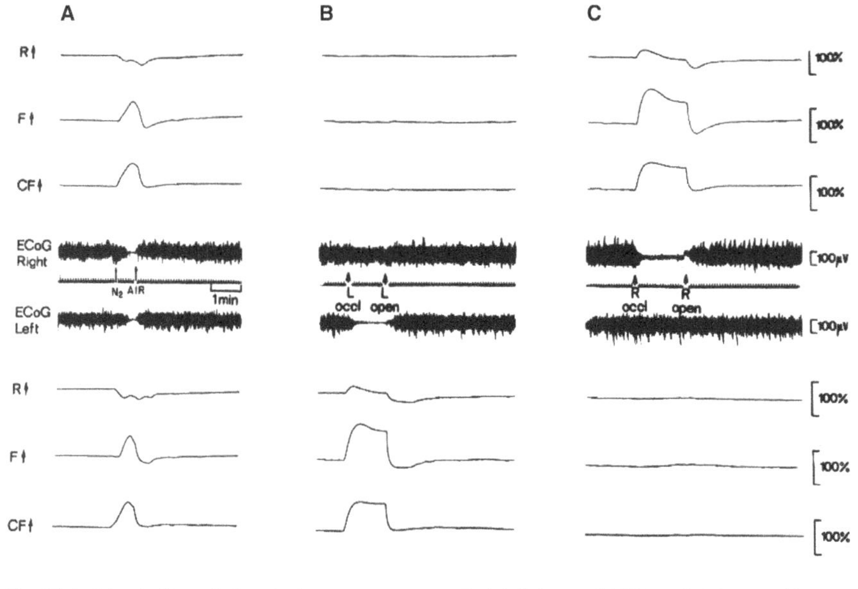


**Fig. 10.1** Metabolic and electrical responses to anoxia (**a**), left carotid (**b**), and right carotid artery occlusion (**c**) measured bilaterally from gerbil brain [18]

in the reflected light intensities were small, and as a result, the difference between the uncorrected (F) and the corrected fluorescence (CF) was minimal. The effect of unilateral carotid artery occlusion is shown in Fig. 10.1b (left artery) and Fig. 10.1c (right artery). The responses show that in this specific gerbil, unilateral occlusion led to almost complete ischemia in the ipsilateral hemisphere while the contralateral hemisphere was not affected. The corrected fluorescence trace reached its maximal level (as compared to the anoxia), and then a small recovery was recorded. The ECoG was also depressed, but not as much as found under anoxia. This record shows that the connection between the anterior cerebral arteries was minimal.

### *10.3.2 Four-Site Monitoring In Vivo*

In this set of experiments, we tested the effects of unilateral and bilateral carotid artery occlusion on the changes in NADH measured in vivo in four sites in the brains of seizure-prone and nonseizure-prone gerbils. The results were collected from 16 animals obtained from IBR in New York, and four Tumblebrook Farm normal gerbils were used as a reference group. The results are shown in Fig. 10.2 and Table 10.1. In each gerbil, we implanted four light guides as described in Sect. 10.2, and the location above the brain is shown in Fig. 10.2. Channels 1 and 2 were in the right hemisphere, while Channels 3 and 4 were in the left one. Channels 1 and 3 were placed in the frontal cortex, while Channels 2 and 4 were in the posterior part of the parietal cortex. Each gerbil was exposed to a short $N_2$ cycle as shown in Fig. 10.2a and later on to unilateral and bilateral carotid artery occlusion. The occlusion was repeated a few times in the same gerbil, alternating between the left and the right artery as the first occlusion site. The response to anoxia, which was very similar qualitatively as measured by the percent change in NADH, showed some variability between the four sites. For this reason, all changes in NADH were calculated in relative terms by comparing the effects of occlusion in each channel to 100% of the change observed under bilateral occlusion (complete ischemia) in the same channel. As can be seen from Fig. 10.2b, the effect of right artery occlusion led to partial ischemia in the ipsilateral hemisphere, which became complete after the occlusion of the left artery. The response of the left hemisphere was maximal after left carotid occlusion because the right artery was already occluded. The results obtained during the occlusions were calculated and summarized in Table 10.1. We found that unilateral or bilateral occlusion induced the same changes in the two sites measured from the same hemisphere (1 and 2 as well as 3 and 4). For this reason, the average between each pair was calculated, and for each hemisphere, one number was presented. The results shown in Table 10.1 are presented as a mean, as well as the range, in each group of numbers. The numbers represent the percent change in NADH fluorescence measured under unilateral occlusion in each hemisphere, calculated on the basis of complete ischemia achieved by bilateral carotid occlusion. When the right artery was occluded, the main change in NADH was measured (as expected) in the ipsilateral hemisphere (RH) in all three groups tested.

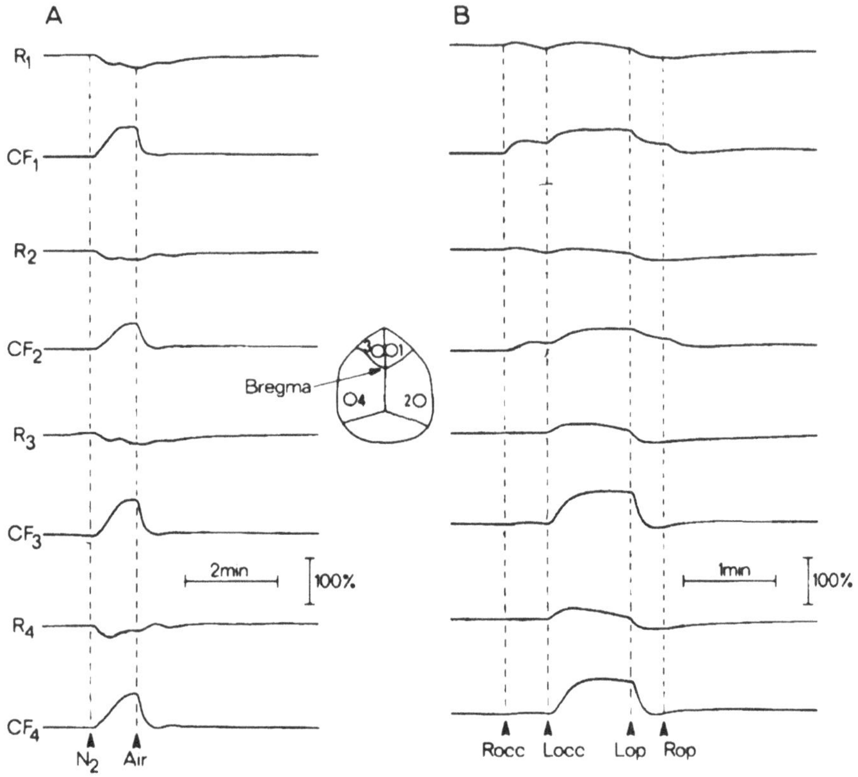


**Fig. 10.2** The effects of anoxia (**a**), partial and complete ischemia (**b**) on NADH fluorescence and reflected light measured from four sites in the gerbil brain (shown in inset) [18]

**Table 10.1** Metabolic responses (NADH % increase) to unilateral carotid artery occlusion in three groups of Mongolian gerbils

| | NSP ($n = 9$) | | SP ($n = 7$) | | Tumblebrook Farms ($n = 4$) | |
|---|---|---|---|---|---|---|
| | RH | LH | RH | LH | RH | LH |
| Right occlusion range | 22–78 | 0–14 | 30–80 | 0–28 | 22–69 | 0–10 |
| Mean ± S.E.M. | 45 ± 5 | 5 ± 1* | 56 ± 4 | 7 ± 3* | 41 ± 5 | 2 ± 1* |
| Left occlusion range | 0–16 | 17–82 | 0–43 | 20–82 | 0–11 | 21–93 |
| Mean ± S.E.M. | 6 ± 2* | 51 ± 6 | 8 ± 3* | 49 ± 5 | 1 ± 1* | 48 ± 10 |

Differences between responses of right and left hemispheres to unilateral occlusion (left or right carotid) were calculated using a Student's $t$-test. All comparisons showed significant differences. SP and NSP are the seizure prone and nonseizure prone from IBR [18]
$*p < 0.001$

The same phenomenon was found when the left carotid artery was occluded. The hemisphere contralateral to the occluded artery showed a small change in comparison to the ipsilateral (average of 1–7% of the maximum). The differences between the right and left hemisphere responses to unilateral occlusion were found to be statistically significant in all three groups tested. The response of the ipsilateral hemisphere to unilateral occlusion was around 50% of the maximal increase under bilateral occlusion in all groups tested. As can be seen from Table 10.1, the minimal value was 17% and the maximal increase in NADH due to unilateral occlusion was 93%, suggesting that the two anterior cerebral arteries exhibited a broad range of communication. According to the results shown in Table 10.1, no significant difference was found between the seizure-prone and nonseizure-prone gerbils, as well as between them and the normal strain of gerbils obtained from Tumblebrook Farms.

### *10.3.3 In Vivo Frozen Brain Comparison*

The aim of the presented sets of experiments was to correlate the changes in NADH redox state measured in vivo under unilateral carotid occlusion with those measured from the same brains after funnel freezing with liquid $N_2$. A total of 24 gerbils were used, 12 of each group (nonseizure prone and seizure prone) divided into two subgroups (left or right occlusion). The animals were held in the head holder after isolation of the common carotid arteries in the neck. Two light guides were held above the center of the parietal cortex with a micromanipulator, enabling fast removal before freezing the brain. The results obtained in vivo are presented in Table 10.2 showing the individual results, as well as the mean for each group and S.E.M. The numbers shown in Table 10.2 are the percent change increase in NADH as compared to the normoxic level. The difference between the two hemispheres in response to unilateral occlusion was not statistically significant ($p < 0.35$ for SP and $p < 0.45$ for NSP). The frozen brains were scanned in two-dimension (2-D) for the redox state of the tissue by measuring the fluorescence intensities of the NADH and FP. The ratio of the two signals was also calculated to give the MR signal [20]. Four samples of brain scans are shown in Figs. 10.3, 10.4, 10.5, and 10.6, representing each one of the four groups of gerbils shown in Table 10.2.

The results obtained in the frozen brain studies confirmed and correlated with those found in the in vivo stage but showed a few interesting findings that can be summarized in the following way.

1. The in vivo NADH increase during the ischemia was correlated to the pyridine nucleotide (PN) changes measured after freezing. High NADH levels are shown as bright areas in the 2-D scan.
2. In most brains, the FP fluorescence was also correlated to the level of ischemia: the ischemic area is the darker one (less fluorescence of the oxidized FP form).
3. The blood supply to the two olfactory bulbs comes through the same artery, as can be seen from the identical response to ischemia. The redox state of the olfactory bulbs was different from that of the rest of the cortex.
4. In many of the brains tested, the borderline of the ischemic area was not the midline as expected from the anatomy of blood vessels to the cortex as reported

**Table 10.2** The effect of unilateral carotid artery occlusion on the percentage increase in redox state in ipsilateral hemisphere

| Strain | Nonseizure prone (NSP) | | | | | | Seizure prone (SP) | | | | | |
|---|---|---|---|---|---|---|---|---|---|---|---|---|
| | Right occlusion (RH) | | | Left occlusion (LH) | | | Right occlusion (RH) | | | Left occlusion (LH) | | |
| Animal no. | In vivo | % PN frozen | % MR | In vivo | % PN frozen | % MR | In vivo | % PN frozen | % MR | In vivo | % PN frozen | % MR |
| 1 | 26 | 29 | −25 | 7 | 19 | −17 | 23 | 21 | −16 | 19 | 24 | −18 |
| 2 | 40 | 68 | −35 | 15 | 24 | −11 | 14 | −5 | −9 | 30 | 48 | −31 |
| 3 | 25 | 18 | −20 | 24 | 16 | −7 | 21 | 9 | −1 | 14 | 31 | −22 |
| 4 | 30 | 76 | −49 | 28 | 15 | −15 | 31 | 14 | −9 | 7 | 6 | −2 |
| 5 | 12 | 9 | −10 | 26 | 13 | −6 | 22 | 2 | −7 | 33 | 18 | −18 |
| 6 | 25 | – | – | 40 | – | – | 18 | 40 | −24 | 7 | 1 | −6 |
| Mean ± S.E.M. | 26 ± 4 | 40 ± 14 | −28 ± 7 | 23 ± 5 | 17 ± 2 | −11 ± 2 | 22 ± 2 | 14 ± 6 | −11 ± 3 | 18 ± 4 | 20 ± 7 | −16 ± 4 |

Four groups of gerbils were used: two seizure prone and two nonseizure prone (total: 24 gerbils). NADH measurements were done in vivo before freezing the brain for 2-D scanning of FP/PN fluorescence to give PN and MR values [18]

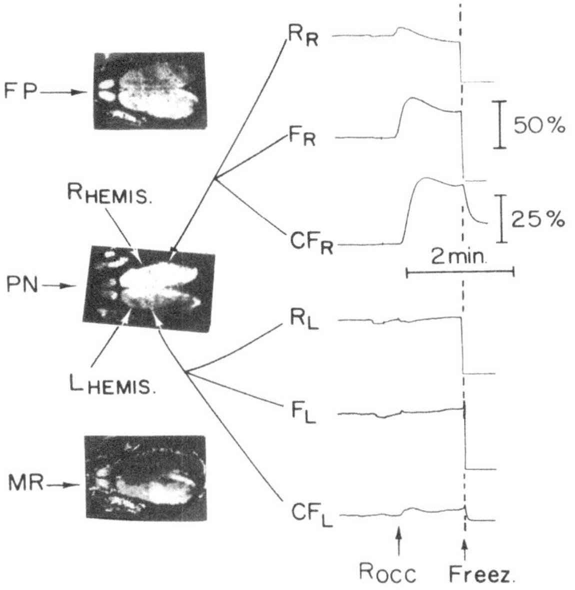


**Fig. 10.3** Correlation between in vivo and frozen brain metabolic responses under right carotid artery occlusion in the nonseizure-prone (NSP) gerbil. An ischemic area was also detected in the midline region of the left hemisphere [18]

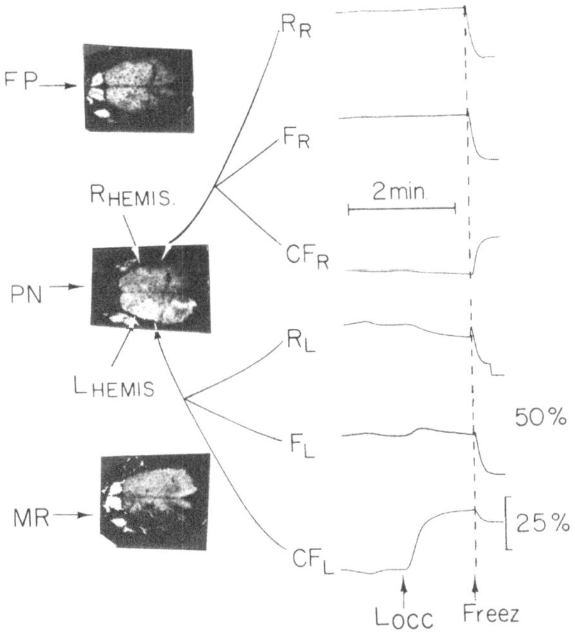


**Fig. 10.4** Responses of nonseizure-prone gerbil to left carotid artery occlusion. Results are similar to those shown in Fig. 10.3 [18]

[2, 7, 13, 14, 17]. A narrow area of the ischemic or normoxic area was found in the contralateral hemisphere, as can be seen in Figs. 10.3 and 10.4. This change was also found in the seizure-prone gerbils (not shown in the two animals pre-

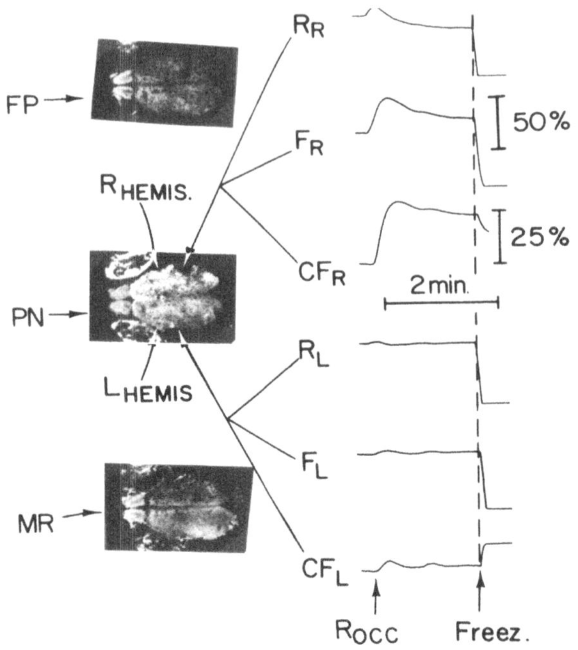


**Fig. 10.5** Effects of right carotid artery occlusion on NADH in vivo and NADH and FP in frozen brain measured from the brain of a seizure-prone (SP) gerbil [18]

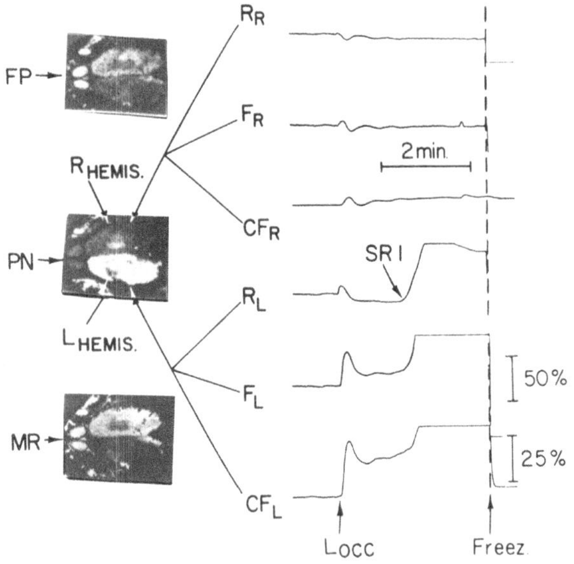


**Fig. 10.6** Unilateral ischemia induced by left carotid artery occlusion in seizure-prone (SP) gerbil. In this gerbil, the olfactory bulbs obtain the blood from the same artery and were not affected by the unilateral occlusion. During the occlusion, a large increase in reflectance was recorded [18]

sented in Figs. 10.5 and 10.6). The unexpected results described in the last paragraph led us to study this phenomenon in more detail and to try to verify it using other approaches.

### 10.3.4 Interhemispheric Transfer of Blood in the Gerbil Brain

In order to clarify the interhemispheric transfer paradox, we used a four-channel light guide (1 mm diameter for each common part) held in a special micromanipulator and placed above the two hemispheres, as shown in the schematic part of Fig. 10.4. Unilateral carotid occlusion (left or right) as well as bilateral occlusion were used, and the NADH responses in vivo were recorded. In gerbils that exhibited the interhemispheric transfer, the results were similar to those presented in Fig. 10.7 (left side), namely, that during unilateral occlusion three out of four areas were affected and NADH levels were higher in comparison to the fourth site. In a few gerbils, one area was affected and the other three were not affected by the unilateral occlusion. In order to obtain the frozen brain data, unilateral occlusion was done, the light guides were removed quickly, and the brain was frozen by the funnel technique described previously. The 2-D distribution of the PN redox state in two slices of the brain is shown below and above the scheme in Fig. 10.7. The verification of the metabolic results obtained was also done by using a different approach. In a normal gerbil, we occluded the left carotid artery, and 0.3 mL of black ink was injected through the tail vein. After 5 min, the brain was perfused through the heart

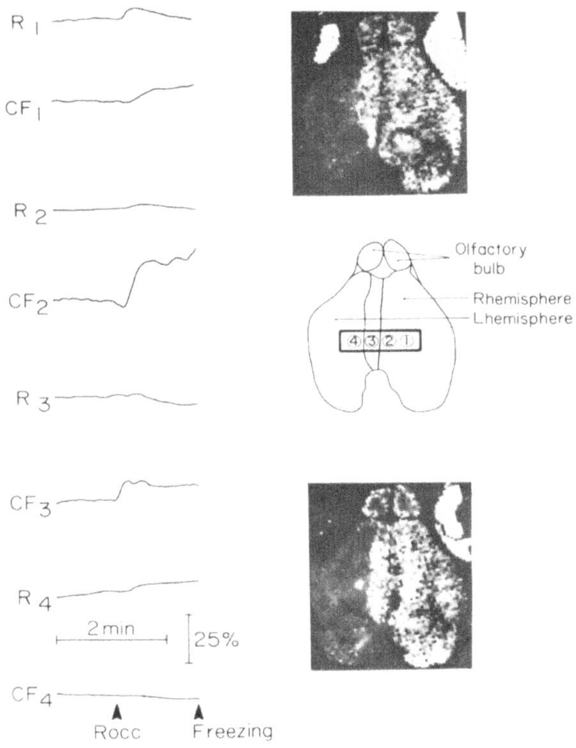


**Fig. 10.7** The effects of right carotid occlusion on NADH fluorescence measured in vivo and in the frozen brain of a normal (Tumblebrook Farm) gerbil. Site 3, located above the midline area of the left hemisphere, did respond to the contralateral carotid occlusion. The two scans shown are of the PN and represent different depths in the brain [18]

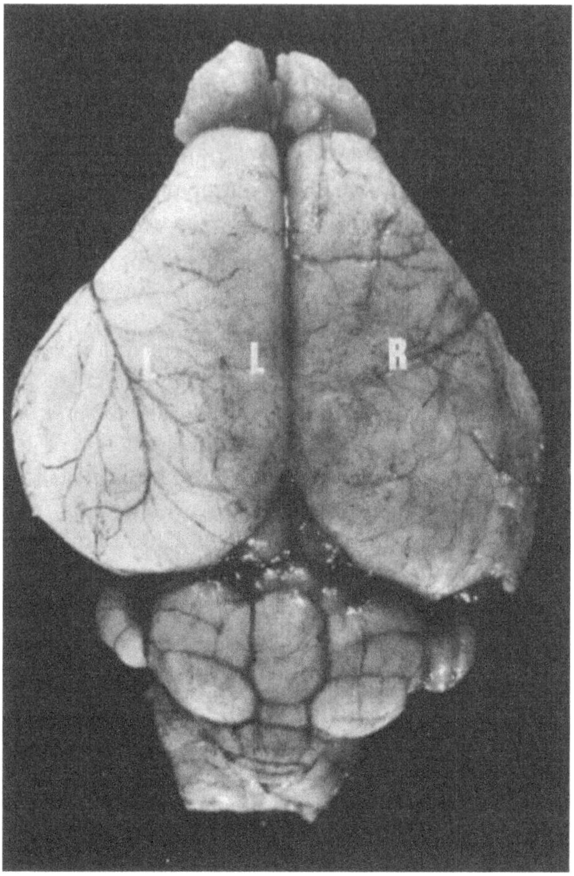


**Fig. 10.8** The distribution of black ink in the two hemispheres of a left carotid artery occluded regular gerbil (Tumblebrook Farm). The blood was removed by regular brain perfusion through the left ventricle of the heart [18]

and removed. Figure 10.8 shows a dorsal view of a gerbil brain after left artery occlusion and injection of black ink. The distribution of the ink is similar to the metabolic responses described in Fig. 10.7.

## 10.4 Discussion

The basis for this study was the recent finding obtained by Donadio et al. [14] regarding the correlation between the two unusual phenomena appearing in the Mongolian gerbil, namely, stroke sensitivity and seizure sensitivity. For approximately the past 15 years, the gerbil has been used as a stroke model. However, very little was done in order to determine the exact pattern of blood vessel anatomy as well as the concomitant physiological and biochemical consequences. Many publications (almost 100) published in the *Gerbil Digest* in 1980 [2] concerned the usage of the gerbil as a stroke model. One aim of this study was to reevaluate the correlation between the potential development of seizures (by behavioral stimulation) and the blood supply to the brain as evaluated by the mitochondrial redox state. In order to answer this question, we used two experimental approaches based on the intracellular signals representing the best evaluation technique for blood delivery to the tissue. Using the two colonies of gerbils obtained from IBR in New York, which were also subjects in this study, we were able to develop seizures by manual stimulation of seizure-prone gerbils while monitoring the metabolic, ionic, and electrical activities. The same stimulation did not cause seizures in the nonseizure-prone gerbils, proving that our operation and the monitoring systems are not affecting the basic phenomenon under investigation. In the first set of experiments, we compared the intramitochondrial responses (NADH fluorescence) to unilateral or/and bilateral carotid artery occlusion (Fig. 10.2 and Table 10.1). The results obtained proved again that bilateral carotid occlusion led to complete ischemia, while unilateral occlusion varied in its effect depending upon the communication in the anterior part of the circle of Willis [2, 7, 13, 14, 17]. Under unilateral occlusion, the change in NADH increase ranged between 17% and 93% of the maximal change obtained under bilateral occlusion, meaning that the communication between the two hemispheres varied. Also, we did not find any difference between the two measurement sites in each hemisphere (Fig. 10.2), namely, the anterior and posterior areas, or between the two strains obtained from the IBR (SP and NSP). In the second part of the study, we developed a new approach to study the correlation between the blood supply to the brain and the propensity of the gerbil to undergo seizures. In the same brain, we monitored bilaterally the response to unilateral carotid occlusion in vivo, and thereafter the brain was frozen for a more detailed study of the 2-D distribution of the redox states. As was discussed by Chance et al. [20], the frozen brain results presented the true redox state of the tissue since the FP signal is coming only from the mitochondria and also the NADH signal is enhanced by the low temperature. Also, the heterogeneity of the tissue in response to ischemia is evaluated by the 2-D scans at various depths. As demonstrated in Figs. 10.3, 10.4, 10.5, and 10.6, the in vivo responses to unilateral carotid occlusion were correlated qualitatively to those obtained after freezing and scanning the brain at low temperatures. The quantitative analysis of the results obtained showed a significant correlation between the changes in NADH measured in vivo and those measured after freezing (PN and MR). These results support the idea that freezing a trapped brain after ischemia can provide good information regarding its redox state. The unexpected result originally only in the frozen brain was the paradoxical crossing over of blood between the two hemispheres near the sagittal sinus area. Later on, we correlated this response during in vivo measurements to those obtained in the frozen brain, and indeed, this phenomenon proved to be a real one. According to the anatomy of the blood vessels in the gerbil brain [7, 13, 14], we are unable to explain the results. Also, it is very difficult to explain our findings by the diaschisis phenomenon described for cerebral blood flow in animals and the human brain [21]. In our animal model—the gerbil—unilateral carotid occlusion resulted (in most gerbils) in a contralateral midline zone of ischemic or non-ischemic tissue depending on the individual animal. Also in the diaschisis phenomenon, the reduction of CBF under unilateral ischemia in the contralateral hemisphere was not spatially specific, while in our case, it was found only in the 1 mm width parallel to the midline zone. When 2-deoxyglucose uptake was measured under unilateral ischemia [22], a narrow zone near the midline behaved as found in the non-ischemic hemisphere (Fig. 4, of Ref. [22]), although they did not mention it. We found support for the metabolic results by using dye injection (Fig. 10.8), and the same overlapping was found. In order to understand the basic mechanism underlying this phenomenon, more studies must be done. Regarding the correlation between the blood supply to the brain and seizure propensity, our results were unable to explain the differences between the findings published by Schonfeld and Glick [13] and those described recently by Donadio et al. [14] about the configuration of the anterior cerebral circulation in the gerbil. Nevertheless, we can confirm the results published by Donadio et al. [14], as was shown by the in vivo results as well as the frozen brain redox state (Tables 10.1 and 10.2). The percent increase in NADH level due to unilateral carotid artery occlusion as compared to the maximal increase was not different in the seizure-prone and nonseizure-prone animals. In addition, no significant difference was observed between occlusion of the left or right artery. These results indicate that the development of seizures is not affected by the oxygen supply to the two hemispheres under normoxic conditions. We cannot provide information about the effect of ischemia on the development of seizures in the gerbil as was discussed by Donadio et al. [14].

## References


1. Kobayashi M, Lust WD, Passonneau JV (1977) Concentrations of energy metabolites and cyclic nucleotides during and after bilateral ischemia in the gerbil cerebral cortex. J Neurochem 29:53–59
2. Robinson DG (1980) The gerbil stroke model. The Gerbil Digest 7:1–4
3. Levine S, Payan H (1966) Effects of ischemia and other procedures on the brain and retina of the gerbil (*Meriones unguiculatus*). Exp Neurol 16:255–262
4. Levine S, Sohn D (1969) Cerebral ischemia in infant and adult gerbils. Relation to incomplete circle of Willis. Arch Pathol 87:315–317
5. Kahn K (1972) The natural course of experimental cerebral infarction in the gerbil. Neurology 22:510–515
6. Levy DE, Brierley JB (1974) Communications between vertebro-basilar and carotid arterial circulations in the gerbil. Exp Neurol 45:503–508
7. Berry K, Wisniewski HM, Svarzbein L, Baez S (1975) On the relationship of brain vasculature to production of neurological deficit and morphological changes following acute unilateral common carotid artery ligation in gerbils. J Neurol Sci 25:75–92
8. Loskota WJ, Lomax P (1975) The Mongolian gerbil (*Meriones unguiculatus*) as a model for the study of the epilepsies: EEG records of seizures. Electroencephalogr Clin Neurophysiol 38:597–604
9. Kaplan H (1975) What triggers seizures in the gerbil, Meriones unguiculatus? Life Sci 17:693–698
10. Thiessen DD, Lindsey G, Friend HC (1968) Spontaneous seizures in the Mongolian gerbil. Psychon Sci 11:227
11. Kaplan H (1981) Effects of fostering on seizure activity in the Mongolian gerbil. Dev Psychobiol 14:565–570

12. Cohn R (1979) Convulsive activity in gerbils subjected to cerebral ischemia. Exp Neurol 65:391–397
13. Schonfeld AR, Glick SD (1979) Cerebrovascular abnormalities associated with seizure susceptibility in the Mongolian gerbil. Brain Res 173:147–151
14. Donadio MF, Kozlowski PB, Kaplan H, Wisniewski HM, Majkowski J (1982) Brain vasculature and induced ischemia in seizure-prone and non-seizure-prone gerbils. Brain Res 234:263–273
15. Mayevsky A (1978) Pyridine nucleotide oxidation reduction state of the cerebral cortex in the awake gerbil. J Neurosci Res 3:369–374
16. Mayevsky A, Chance B (1983) Multisite measurements of NADH redox state from cerebral cortex of the awake animal. Adv Exp Med Biol 159:143–155
17. Mayevsky A, Bar-Sagie D, Levy N, Amosi M (1978) The effects of ischemia on the metabolic response of the cerebral cortex of the gerbil. (Abstract). Fed Proc 37:97
18. Mayevsky A, Zarchin N, Kaplan H, Haveri J, Haselgrove J, Chance B (1983) Brain metabolic responses to ischemia in the Mongolian gerbil: *in vivo* and freeze trapped redox state scanning. Brain Res 276:95–107
19. Mayevsky A, Bar-Sagie D (1978) The interrelation between CBF, energy metabolism and ECoG in a new awake brain model. Adv Exp Med Biol 92:761–768
20. Chance B, Schoener B, Oshino R, Itshak F, Nakase Y (1979) Oxidation-reduction ratio studies of mitochondria in freeze-trapped samples. NADH and flavoprotein fluorescence signals. J Biol Chem 254:4764–4771
21. Reivich M, Ginsberg M, Slater R, Jones S, Kovach A, Greenberg J, Goldberg H (1978) Alterations in regional cerebral hemodynamics and metabolism produced by focal cerebral ischemia. Eur Neurol 17(Suppl 1):9–16
22. Levy DE, Duffy TE (1977) Cerebral energy metabolism during transient ischemia and recovery in the gerbil. J Neurochem 28:63–70

# Chapter 11
# Discussion and Conclusions

**Abstract** This book is devoted to a unique experimental animal, the Mongolian gerbil that was adopted in the early 1960s as an experimental animal in studying brain ischemia and epileptic seizures. The uniqueness of this book is that it emphasizes the relationship between brain mitochondrial function, vasculature, and brain pathology. The anatomy of the blood vessels supplying the blood to the brain was compared to the responses of the brain mitochondria to ischemia induced by the ligation of one or two carotid arteries. Qualitative and quantitative analyses of the data accumulated were performed and discussed.

The advantages of the gerbil as an animal model were discussed in various parts of this book. Also, the sensitivity of the gerbil to experimental manipulation was discussed.

**Keywords** Redox state mapping · Frozen brain · Brain circulation · Brain ischemia · Seizures

## 11.1 Introduction

Monitoring of brain mitochondrial NADH redox state in vivo was initiated in Philadelphia at the University of Pennsylvania in the late 1950s and early 1960s. Prof. Britton Chance and his team were able to develop monitoring systems that were able to use optical technology that was able to continuously monitor the metabolic state of brain mitochondrial NADH together with brain EEG or ECoG [1]. The availability of optical fibers that could transmit the UV light in the early 1970s enabled me to monitor the brains of rats in the unanesthetized animal [2, 3]. In 1978, I applied the developed technology to monitor rats exposed to ischemia by ligation of 1 or 2 carotid arteries in the rat. Due to the cerebral blood flow via the circle of Willis in the rat, the level of ischemia developed was very low.

 
A. Mayevsky, *The Mongolian Gerbil Brain*,
https://doi.org/10.1007/978-3-031-69549-0_11

The availability of the Mongolian Gerbil during the late 1960s and early 1970s enabled me to use the monitoring of mitochondrial NADH in the brain of this unique animal [4]. In 1980, we used for the first time a multiparametric monitoring system in the brain of the gerbil exposed to partial and complete ischemia or to cortical spreading depression (details are in paper [5]).

## 11.2 The Uniqueness of the Gerbil Brain in Stroke Studies

In order to monitor the effects of brain ischemia, it is important to use an animal model that will be easy and simple to prepare as well as create a wide range of ischemia after occlusion of the related blood vessels. Figure 11.1 presents the arterial blood supply to the brain of a rat (a) [6] and a gerbil (b) [7]. The "circle of Willis" in the rat is complete, while in the gerbil, it is variable between different animals. The meaning of this difference in the architecture of blood supply to the brain between the rat and gerbil is that the gerbil was adopted as a model in ischemia studies in the middle of the 1960s. The next three figures demonstrate the range of ischemia obtained in models of rats.

Figure 11.2 shows the response of NADH fluorescence and ECoG to carotid artery ligation. After ligation of the right artery, an increase of 25% in NADH fluorescence was recorded (second and third traces from the top). The effect on 366 nm

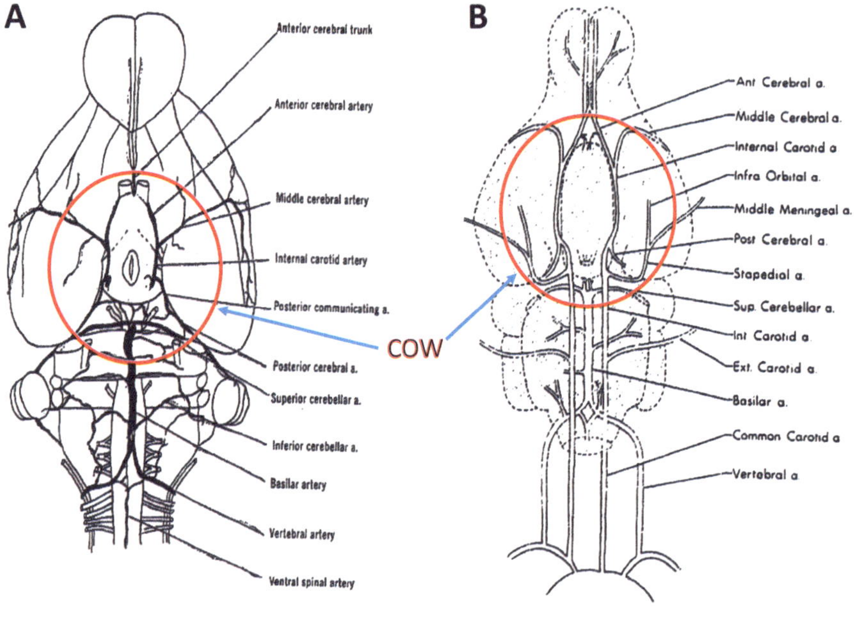


**Fig. 11.1** Ventral view of the rat's brain (**a**) [6] and gerbil (**b**) [7] and its arterial system, including the Circle of Willis

reflectance was very small. The ECoG of the right hemisphere showed a short transient depression that recovered very quickly to normal (30 s). The ligation of the left artery (while the right artery remained closed) resulted in an increase in NADH fluorescence, which remained higher compared to that in the pre-ligation period. A gradual increase in NADH fluorescence (about 10%) was recorded during the time interval of ligation of the right and left arteries. This increase in NADH, however, was not large enough to have an immediate effect on the ECoG, perhaps due to the compensatory blood flow received by the brain from the vertebral arteries through the circle of Willis [6].

Another example is presented in the next figure. The effects of ischemia on the response to CSD are dependent upon the time interval between the ligation of the carotid arteries and the induction of the CSD wave. The effects of acute bilateral ligation in the rat on the response to CSD are presented in Fig. 11.3 [9]. Part (a) shows the response of the NADH to spreading depression and the effect of the ligation on the steady state level of NADH. Parts (b)–(d) show the responses of NADH to spreading depression elicited 15 min, 90 min, and 18 h, respectively, after the ligation.

In this particular rat, the changes of the reflectance (upper trace in each part) were minimal during the spreading depression cycles. In Part (d), one can see very clearly the very long biphasic cycle of the reflectance. The effect of the ligation on the oxidation-reduction state of NADH was transient, namely, a small increase was recorded during the ligation, but within 2 min, it recovered to the baseline. The main effect of the ligation is on the response of the brain to spreading depression, which increases oxygen utilization. During the first 60 min after ligation, the oxidation cycle can be detected as a response to SD, but the amplitude of the oxidation (decrease) was diminished and the duration of the cycle became much longer (almost 4 min in (b) and only about 2 min in (a)). The qualitative change in the

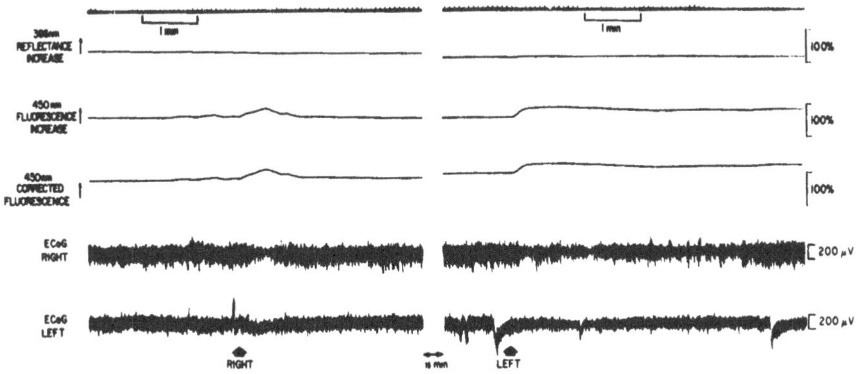


**Fig. 11.2** Effects of bilateral carotid artery ligation on NADH levels and ECoG. The arrows (right and left) at the bottom show the time of ligation [8]

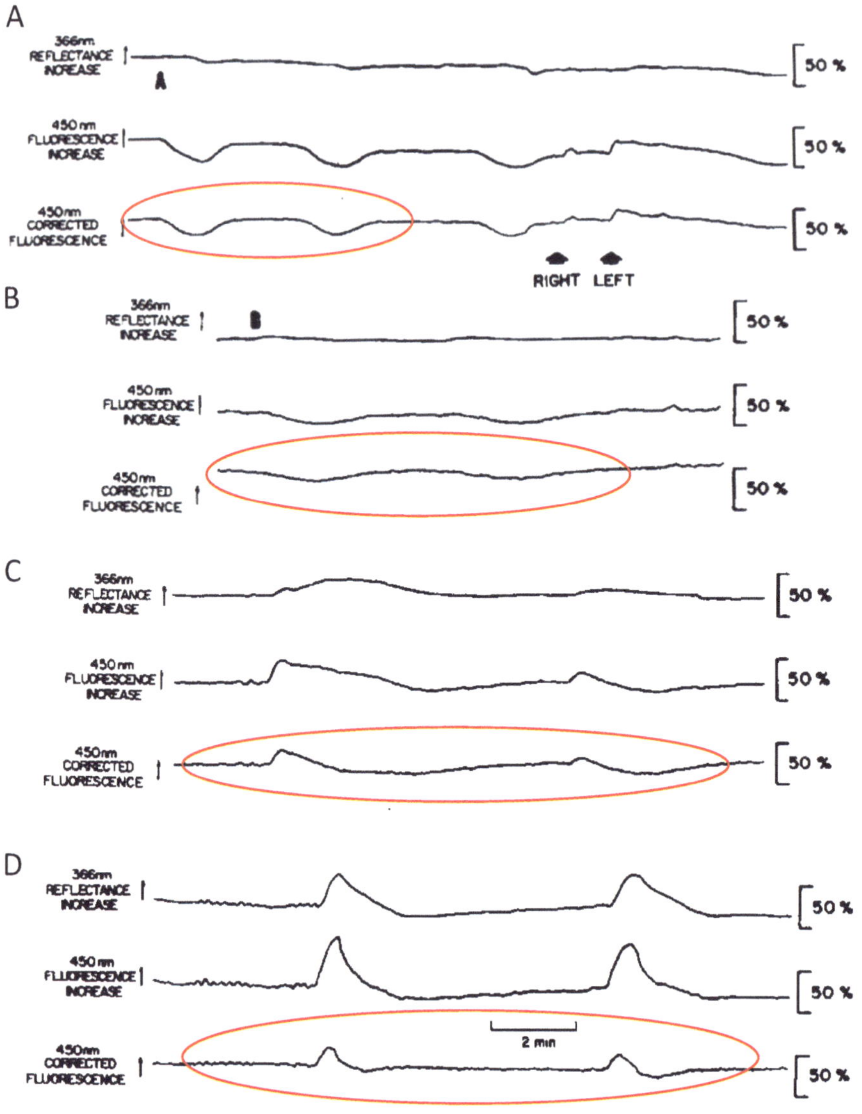


**Fig. 11.3** (**a**) The effect of acute bilateral carotids occlusion (Right, Left) on the metabolic response to cortical spreading depression (SD). Parts (**b**)–(**d**) were measured 15 min, 90 min, and 18 h after occlusion, respectively [9]

response of the brain to SD started several hours after the ligation. In Part (c), one can see the response to SD 90 min after the ligation. The uncorrected trace, as well as the corrected one, shows a biphasic response of NADH to CSD. At the beginning of the cycle, an increase in NADH was recorded, while afterwards the NADH decreased below the baseline level. In Part (d) (18 h after ligation), the response of the brain was completely opposite to what was measured before the ligation (Part a). It seems that instead of an "oxidation cycle" as a response to SD, we found a "reduction cycle." In this animal, the correction of the NADH for hemodynamic changes is not necessary, and the uncorrected trace also shows the same results.

As presented in Figs. 11.2 and 11.3, the level of ischemia created by the occlusion of the two carotid arteries is in the low range. The only way to create severe brain ischemia is to use the three to four-vessel occlusion model [10]. Using this model, we were able to create various levels of ischemia while monitoring the brain NADH redox state together with ECoG [11], as seen in Fig. 11.4. As seen, a typical record is obtained after occlusion of the right carotid artery ($R_{occl}$) followed by the left carotid occlusion ($L_{occl}$). As can be seen, the right hemisphere became more reduced (increased $CF_R$) and partially re-oxidized later via the anterior circulation and the left carotid artery (still open). The signals of the left hemisphere remained unchanged under right carotid occlusion. Bilateral occlusion induced a large increase in the CF signals of the two hemispheres, which recovered after reopening of the blood vessels (L+ R open). The electrical activity (ECoG) disappeared 30 s after occlusion of the left carotid artery. The reflectance traces ($R_R$ and $R_L$) show an initial small transient increase followed by a decrease due to hemodynamic changes. Four minutes after the occlusion, an increase in the reflectance was recorded and has been described by us as a secondary reflectance increase [12]. This change in reflectance is not always corrected by the subtraction technique, and a transient apparent increase of NADH can be seen in the corrected fluorescence trace (CF). During the reperfusion stage, a large decrease in the reflectance was recorded (increased blood volume), which gradually returned to the pre-ischemic baseline. This large change in reflectance is due to massive vasoconstriction, which occurred under the ion shifts.

The great disadvantage of the four-vessel occlusion rat model is the need for the permanent coagulation of the two vertebral arteries, which is a time-consuming procedure, as well as the irreversibility of the coagulated vertebral arteries.

In comparison of the responses to ischemia in the rat brain is presented in Figs. 11.2 and 11.3. Figure 11.5 presents the responses of two different gerbils exposed to carotid artery ligations followed by 100% nitrogen exposure. In the gerbil presented in Part (a), the effect of unilateral and bilateral carotid occlusion led to relatively small changes in NADH levels (CF) in the two hemispheres. When nitrogen was applied while the carotids were occluded, the increase in NADH levels in the two hemispheres was very large. Those results are very similar to the responses of a regular rat. The responses obtained in the gerbil presented in Fig. 11.5b are completely different. As seen, the occlusion of the right carotid artery led to a small increase in NADH in both hemispheres, but after occlusion of the left carotid artery, while the right one remains occluded, the level of NADH reached its maximal level. Application of nitrogen at this stage did not affect the NADH levels in both hemispheres.

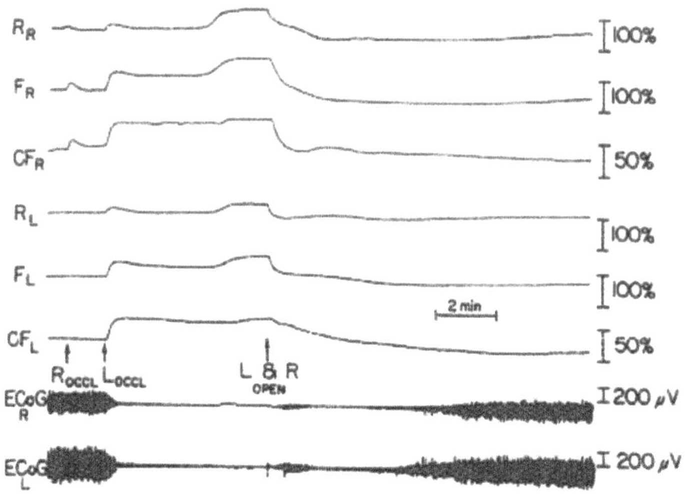


**Fig. 11.4** Metabolic and electrical responses to severe ischemia induced by four-vessel occlusion in the rat. $R_R$, $F_R$, and $CF_R$, reflectance, fluorescence, and corrected fluorescence measured from the right hemisphere; $R_L$, $F_L$, and $CF_L$, the same three signals measured from the left hemisphere. $ECoG_R$ and $ECoG_L$, electrocorticogram monitored from the right and left hemisphere, respectively [11]

## 11.3 Responses of the Gerbil Brain to CSD

Activation of energy metabolism in the brain may occur under various physiological or pathological conditions. If (under those conditions) oxygen is not a limiting factor, then the NADH redox state will shift to a more oxidized level, defined by Chance and Williams (1955) [14] to be a State 4 to State 3 transition. The main factor that increases energy metabolism in vivo is the activation of $Na^+$-$K^+$-ATPase system. This occurs when extracellular $K^+$ is elevated during the activation of the brain [15, 16]. The development of fiber optic fluorometry enabled us to use an awake animal exposed to SD or epileptic activity. In the awake rat exposed to SD, an oxidation of NADH was recorded simultaneously with a biphasic change in the reflected light [3]. The repetitive nature of the SD phenomenon and the metabolic responses to it were recorded when higher concentrations of KCl (0.3–0.4 M) were used to stimulate the cortex [17]. The correlation between the behavior of the extracellular $K^+$ and the NADH redox state was first demonstrated using light guide fluorometry [18] and later using the $K^+$ surface electrode [5]. In the bilaterally carotid-ligated rat, the response to spreading depression (State 4 to State 3 transition) showed peculiar kinetics up to 14 days after ligation. We found that under partial ischemia, NADH showed a "reduction cycle." We speculated that due to the extra energy needed for pumping the $K^+$ into the cells, an imbalance in oxygen suppy-demand was established and NADH became more reduced.

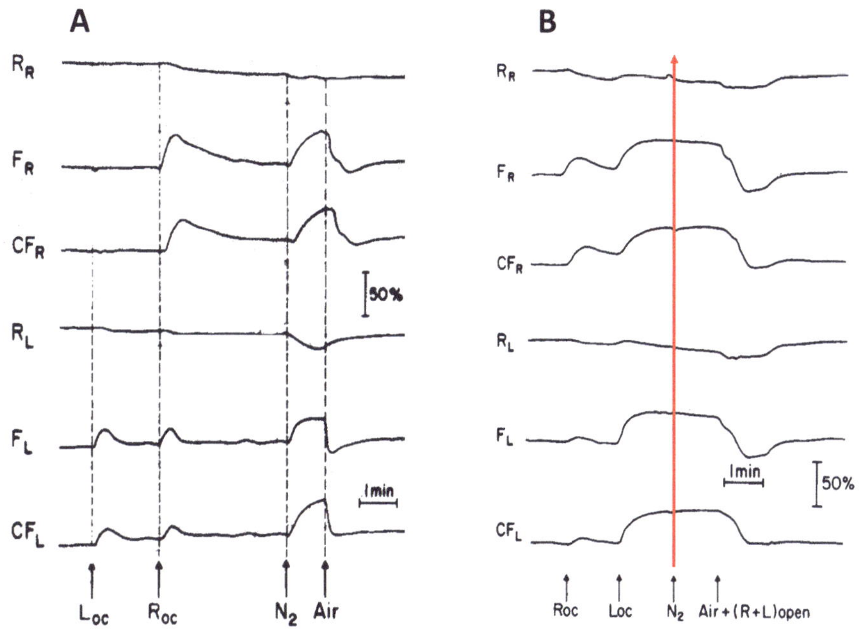


**Fig. 11.5** (**a**) Metabolic and hemodynamic responses to unilateral and bilateral carotid artery occlusion ($L_{oc}$, $R_{oc}$) and anoxia ($N_2$) measured from a gerbil [13]. (**b**) Effects of ischemia and anoxia on the metabolic and hemodynamic states of another type of gerbil brain [13]. $R_R$ and $R_L$, reflectance at 366 nm of the right and left hemispheres, respectively; $F_R$, $F_L$, $CF_R$, and $CF_L$, 450 nm fluorescence (F) and corrected fluorescence (CF) measured from the right and left hemispheres

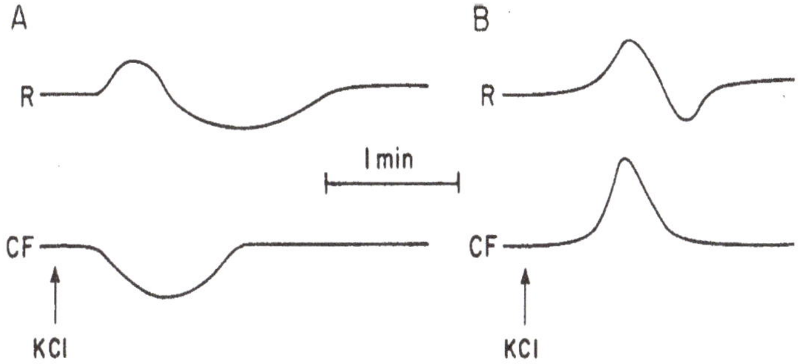


**Fig. 11.6** Metabolic responses to cortical spreading depression (induced by KCl) in the normoxic (**a**) and partially ischemic, hypoxic, or anesthetized gerbil brain (**b**) [19]

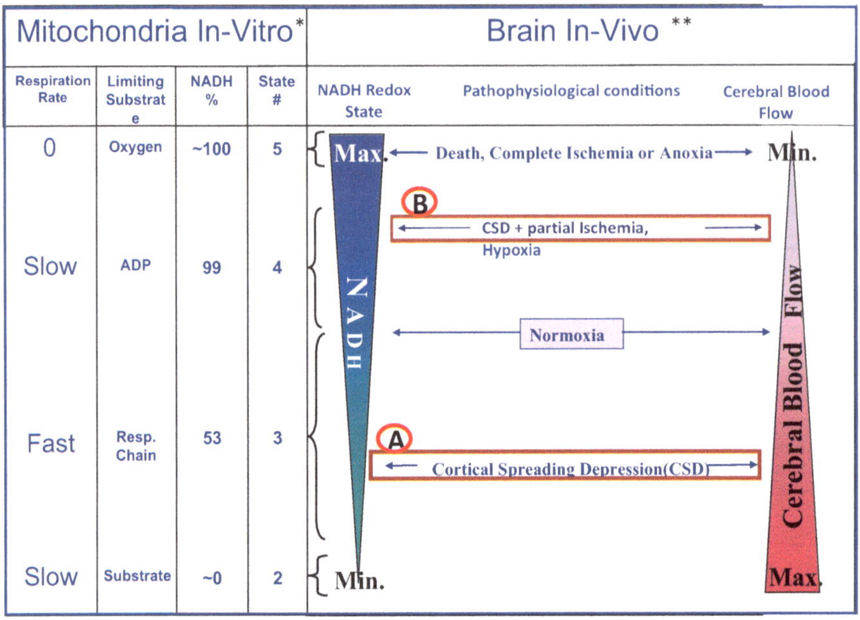


**Fig. 11.7** Comparison between the classical definition of Chance and Williams regarding the metabolic state of isolated mitochondria and the metabolic state of the in vivo brain [21] (*according to Chance and Williams (1955) [14]; **according to Mayevsky (1984) [22]); CSD, cortical spreading depression. The lower red rectangle—**A** shows the response to CSD in normoxia; the upper red rectangle—**B** shows the response to CSD under partial ischemia or mild hypoxia

Figure 11.6 shows the typical response to SD in the normoxic brain (a) as compared to the one recorded in the partial ischemic brain (b). This change in the direction of the NADH response to SD (reduction or oxidation) can be used as an indicator of the intactness of the microcirculatory system [20].

### *11.3.1 NADH Responses to CSD In Vivo Versus Isolated Mitochondria In Vitro*

Figure 11.7 presents the comparison between the mitochondrial metabolic states in vitro and the states in the brain in vivo. We created these in vivo responses of the brain during real-time measurements of NADH and CBF in most of the perturbation induced. We concluded from many of the experiments performed in our laboratory that the resting states of the brain in vivo are around 3.6 and not State 4 as determined by Chance and his team for the mitochondria in vitro [14].

### 11.3.2 Cortical Spreading Depression (CSD) Under Pathophysiological Conditions

During our long-term period in studying CSD, it was clear that the development of one or more waves of this unique event depends on the appropriate stimulation given to the cerebral cortex. Once the stimulus intensity is appropriate, the wave is developed and propagates to the other areas of the stimulated hemisphere. Nevertheless, the recovery of the brain tissue is dependent upon the availability of energy (ATP) to restore the ion homeostasis to normal.

Mitochondrial function and its metabolic states under in vivo conditions are completely different from the definition made for isolated mitochondria by Chance and Williams in 1955 [14], as seen in the left side of Fig. 11.7. In order to compare the in vitro and the in vivo states, only four states (States 2–5) are presented. The "resting" state of the mitochondria in vitro was defined as "State 4," where NADH was 99% in the reduced form and ADP was the rate-limiting substance. If ADP is added to a suspension, oxygen consumption will increase, and the rate limit will be determined by the activity of the respiratory chain. During this State 3, or the "active state," the NADH redox state will decrease or become more oxidized (about 50%). When the "resting" mitochondria are deprived of oxygen, the activity of the respiratory chain will stop and NADH will reach its maximum redox state (State 5).

A definitive description of the mitochondrial metabolic state has never been given for in vivo conditions. Therefore, we described the in vivo mitochondrial conditions as recorded by NADH fluorescence in a representative tissue or organ—for example, the brain. While the range between minimal NADH (~0) and its maximal level was determined in vitro, it is almost impossible to determine it in the intact brain or other organs in vivo. For example, State 2, with a substrate-free medium, could not be achieved in vivo since the tissue would die due to the lack of substrate. On the other hand, the maximal level of NADH (State 5) could be monitored in vivo under complete deprivation of oxygen by anoxia or complete ischemia. We used the changes in NADH levels monitored in vivo to create a new scale ranging from a maximal definite point to the minimal level as shown in the right side of Fig. 11.7. Details of this approach have been published [22].

As seen in Fig. 11.7, the maximal NADH level is achieved under complete $O_2$ deprivation that can be induced both under in vitro and in vivo conditions. This signifies that this definitive point can be used to determine State 5 in vivo as well. The problem is to determine the metabolic state of a tissue in an in vivo situation. If we adopt the in vitro value of a resting state (State 4), this would signify that the increase in NADH during State 5, induced by anoxia (0% $O_2$), would be only 1%. According to all in vivo studies, this is not the case, and during anoxia, the increase in NADH is larger than the decrease. Therefore, we concluded that, under in vivo conditions, the "resting" metabolic state of the brain is found between States 4 and 3 rather than in State 4 as defined in vitro [22]. In order to determine the maximal levels of NADH in vivo, it is almost impossible to use cyanide; therefore, we used 100% nitrogen until the cessation of respiration and the recovery to air were monitored. It is almost impossible to achieve the minimal level of NADH (maximum oxidation) in vivo as compared to the level determined in State 2 when substrate availability is minimal.

Between 1973 and 1980, my laboratory was equipped and able to monitor, under CSD, only mitochondrial NADH regarding the direct main parameter that represents the cellular energy metabolism process. In 1980, we added the measurement of tissue $pO_2$ using oxygen electrode to the NADH measurement, but it was clear that this is not the ultimate best parameter to measure in the brain due to its dependence on cerebral blood flow. Therefore, after seeking a better parameter to add to the NADH, we adopted the measurement of CBF using the newly developed technology based on the laser Doppler flowmetry in 1990–1991 [23–25].

The main issue that came up was, could we interpret the in vivo data according to the concept developed for the in vitro conditions? A good example to demonstrate the dilemma is presented in Fig. 11.7. The lower red rectangle (a) shows that when CSD developed in the normoxic brain, the NADH was oxidized (decreased fluorescence), and a large increase in CBF was recorded. But when the CSD developed under partial ischemia, for example, as seen in the upper red rectangle (b), NADH was elevated, namely, more reduced due to the decrease in CBF and a limitation in oxygen due to the partial ischemic condition. According to Chance and Williams [14, 26], ATP synthesis is not possible if the tissue is in State 4 due to the unavailable ADP needed. This is one of the main reasons that we concluded that the "resting" brain in vivo is in States 3.5–3.6 and not in State 4. The fact that the brain tissue is recovered from a wave of CSD even under partial ischemia (Fig. 11.6).

Schematically, the differences in the responses to CSD in the normoxic and ischemic brains are demonstrated in Fig. 11.8.

The metabolic responses to the Leão spreading depression [27] phenomenon (CSD) have been studied by various investigators in different animal models. Few investigators showed that CSD caused an increase in oxygen utilization as evaluated by measuring $pO_2$ change in the cortex [28]. In order to supply the need for extra $O_2$, hemodynamic changes take place, and indeed, some researchers showed blood flow changes during cortical CSD [29–31]. Using fluorometric/reflectometric techniques for monitoring mitochondrial signals, the same conclusion was drawn, namely, that CSD caused an increase in energy utilization. In 1973, Rosenthal and Somjen [32] showed an oxidation of NADH due to CSD in the cerveau isole cat preparation. Using the awake rat model, we also showed oxidation of NADH due to CSD [3, 18]. We were able to correlate the NADH oxidation cycle with the pumping activity of the extracellular $K^+$ in the recovery phase of CSD in rats [33]. The effects of pentobarbital and ouabain on the metabolic responses to CSD in the cat brain were described by LaManna and Rosenthal [34]. We showed in the awake rat the effects of hypoxia and partial ischemia [9] on the responses to CSD.

As shown in the past, during CSD, there is a large, passive efflux of $K^+$ into the extracellular space [33], regardless of energy availability. The pumping back of the $K^+$ by the ATPase into the cells is an energy-dependent process and increases the energy demand of the tissue [5, 32, 33, 35]. As a result, the mitochondria are activated (toward State 3) in order to produce the extra ATP needed, and oxygen utilization is increased [16, 36]. In the normoxic brain, the increased blood flow will

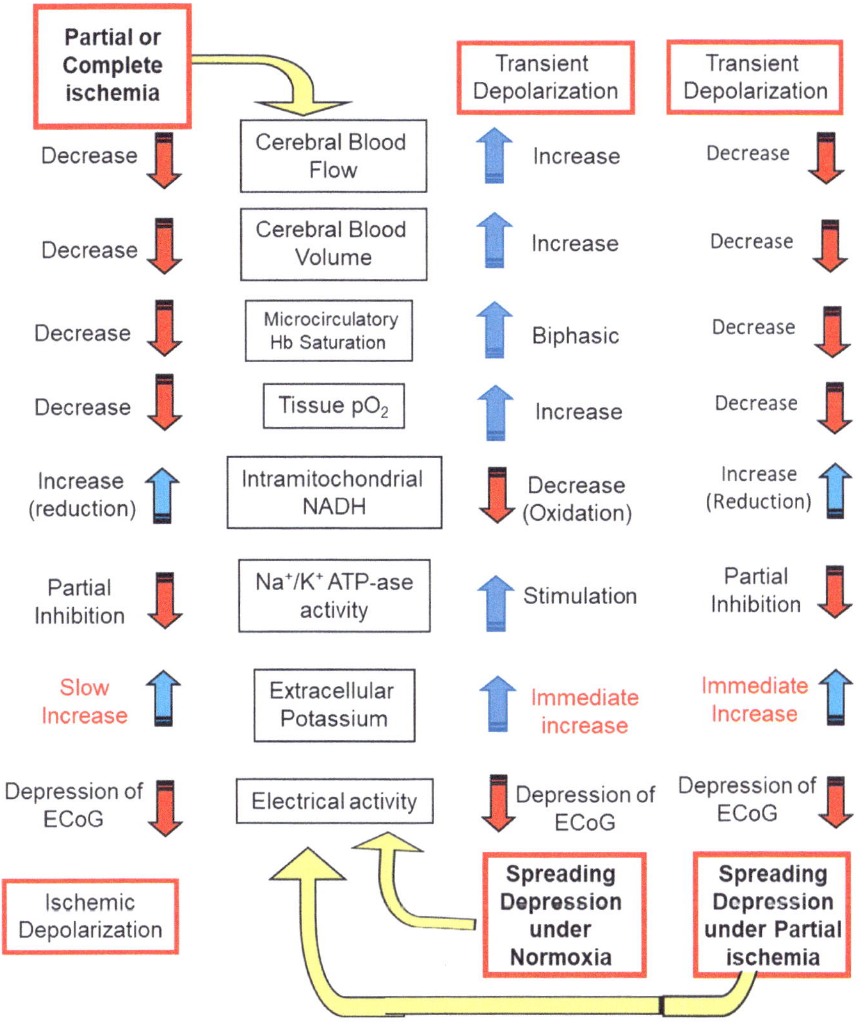


**Fig. 11.8** Schematic comparison between two responses to CSD induced in the normoxic brain (right side) and in the partial ischemic brain (left side). Also, the response to partial or complete ischemia without CSD is presented on the left side

compensate for the extra $O_2$ needed and will keep the balance of oxygen in the normal range so that oxidation of NADH is observed. When oxygen delivery to the tissue is limited, an increase in the energy demand during CSD will lead to an imbalance between supply and demand, resulting in a reduction of NADH. In severe oxygen imbalance, the ATP production will be so slow (anoxia or complete ischemia) that the $K^+$ will not return into the cells during the CSD episode and NADH will be fully reduced.

### 11.3.3 Modeling of Brain Energy Metabolism and Function

#### 11.3.3.1 Introduction

In order to test the validity of the two types of NADH responses to CSD induced in the normoxic and ischemic brain, for example, we used another technique. We constructed and tested a mathematical model [37], capable of simulating changes in brain energy metabolism that develop in real time under various pathophysiological conditions.

The goal of our study was to develop and test a mathematical model that permits the exploration of brain tissue metabolism in real time, based on several observable and hidden parameters computed via the model. Such a model has the potential to serve as a tool in simulating the effects of new pathophysiological conditions on the brain responses. A sufficiently detailed model of this kind, when properly tuned, can be used to assess the brain tissue functioning even when not all the parameters are monitored in an actual experiment.

#### 11.3.3.2 The Mathematical Model

In the framework of our research, we refer to the tissue as a whole, that is., as a self-contained system with inputs and outputs. This simplification enables us to utilize the so-called integral- or point-model approach when the physiological variables are regarded as functions of time only (their interdependence apart) rather than as space-distributed variables. Hence, the "diffusion" terms and the spatial derivatives that appear in the space-distributed models [38, 39] are omitted, and we arrive at a classical dynamic system, whose states are described by ordinary differential equations of the kinetic type. Furthermore, using this approach, we can tune the model, maximizing the fit to the actual clinical and experimental data with little risk of overfitting. In the course of data fitting, different kinds of nonlinearities in the right-hand parts of the differential equations were tested. The main objective was to keep the model physiologically meaningful, simple as possible, and universal in the sense that the same values of the model parameters would be used to simulate the normal and different pathological conditions of the brain tissue. The goodness-of-fit was examined and confirmed in numerous independent runs using various laboratory and clinical data published earlier [40–44]. The mathematical model manifests many of the essential characteristics of the experimental data with regard to certain observable variables, providing a quantitative fit, which is indicative of the model's reliability. We have tested the ability of the model to correctly reproduce the dynamics of a reduced number of the physiological variables when the dynamics of the remaining variables are inputted in the model. This property of our model may be used in clinical practice to complement real-time measurements. Our model is based on the assumption that the tissue functional state is fully determined by the following metabolic variables:

1. Concentration of extracellular potassium (K)
2. Rate of potassium reuptake (R)
3. Blood flow (F)
4. Oxygen (O)
5. NADH (N)
6. Partial impairment (P)

Such a formulation of the model was selected because quantitative data on the rates of change of the model variables are scarce. In clinical conditions, only some of the above tissue parameters, such as the blood flow, the concentration of extracellular potassium, and the level of NADH, can be measured. Other, hidden variables (the rate of potassium reuptake, partial impairment, and the level of oxygen) are either not measurable or not measured at the present. Their meaning of the model variables was described [37].

#### 11.3.3.3 Results

**Spreading depression** The pathological state of spreading depression can be induced by washing the cerebral cortex with a diluted solution of KCl (the "potassium injection"). This is modeled by putting $K_{inj} > 0$ within the time interval $t_1 < t < t_2$. When spreading depression is induced, the increase in energy requirement leads to the activation of $Na^+$-$K^+$-ATPase in order to restore the normal potassium distribution and causes an increase in blood flow and a decrease in NADH (Fig. 11.9).

**Spreading depression after partial ischemia** In this case, the blood flow decreases mildly ($F_{max} = 0.5$ within the time interval from $t_1$ to $t_2$), but within the ischemic period, an infusion of potassium is modeled; that is, $K_{inj}$ is assumed to be positive within a relatively short time interval contained in the interval $t_1 < t < t_2$ during which the main events occur. Accordingly, the blood flow slightly decreases as a result of the infusion and then increases to the new resting level (Fig. 11.10).

In the cases of global ischemia and CSD, the temporal behavior of the model variables was compared with the corresponding published clinical and experimental data [43, 45]. In order to compare the simulated and experimental dynamics, the scales were adjusted. The comparison (Figs. 11.11 and 11.12) testifies to the qualitative agreement between the computed and experimental values.

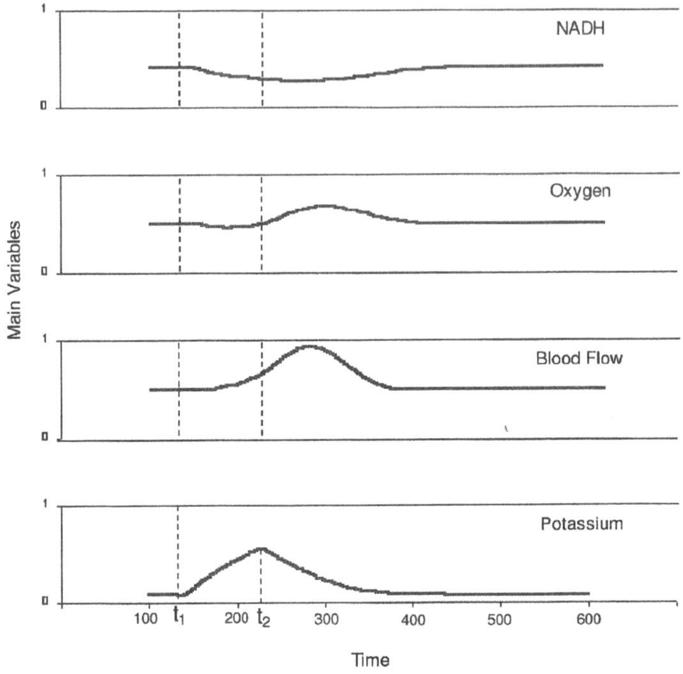


**Fig. 11.9** Simulated spreading depression, induced by an injection of potassium when it is at the resting level. The duration of injection is from $t_1$ to $t_2$ [37]

#### 11.3.3.4 Model Discussion

Our results demonstrate the potential of the computational mathematical model of tissue metabolism to describe the pathogenesis of tissue sustains in various clinical and experimental situations. The model simulates five pathophysiological tissue conditions, shown in Figures 1–5 of our previous publication [37] and Figs. 11.9, 11.10, 11.11, and 11.12 in the current chapter. This includes:

1. Global ischemia,
2. Partial ischemia,
3. Partial ischemia without depolarization,
4. Spreading depression,
5. Spreading depression induced after partial ischemia.

Note that the various states of tissue activity are described using the same values of constants [37]. This indicates the model's adequacy for the processes occurring in the tissue. The considerable qualitative and quantitative agreement with the experimental/clinical data (Figs. 11.11 and 11.12) as tests to the model's reliability. The number of variables routinely monitored in clinical situations is limited. Using our model in clinical and laboratory practice, the number of measured parameters can

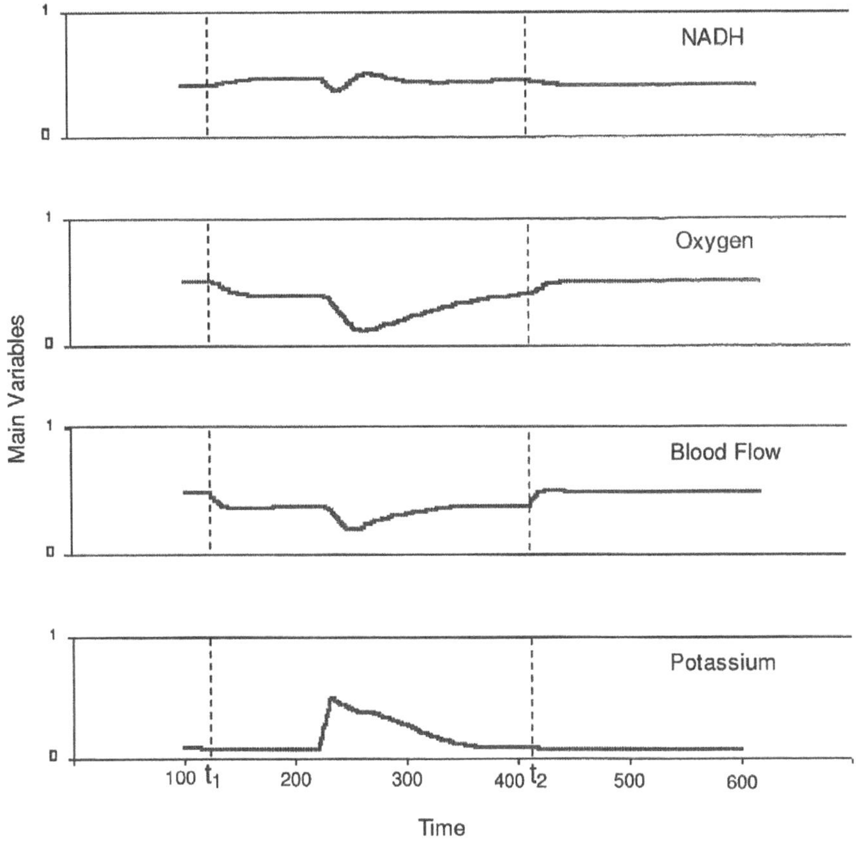


**Fig. 11.10** Simulated spreading depression induced after Partial Ischemia, resulting from a mild decrease in blood flow with a short injection of potassium [37]

be minimized, since the nonmeasured variables can be evaluated by the model in real time. Moreover, in principle, the comparison of the computed and clinical data can be used to test the tissue state and predict the recovery process. When utilized along with real-time clinical measurements, the model may provide the physician with a means for early diagnosis of pathological processes at their early stages.

A wide range of pathological events may ensue due to the limitation of oxygen supply to the tissue by a restriction of blood flow (ischemia) or by a lowering of the $O_2$ level in the inspired air (hypoxia). As a result of oxygen restriction, mitochondrial activity is inhibited, ATP production decreases, and ionic homeostasis is disturbed, leading to extracellular $K^+$ accumulation. A high correlation has been found between changes in the cerebral blood flow and mitochondrial NADH redox state during ischemia [19, 44], as presented in Fig. 11.13. On the other hand, high levels of extracellular potassium produced in CSD lead to an increase in the tissue metabolic demands, as indicated by a rise in $O_2$ consumption [47]. This process is

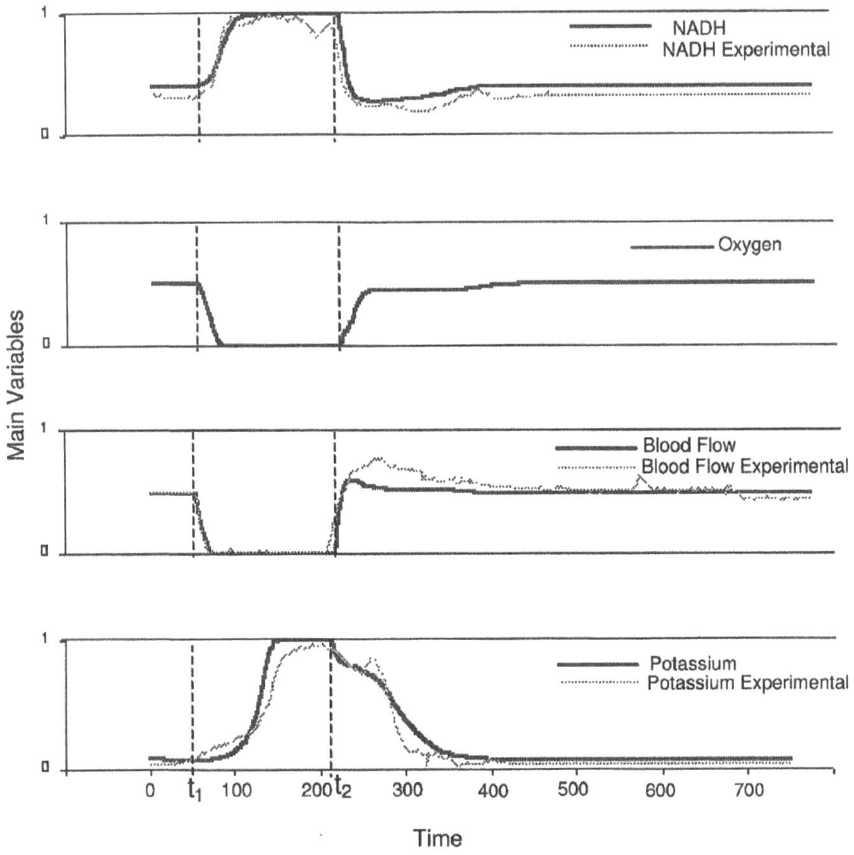


**Fig. 11.11** The comparison of the temporal behavior of the model variables to the corresponding experimental data for global ischemia. The simulated dynamics (shown in Fig. 1 of the original publication [37]) overlap with the experimental data from [46]

usually accompanied by an increase in blood flow; as a result, the activation of mitochondrial activity leads to NADH oxidation, as demonstrated in Fig. 11.14.

## *11.3.4 Correlations Between Parameters in Pathophysiological Events*

### 11.3.4.1 The First Study [46]

In order to study the correlation between mitochondrial activity and the hemodynamic responses under various conditions, the analog signals recorded were analyzed quantitatively. The resting levels of the signal before the perturbation were

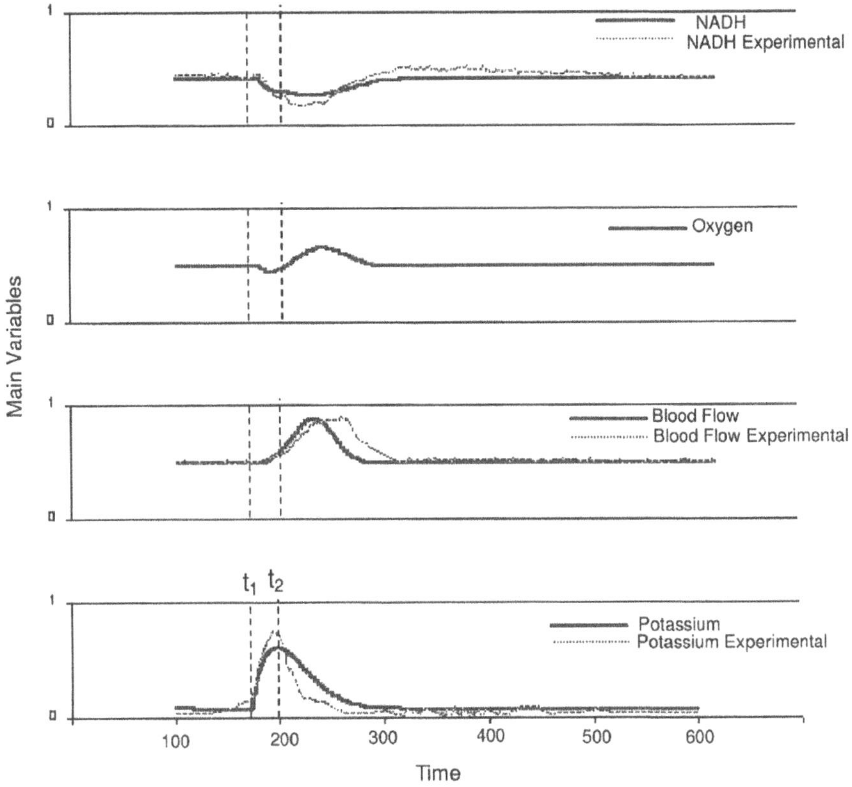


**Fig. 11.12** The comparison of the temporal behavior of the model variables to the corresponding clinical data for spreading depression from [37, 41]

considered a 100% value, and the change during the perturbation was calculated in percent change relative to the resting level.

Figure 11.15 shows the relationship between the CBF (measured by LDF) and mitochondrial redox state (NADH-CF) as well as between the reflectance and blood volume measured under ischemia, hypoxia, and spreading depression in the gerbil brain. Since various gerbils have a different anatomy of the anterior part of the circle of Willis, the level of ischemia created (Part a) covers the range from 10% to 100% ischemia. Also, during the graded hypoxia, a large increase in CBF up to 400% was noted simultaneously with the increase in NADH. During SD, the NADH became oxidized while the flow was increased by up to 400%. The correlation shown in Part (b) presents the changes in blood volume measured by the laser Doppler flowmeter versus the changes in intensity of the reflected light (366 nm) measured by the NADH fluorometer. The correlations between the three different parameters measured by the LDF are shown in Fig. 11.16. A very clear positive correlation can be seen between the two pairs of parameters. The calculations of the correlation coefficients (R) reveal that most of the values were statistically significant.

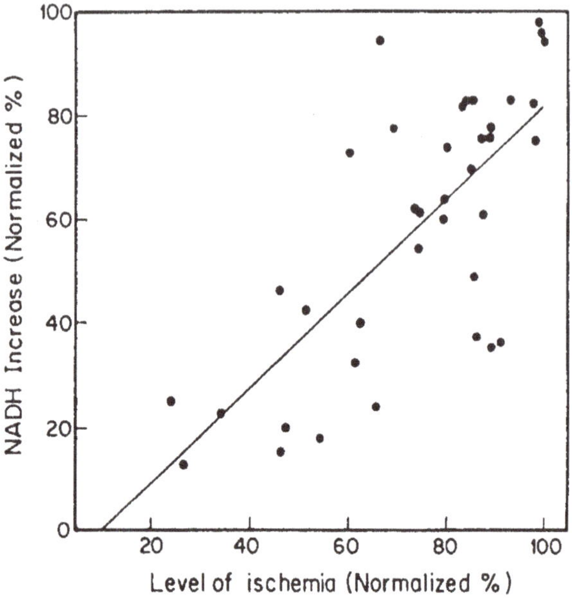


**Fig. 11.13** The correlation between the level of ischemia (carotid artery occlusion) and the NADH response in the Mongolian gerbil. Each point represents one occlusion, and the values were normalized in the 0–100% range of each parameter [19]

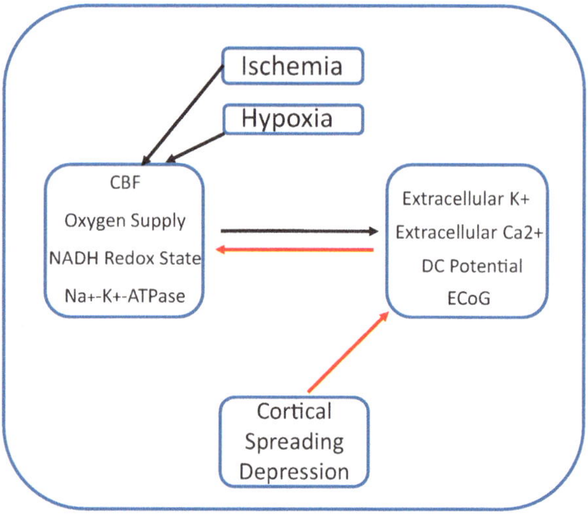


**Fig. 11.14** Schematic presentation or the various changes taking place in the brain under ischemia, hypoxia, or spreading depression [48] (modified)

In Table 11.1, two sets of R values are presented. In the upper right side, the values of each individual treatment, and in the lower left, the pooled values of all 92 pairs of numbers. Under spreading depression, most of the R values were not statistically significant except NADH and CBF, as well as CBF velocity. The interrelation between energy supply and the ability of the brain to recover from spreading depression is shown in Fig. 5.17. Line A shows the initial effect of $K^+_{(2)}$ electrode, followed by the other sensors. Due to the occlusion ($R_{occl}$), a small decrease in CBF

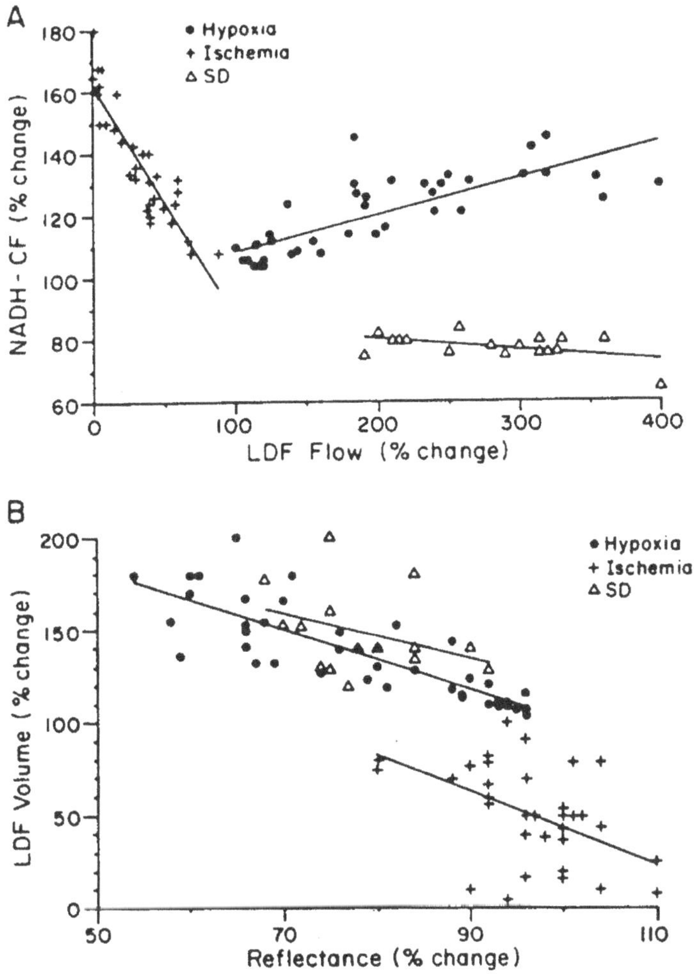


**Fig. 11.15** Correlations between parameters measured by the laser Doppler flowmeter (flow and volume) and those monitored by the fluorometer reflectometer (NADH fluorescence (**a**) and reflected light (**b**)). The gerbils were exposed to ischemia, hypoxia, and spreading depression [46]

was recorded (LDF) simultaneously with an increase in NADH redox state (CF). The propagation of the SD wave was blocked by the ischemic episode, and recovery processes started only after the reopening of the occluded artery ($R_{open}$).

In order to correlate the two parameters, namely, the NADH redox state and CBF, it was necessary to integrate the probes so that both of them would "sample" approximately the same tissue volume. This was achieved by developing the meta-flow probe consisting of two sets of optical fibers connecting the surface of the brain to the NADH fluorometer and to the laser Doppler flowmeter. As seen in Fig. 3.23 or 3.25 in Chap. 3, the three fibers of the LDF are mixed between the NADH fibers,

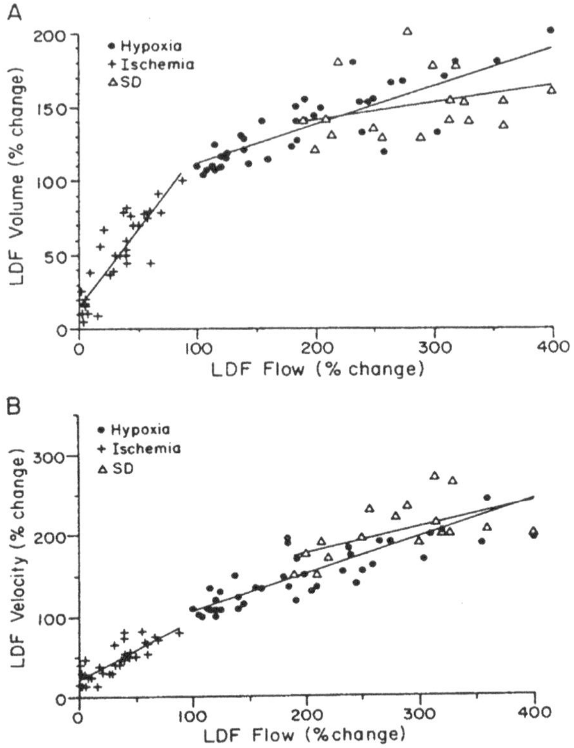


**Fig. 11.16** The relationship between all three parameters measured by the laser Doppler flowmeter. (**a**) LDF flow and LDF Volume, (**b**) LDF flow and LDF Velocity [46]

and since the light wavelength of the LDF is longer (>600 nm), it will penetrate better than the NADH fibers.

The three kinds of perturbation used, namely, ischemia, hypoxia, and spreading depression, cover a wide range of clinical pathological situations; thus, the information regarding the correlation between flow and NADH redox state is very significant. As seen in Fig. 11.15a, the relationship of the CBF and NADH redox states was different under each of the three perturbations used. Under ischemia, the decrease in CBF is expected by definition (decrease in blood supply), and indeed the CF signal increased according to the decrease in CBF. The response to hypoxia was characterized by an increase in CBF (due to autoregulatory response), and the NADH also increased. The response to spreading depression was completely different. The SD wave led to an increase in ion pumping activity, which stimulated the energy consumption [49]. Under those conditions, the CBF is stimulated in order to compensate for the extra oxygen consumed [47], as detected by the oxidation of the NADH (decrease of CF signal). As shown in Fig. 11.15b, the changes in the reflectance trace were correlated to the blood volume changes measured by the LDF. The correlation was very clear under hypoxia and not so clear under SD.

According to the manufacturer of LDF (TSI Inc.), the volume parameter represents only the volume of the flowing fraction of the blood and not the total volume; therefore, the meaning of it is not clear as yet. As seen in Fig. 11.16, all three

**Table 11.1** The correlation coefficient values calculated between the various metabolic and hemodynamic parameters monitored during hypoxia, ischemia, and spreading depression [46]

| | Flow | Volume | Velocity | Reflectance | Corrected fluorescence |
|---|---|---|---|---|---|
| Flow | | 0.667<br>0.753<br>0.095 (N.S) | 0.592<br>0.676<br>0.319 ($p < 0.02$) | 0.742<br>0.158 ($p < 0.02$)<br>0.176 (N.S) | 0.592<br>0.808<br>0.241 ($p < 0.05$) |
| Volume | 0.799 | | 0.490<br>0.561<br>0.005 (N.S) | 0.684<br>0.259 ($p < 0.002$)<br>0.156 (N.S) | 0.653<br>0.682<br>0.057 (N.S) |
| Velocity | 0.862 | 0.762 | | 0.598<br>0.148 ($p < 0.05$)<br>0.019 (N.S) | 0.606<br>0.665<br>0.010 (N.S) |
| Reflectance | 0.706 | 0.678 | 0.621 | | 0.822<br>0.091 (N.S)<br>0.100 (N.S) |
| Corrected fluorescence | 0.307 | 0.339 | 0.328 | 0.067 ($p < 0.01$) | |

The calculations were done on 41 pairs of hypoxia (upper number), 34 pairs of ischemia (middle number), and 17 pairs of spreading depression (lower number)
The values appearing as single numbers per box were calculated by using all 92 pairs collected in the three kinds of perturbations. The level of significance was $p < 0.0001$ for all correlations except in cases that it was lower and are presented in the parentheses

parameters measured by the LDF, namely, flow, volume, and velocity, were correlated very well with each other. Recently, we started using the LDF produced by Perimed Inc., and the information obtained regarding the volume is different. In the Perimed instrument, the total backscattering light is used to evaluate the changes in blood volume, and therefore, it may be more adequate for comparison with the reflectance changes.

#### 11.3.4.2 The Second Study [50]

In order to understand the relationship between blood oxygenation and mitochondrial redox state, we quantified the traces of $HbO_2$, Hb, R, and CF monitored in a few hypoxic episodes (Fig. 11.17). As seen, a very significant correlation was found between the decrease in oxygenation of $HbO_2$ and the relative increase in NADH redox state (CF). Also, the two parameters representing the blood volume changes, namely, the Hb relative concentration and the reflected light changes (R), correlated very well (Part b). The same kind of correlation was found when the gerbils were exposed to other oxygen deprivation perturbations (anoxia or ischemia). The last subject of this study to be discussed is the technique to measure and calculate the $O_2$ consumption rate in the same animal model. To do this kind of measurement in the

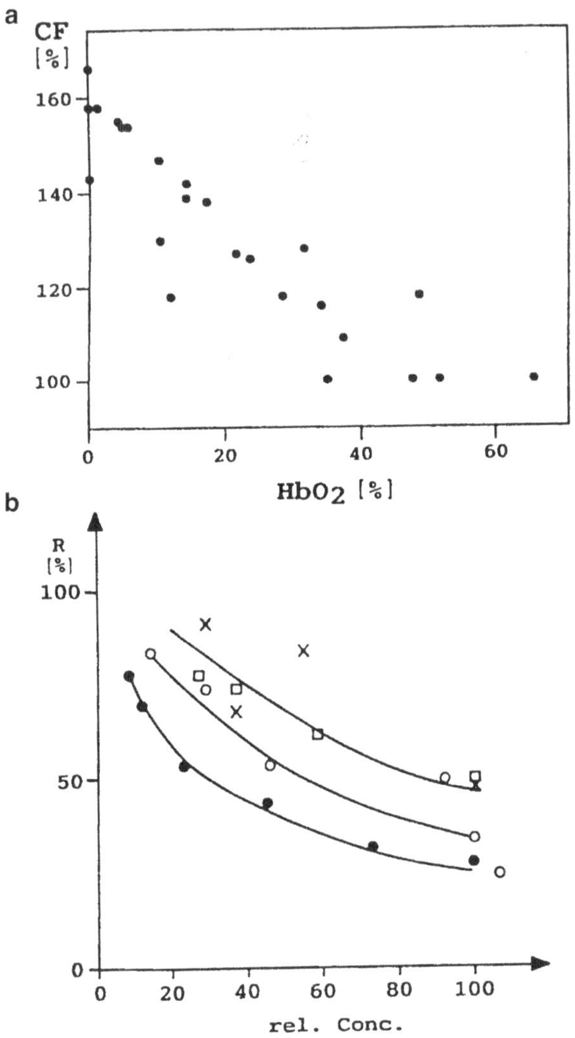


**Fig. 11.17** The correlation between NADH fluorescence (CF) and $HbO_2$ (**a**) as well as between 366 nm reflectance (R) and relative concentration of Hb (blood volume) shown in (**b**). The data were collected from gerbils exposed to hypoxia. The various symbols represent different animals [50]

gerbil, we exposed the brain to a short bilateral carotid artery occlusion (2–5 s) while recording the deoxygenation of the $HbO_2$ with fast scanning of the 500–600 nm range using the EMPHO unit. The response to this $O_2$ consumption test, shown in Figs. 4.12, 4.60, and 5.13 and labeled by T, is based on the principle that the rate of deoxygenation of $HbO_2$ during the short occlusion is proportional to the amount of $O_2$ used (consumed) by the tissue at that time. Figure 11.18 shows the responses recorded under normoxic (top) and hypoxic (bottom) conditions. The $O_2$ consumption was calculated for the various tests performed during the few hypoxia episodes and is presented in Fig. 11.19. The results were calculated in terms of *dc*/*dt* versus the initial content of $HbO_2$ ($C_{in}$) when the test was performed. As seen in Fig. 11.19,

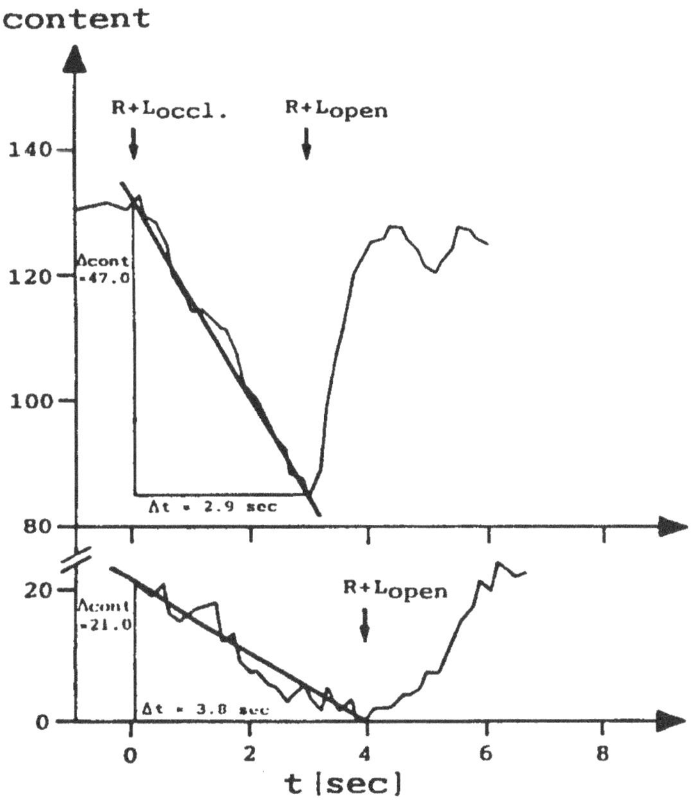


**Fig. 11.18** The effect of bilateral carotid occlusion on tissue $HbO_2$ level during normoxia (upper part) and hypoxia (lower section). The calculated values of content over time ($t$) provide the relative $O_2$ consumption rate [50]

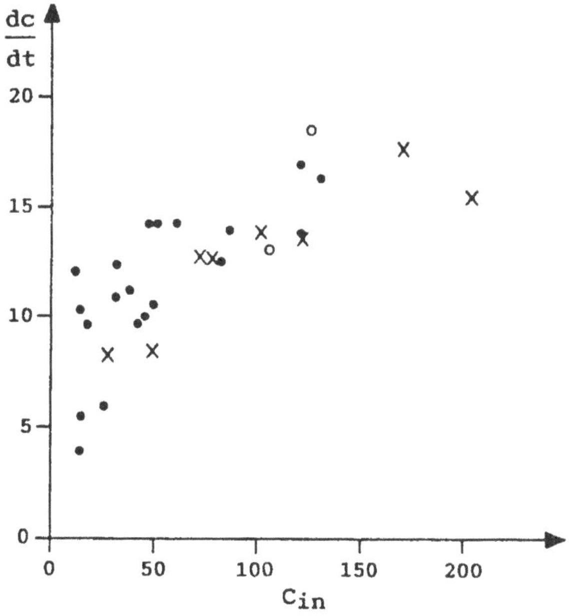


**Fig. 11.19** The relationship between relative $O_2$ content in the brain and the rate of $O_2$ consumption in normoxic gerbils exposed to hypoxia. The various symbols represent different animals [50]

the rate of $O_2$ consumption decreased as the animal became hypoxic and the $O_2$ content in blood decreased. A few tests of $O_2$ consumption were performed under SD wave conditions, and a large increase was recorded, although more studies need to be carried out in order to evaluate this statistically.

## References

1. Chance B, Cohen P, Jobsis F, Schoener B (1962) Intracellular oxidation-reduction states *in vivo*. Science 137:499–508
2. Chance B, Oshino N, Sugano T, Mayevsky A (1973) Basic principles of tissue oxygen determination from mitochondrial signals. In: Oxygen transport to tissue, Advances in experimental medicine and biology, vol 37A. Springer, Boston, pp 277–292
3. Mayevsky A, Chance B (1973) In: Bicher HI, Bruley DF (eds) A new long-term method for the measurement of NADH fluorescence in intact rat brain with implanted cannula. Plenum Press, New York, pp 239–244
4. Mayevsky A (1978) Pyridine nucleotide oxidation reduction state of the cerebral cortex in the awake gerbil. J Neurosci Res 3:369–374
5. Mayevsky A, Crowe W, Mela L (1980) The interrelation between brain oxidative metabolism and extracellular potassium in the unanesthetized gerbil. Neurol Res 1:213–226
6. Zeman W, Innes JRM (1963) Craigie's neuroanatomy of the rat. Academic Press, New York
7. Berry K, Wisniewski HM, Svarzbein L, Baez S (1975) On the relationship of brain vasculature to production of neurological deficit and morphological changes following acute unilateral common carotid artery ligation in gerbils. J Neurol Sci 25:75–92
8. Mayevsky A (1976) Brain energy metabolism of the conscious rat exposed to various physiological and pathological situations. Brain Res 113:327–338
9. Mayevsky A (1978) Ischemia in the brain: the effects of carotid artery ligation and decapitation on the energy state of the awake and anesthetized rat. Brain Res 140:217–230
10. Pulsinelli WA, Brierley JB (1979) A new model of bilateral hemispheric ischemia in the unanesthetized rat. Stroke 10:267–272
11. Mayevsky A, Duckrow RB, Yoles E, Zarchin N, Kanshansky D (1990) Brain mitochondrial redox state, tissue hemodynamic and extracellular ion responses to four-vessel occlusion and spreading depression in the rat. Neurol Res 12:243–248
12. Friedli CM, Sclarsky DS, Mayevsky A (1982) Multiprobe monitoring of ionic, metabolic, and electrical activities in the awake brain. Am J Phys 243:R462–R469
13. Mayevsky A, Breuer Z (1992) Brain vasculature and mitochondrial responses to ischemia in gerbils. I. Basic anatomical patterns and biochemical correlates. Brain Res 58:242–250
14. Chance B, Williams GR (1955) Respiratory enzymes in oxidative phosphorylation (III-The steady state). J Biol Chem 217:409–427
15. Rosenthal M, Jobsis FF (1971) Intracellular redox changes in functioning cerebral cortex. II. Effects of direct cortical stimulation. J Neurophysiol 34:750–762
16. Lothman E, LaManna J, Cordingley G, Rosenthal M, Somjen G (1975) Responses of electrical potential potassium levels, and oxidative metabolic activity of the cerebral neocortex of cats. Brain Res 88:15–36
17. Mayevsky A, Cohen S (1990) Brain ischemic depolarization and vasospasm in the Mongolian gerbil: the dependence on energy depletion levels. Adv Exp Med Biol 277:315–322
18. Mayevsky A, Chance B (1974) Repetitive patterns of metabolic changes during cortical spreading depression of the awake rat. Brain Res 65:529–533
19. Mayevsky A (1993) Biochemical and physiological activities of the brain as *in vivo* markers of brain pathology. In: Bernstein EF, Callow AD, Nicolaides AN, Shifrin EG (eds) Cerebral revascularization. Med-Orion Pub, pp 51–69

20. Mayevsky A, Zarchin N, Friedli CM (1982) Factors affecting the oxygen balance in the awake cerebral cortex exposed to spreading depression. Brain Res 236:93–105
21. Mayevsky A, Meilin S, Manor T, Zarchin N, Sonn J (1999) Optical monitoring of NADH redox state and blood flow as indicators of brain energy balance. Adv Exp Med Biol 471:133–140
22. Mayevsky A (1984) Brain NADH redox state monitored *in vivo* by fiber optic surface fluorometry. Brain Res Rev 7:49–68
23. Mayevsky A, Yoles E, Zarchin N, Kaushansky D (1990) Brain vascular ionic and metabolic responses to ischemia in the Mongolian gerbil. J Basic Clin Physiol Pharmacol 1:207–220
24. Mayevsky A, Ziv I (1991) Oscillations of cortical oxidative metabolism and microcirculation in the ischaemic brain. Neurol Res 13:39–47
25. Mayevsky A, Flamm ES, Pennie W, Chance B (1991) A fiber optic based multiprobes system for intraoperative monitoring of brain functions. SPIE 1431:303–313
26. Chance B, Williams GR (1956) The respiratory chain and oxidative phosphorylation. In: Nord FF (ed) Advances in enzymology. Interscience Publisher, Inc., New York, pp 65–134
27. Leão AAP (1944) Spreading depression of activity in cerebral cortex. J Neurophysiol 7:359–390
28. Mayevsky A, Lebourdais S, Chance B (1980) The interrelation between brain $PO_2$ and NADH oxidation-reduction state in the gerbil. J Neurosci Res 5:173–182
29. Buresova O (1957) Changes in cerebral circulationin rats during spreading EEG depression. Physiol Bohemoslov 6:1–11
30. Leão AAP (1944) Pial circulation and spreading depression of activity in the cerebral cortex. J Neurophysiol 7:391–396
31. Van Harreveld A, Ochs S (1957) Electrical and vascular concomitants of spreading depression. Am J Phys 189:159–166
32. Rosenthal M, Somjen G (1973) Spreading depression, sustained potential shifts, and metabolic activity of cerebral cortex of cats. J Neurophysiol 36:739–749
33. Mayevsky A, Zeuthen T, Chance B (1974) Measurements of extracellular potassium, ECoG and pyridine nucleotide levels during cortical spreading depression in rats. Brain Res 76:347–349
34. LaManna JC, Rosenthal M (1975) Effect of ouabain and phenobarbital on oxidative metabolic activity associated with spreading cortical depression in cats. Brain Res 88:145–149
35. Crowe W, Mayevsky A, Mela L (1979) The dynamics of $K^+$ leakage and recovery in cerebral ischemia. In: Leter A, Saba T, Mela L (eds) Advances in shock research. Alan R. Liss, New York, pp 221–232
36. Somjen GG, Rosenthal M, Cordingley G, LaManna J, Lothman E (1976) Potassium, neuroglia, and oxidative metabolism in central gray matter. Fed Proc 35:1266–1271
37. Vatov L, Kizner Z, Ruppin E, Meilin S, Manor T, Mayevsky A (2006) Modeling brain energy metabolism and function: a multiparametric monitoring approach. Bull Math Biol 68:275–291
38. Revett K, Ruppin E, Goodall S, Reggia JA (1998) Spreading depression in focal ischemia: a computational study. J Cereb Blood Flow Metab 18:998–1007
39. Ruppin E, Ofer E, Reggia JA, Revett K, Goodall S (1999) Pathogenic mechanisms in ischemic damage: a computational study. Comput Biol Med 29:39–59
40. Mayevsky A, Zarchin N, Tannenbaum B (1984) Brain responses to experimental oxygen deficiency in the Mongolian gerbil. Adv Exp Med Biol 180:191–202
41. Mayevsky A, Doron A, Manor T, Meilin S, Zarchin N, Ouaknine GE (1996) Cortical spreading depression recorded from the human brain using a multiparametric monitoring system. Brain Res 740:268–274
42. Mayevsky A, Meilin S, Manor T, Ornstein E, Zarchin N, Sonn J (1998) Multiparametric monitoring of brain oxygen balance under experimental and clinical conditions. Neurol Res 20:S76–S80
43. Meilin S, Zarchin N, Mayevsky A (1999) Inter-relation between hemodynamic, metabolic, ionic and electrical activities during ischemia and reperfusion in the gerbil brain. Neurol Res 21:699–704
44. Mayevsky A (1990) Level of ischemia and brain functions in the Mongolian gerbil *in vivo*. Brain Res 524:1–9

45. Mayevsky A, Manor T, Meilin S, Doron A, Ouaknine GE (1998) Real-time multiparametric monitoring of the injured human cerebral cortex – a new approach. In: Marmarou R, Bullock R (eds) Acta Neurochir. Suppl (Wien.). ICPX, pp 78–81
46. Mayevsky A (1992) Cerebral blood flow and brain mitochondrial redox state responses to various perturbations in gerbils. Adv Exp Med Biol 317:707–716
47. Mayevsky A, Weiss HR (1991) Cerebral blood flow and oxygen consumption in cortical spreading depression. J Cereb Blood Flow Metab 11:829–836
48. Mayevsky A (1992) Interrelation between intracellular redox state and ion homeostasis in the brain *in vivo*. In: Frank K, Kessler M (eds) Quantitative spectroscopy in tissues. Verlasgruppe GmbH, pp 155–168
49. Erecinska M, Silver IA (1989) ATP and brain function. J Cereb Blood Flow Metab 9:2–19
50. Mayevsky A, Frank K, Muck M, Nioka S, Kessler M, Chance B (1992) Multiparametric evaluation of brain functions in the Mongolian gerbil *in vivo*. J Basic Clin Physiol Pharmacol 3:323–342

# Index




A. Mayevsky, *The Mongolian Gerbil Brain*,
https://doi.org/10.1007/978-3-031-69549-0

The manufacturer's authorised representative in the EU is Springer Nature Customer Service Centre GmbH, Europaplatz 3, 69115 Heidelberg, Germany. If you have any concerns regarding our products, please contact ProductSafety@springernature.com

Printed and bound by CPI Group (UK) Ltd, Croydon, CR0 4YY
01/10/2026
02225904-0011